THE ATMOSPHERE

FOURTH EDITION

THE ATMOSPHERE

AN INTRODUCTION TO METEOROLOGY

FREDERICK K. LUTGENS
EDWARD J. TARBUCK

Illinois Central College

PRENTICE HALL
Englewood Cliffs, New Jersey 07632

Library of Congress Cataloging-in-Publication Data

Lutgens, Frederick K.
   The atmosphere : an introduction to meteorology.

   Includes index.
   1. Atmosphere. 2. Meteorology. 3. Weather.
I. Tarbuck, Edward J. II. Title.
QC861.2.L87 1989   551.5   88–4188
ISBN 0–13–050196–4

Interior design supervision: Janet Schmid
Cover design: Bruce Kenselaar
Manufacturing buyer: Paula Massenaro
Page layout: Robert Wullen
Illustrations: Tasa Graphic Arts Inc., Tijeras, New Mexico

 © 1989, 1986, 1982, 1979 by Prentice-Hall, Inc.
A Division of Simon & Schuster
Englewood Cliffs, New Jersey 07632

Printed in the United States of America

10 9 8 7 6 5 4 3 2

ISBN 0-13-050196-4

Prentice-Hall International (UK) Limited, *London*
Prentice-Hall of Australia Pty. Limited, *Sydney*
Prentice-Hall Canada Inc., *Toronto*
Prentice-Hall Hispanoamericana, S.A., *Mexico*
Prentice-Hall of India Private Limited, *New Delhi*
Prentice-Hall of Japan, Inc., *Tokyo*
Simon & Schuster Asia Pte. Ldt., *Singapore*
Editora Prentice-Hall do Brasil, Ltda., *Rio de Janeiro*

To our parents—our first and best teachers;
and to our wives and children
for their encouragement and patience.

Contents in Brief

Contents

3 Temperature   53

4 Humidity, Condensation, and Atmospheric Stability   82

5 Forms of Condensation and Precipitation   121

6 Air Pressure and Winds   157

13 The Changing Climate   354

14 World Climates   390

Preface

There are few aspects of the physical environment that influence our daily lives more than the phenomena we collectively call weather. Newspapers, magazines, and television stations frequently report a wide range of weather events as major news stories—an obvious reflection of people's interest and curiosity about the atmosphere. Since a meteorology course can take advantage of this interest and curiosity about the weather, it is not surprising that introductory courses dealing with the atmosphere are popular on many college campuses.

The Atmosphere: An Introduction to Meteorology, 4th Ed. is designed to meet the needs of students who enroll in such a course. The book is a current and comprehensive, yet not encyclopedic, survey of the atmosphere intended for an introductory audience. Although a portion of modern meteorology is highly quantitative, it was our intent to produce a nontechnical and nonmathematical text. It is our firm belief that weather phenomena can be understood and appreciated by people who do not have a strong background in mathematics.

Since student use of the text was one of our primary concerns, we made a concentrated effort to write a book that is highly usable—a tool for learning the basic concepts of meteorology. To achieve this goal, the book is written concisely and is well illustrated. The useful learning aids found in the three previous editions have been retained and strengthened in the fourth edition. Important terms are printed in boldface type within the body of the text and also appear in a vocabulary review at the end of each chapter. A glossary is included at the end of the text for easy reference to key terms. Each chapter concludes with review questions and problems that highlight major concepts and aid student review. Information on metric conversions and weather maps is found in the appendices along with other helpful reference material.

In previous editions of *The Atmosphere*, special attention was always given to the quality of the artwork and photographs. This focus has been given even greater emphasis in the fourth edition. Thirty-eight new photographs have been integrated into the body of the text, with 28 in full color. The photos were carefully selected to add realism to the subject and to heighten the interest of the reader. The art program has also received a great deal of attention. Because we believe that carefully planned and executed line art significantly aids student understanding, more than 100 diagrams and maps are either new or substantially redrawn. Moreover, 43 of the most important pieces have been produced in full color. A glance at the temperature maps in Chapter 3, the pressure or precipitation maps in Chapter 7, and the many new diagrams relating to fronts in Chapter 9 serve to illustrate this important improvement.

The fourth edition of *The Atmosphere: An Introduction to Meteorology* represents a thorough revision. However, it should be emphasized that the main focus of the new edition remains the same as that of its predecessors—to foster a basic understanding of the atmospheric environment. In keeping with this aim, the first ten chapters continue to be devoted to a presentation of the major elements and concepts of meteorology. With the goal of achieving greater clarity, many basic discussions were rewritten and/or expanded. Some examples include sections on exploring the atmosphere, temperature measurement, curved flow, jet streams, cyclogenesis, and traveling cyclones and anticyclones. Although the first ten chapters stress fundamentals, applications and topical issues are not ignored. Indeed, the fourth edition contains updated and expanded coverage of the ozone problem, air pollution, acid rain, and intentional weather modification. Coverage of subjects as diverse as indices of human discomfort and the effects of El Niño on global weather also continue to be important parts of the book.

Chapter 11 on weather analysis follows the chapters that deal with basic meteorological principles and serves to reinforce and apply many of the concepts presented in the first ten chapters. In addition to introducing the reader to the nature of modern weather forecasting, the chapter provides a case history of a storm. The several different perspectives and views of the storm as it passes across the United States are designed to increase the reader's understanding of our often complex middle-latitude weather.

Chapter 12 examines a topic about which many beginning students are curious—the atmosphere's varied optical phenomena. The chapter seeks to provide nontechnical yet meaningful insights into many of these beautiful and often complex phenomena. After investigating Chapter 12, such features as rainbows, mirages, and halos will take on new meaning for the beginning student.

The text concludes with two chapters on climate. Chapter 13, "The Changing Climate," explores a topic that is the focus of increasing interest among both the scientific community and the public at large. How are people modifying the climates of cities? Is global climate changing, and, if so, in what ways? Are people causing or contributing to any of these changes? These questions, and in particular, the last question, have received growing attention in the past few years. The number of international conferences, scholarly publications, and popular magazine articles devoted to human impact on global climate testifies to this heightened concern. The thoroughly revised and updated discussions in Chapter 13 reflect the fast-changing nature of this sometimes controversial subject. Finally, Chapter 14 examines climates in a more traditional manner. Here, the reader will investigate the broad diversity of climatic types around the world and the controls that have created them.

A popular feature in previous editions that has been retained in the fourth edition is the inclusion of guest essays. These well written pieces by experts in the field allow the reader to explore some topics from a different viewpoint. In the fourth edition, a new essay has been added to the four that appeared in the third edition. We believe that the American Meteorological Society's statement of concern entitled, "Is the United States Headed for a Hurricane Disaster?" will prove to be of interest to many.

The authors wish to express their thanks to the many individuals, institutions, government agencies, and businesses that provided photographs and illustrations for use in the text. Further, we would like to acknowledge the aid of our students. Their comments continue to help us maintain our focus on readability and under-standing. A special debt of gratitude goes to those colleagues who prepared in-depth prerevision reviews of the third edition of *The Atmosphere*. Their critical comments and thoughtful input helped guide our revision and strengthen the text. We wish to thank

Dr. Philip W. Suckling
University of Georgia

Prof. Alan Anderson
St. Cloud State University

Prof. John G. Gehr
The University of Arkansas

Prof. Robert Rouse
Elmhurst College

Prof. Scott A. Isrod
University of Illinois

We are also indebted to Dennis Tasa of Tasa Graphic Arts, Inc. for his imaginative production of the new line art. The text has greatly benefited from his considerable talent. Our thanks also go to the fine production staff at Prentice-Hall who skillfully transformed our manuscript into a finished product.

Frederick K. Lutgens

Edward J. Tarbuck

1

Introduction to the Atmosphere

WEATHER AND CLIMATE
COMPOSITION OF THE ATMOSPHERE
THE OZONE PROBLEM
ORIGIN OF THE ATMOSPHERE
EXPLORING THE ATMOSPHERE
VERTICAL STRUCTURE OF THE ATMOSPHERE
    Temperature Changes in the Vertical
    Vertical Variations in Composition
    The Ionosphere

A view of the earth from space affords us a unique perspective of our planet (Figure 1–1). At first it may strike us that the earth is a fragile-appearing sphere surrounded by the blackness of space. It is, in fact, just a speck of matter in an infinite universe. As we look more closely, it becomes apparent that the earth is much more than rock and soil. Indeed, the most conspicuous features are not the continents but the swirling clouds suspended above the surface and the vast global ocean. From such a vantage point we can appreciate why the earth's physical environment is traditionally divided into three major parts: the solid earth or **lithosphere;** the water portion of our planet, the **hydrosphere;** and the earth's life-giving gaseous envelope, the **atmosphere.** It is this last portion of our physical environment, the atmosphere, that is the primary focus of this text. It should be emphasized, however, that our environment is highly integrated and is not dominated by rock, water, or air alone. Rather, it is characterized by continuous interactions as air comes in contact with rock, rock with water, and water with air.

Lithosphere

Hydrosphere

Atmosphere

FIGURE 1–1
Earth from space. (Courtesy of NASA)

Biosphere

Moreover, the **biosphere,** the totality of life-forms on our planet, extends into each of the three physical realms and is an equally integral part of the earth.

If, like the moon, the earth had no atmosphere, our planet would not only be lifeless, but many of the processes and interactions that make the surface such a dynamic place could not operate. Without weathering and erosion, the face of our planet might more closely resemble the lunar surface, which has not changed appreciably in nearly 3 billion years.

People's interest in the atmosphere is probably as old as the history of humanity. Certainly there are few other aspects of our physical environment that affect our daily lives more than the phenomena we collectively call the weather. The clothes we wear and the activities we engage in are strongly influenced by the weather. Yet in addition to the day-to-day personal decisions we must make about the weather, there have been, and will continue to be, an increasing number of political decisions to make involving the atmosphere. Answers to questions regarding air pollution and its control, the effects of various chemicals on the atmosphere's ozone layer, the positive or adverse effects of intentional weather modification are all important and, in some cases, perhaps even vital. So there is a need for increased awareness and understanding of our atmosphere and its behavior.

WEATHER AND CLIMATE

Acted on by the combined effects of the earth's motions and energy from the sun, our planet's formless and invisible envelope of air reacts by producing an infinite variety of weather, which, in turn, creates the basic pattern of global climates. Although not identical, weather and climate have much in common.

Weather

Weather is constantly changing, sometimes from hour to hour and at other times from day to day. It is a term that denotes the state of the atmosphere at a given time and place. Whereas changes in the weather are continuous and sometimes seemingly erratic, it is nevertheless possible to arrive at a generalization of these variations. Such a description of aggregate weather conditions is termed **climate.**

Climate

Climate is often defined simply as "average weather," but this is an inaccurate definition. In order to more accurately portray the character of an area, variations and extremes must also be included. Thus climate is the sum of all statistical weather information that helps describe a place or region.

Elements (atmospheric)

The nature of both weather and climate is expressed in terms of the same basic **elements,** those quantities or properties that are measured regularly. The most important are (1) the temperature of the air, (2) the humidity of the air, (3) the type and amount of cloudiness, (4) the type and amount of precipitation, (5) the pressure exerted by the air, and (6) the speed and direction of the wind. These elements constitute the variables from which weather patterns and climatic types are deciphered. Although we shall study these elements separately at first, keep in mind that they are very much interrelated. A change in one of the elements often produces changes in the others.

Before you study the elements of weather and climate in detail, which begins in Chapter 2, you should read the remainder of this chapter as a general introduction to the physical and chemical nature of the atmosphere.

COMPOSITION OF THE ATMOSPHERE

Air

In the days of Aristotle air was believed to be one of four fundamental substances that could not be further subdivided into constituent components. The other three substances were fire, earth (soil), and water. Even today the term **air** is sometimes used as if it were a specific gas, which, of course, it is not. The envelope of air that surrounds our planet is a *mixture* of many discrete gases, each with its own physical properties, in which varying quantities of tiny solid and liquid particles are suspended.

The composition of air is not constant; it varies from time to time and from place to place. If the suspended particles, water vapor, and other variable gases were removed from the atmosphere, however, we would find that its makeup is, in fact, very stable all over the earth up to an altitude of about 80 kilometers. As can be seen in Figure 1–2 and Table 1–1, two elements, nitrogen and oxygen, make up 99 percent of the volume of clean, dry air; most of the remaining 1 percent is accounted for by the inert gaseous element, argon. Although these elements are the most plentiful components of the atmosphere, they are of little or no importance in affecting weather phenomena. Carbon dioxide, however, is present in only minute quantities, but it is nevertheless a meterologically important constituent of air.

Because carbon dioxide is an efficient absorber of radiant energy emitted by the earth and thus influences the flow of energy through the atmosphere, it is of great interest to meteorologists. Although the proportion of carbon dioxide is

FIGURE 1–2
Proportional volume of gases composing dry air. Nitrogen and oxygen obviously dominate.

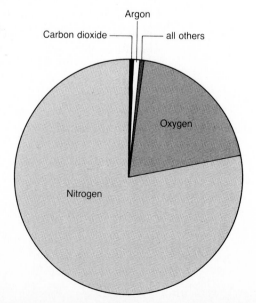

TABLE 1–1 Principal Gases of Dry Air

Constituent	Percent by Volume	Concentration in Parts per Million (ppm)
Nitrogen (N_2)	78.084	—
Oxygen (O_2)	20.946	—
Argon (A)	0.934	—
Carbon dioxide (CO_2)	0.034	—
Neon (Ne)	0.00182	18.2
Helium (He)	0.000524	5.24
Methane (CH_4)	0.00015	1.5
Krypton (Kr)	0.000114	1.14
Hydrogen (H_2)	0.00005	0.5

relatively uniform in the air, its percentage has been rising steadily for more than a century. This rise is attributed to the burning of ever-increasing quantities of fossil fuels, such as coal, oil, and natural gas. Much of this additional carbon dioxide is absorbed by the waters of the ocean or is used by plants, but some (about 50 percent) remains in the air. Estimates project that by sometime in the second half of the next century carbon dioxide levels will be twice as high as they were early in the twentieth century. Although the precise impact of the increased carbon dioxide is difficult to predict, most atmospheric scientists believe that it will bring about a warming of the lower atmosphere and hence produce global climatic change. The role of carbon dioxide in the atmosphere and its possible effect on climate are discussed in more detail in Chapter 13.

Like carbon dioxide, some of the atmosphere's variable components, notably water vapor, ozone, and dust particles, are also present in small amounts; nonetheless, they are important. Water vapor is one of the most variable gases in our atmosphere. In the warm and wet tropics it may account for up to 4 percent, by volume, of the atmosphere, whereas in the air of deserts and polar regions it may constitute but a tiny fraction of 1 percent. The significance of atmospheric moisture to all life is universally recognized and does not require further emphasis here. The roles that it plays in the physical processes of the atmosphere are less well known, however, and are extremely important meteorologically.

In addition to the fact that water vapor is the source of all clouds and precipitation, it has the ability to absorb not only radiant energy emitted by the earth but also some solar energy. Therefore along with carbon dioxide, it is a controlling influence on energy transfer through the atmosphere. Furthermore, water is the only substance that can exist in all three states of matter (solid, liquid, and gas) at the temperatures and pressures that normally exist on earth. As water changes from one state to another, it absorbs or releases heat energy (termed *latent heat*). In this manner, heat absorbed at one location is transported by winds to other locales and released. The released latent heat, in turn, supplies the energy that drives many storms. Chapters 4 and 5 examine the many important roles of water in the air.

The movements of the atmosphere are sufficient to keep a large quantity of solid particles suspended within it. Although visible dust sometimes clouds the sky, these relatively large particles are too heavy to remain for very long in the air. Still, many particles are microscopic and remain suspended for considerable periods of time. They may originate from many sources, both natural and human made, and include sea salts from breaking waves, fine soil blown into the air, smoke and soot from fires, pollen and microorganisms lifted by the wind, ash and dust from volcanic eruptions, and more. Such particles are most numerous in the lower part of the atmosphere near their primary source, the earth's surface. Nevertheless, the upper atmosphere is not free of them, because some dust is carried to great heights by rising currents of air and other particles are contributed by meteoroids that disintegrate as they pass through the earth's envelope of air. From a meteorological standpoint these tiny, often invisible particles can be significant. First, many are water absorbent and consequently act as surfaces on which water vapor may condense, an important function in the formation of clouds and fog. Second, dust may absorb or reflect incoming solar radiation. Thus when dust loading of the atmosphere is great, as it may be following an explosive volcanic eruption, the amount of sunlight reaching the earth's surface can be measurably reduced. Finally, dust in the air contributes to an optical phenomenon we have all observed—the varied hues of red and orange at sunrise and sunset.

Ozone

Another important component of the atmosphere is **ozone,** the triatomic form of oxygen (O_3). Ozone is not the same as the oxygen we breathe, which has two atoms per molecule (O_2). There is little of this gas in the atmosphere (less than 0.00005 percent by volume). Put another way, if all the ozone in the atmosphere were brought down to the earth's surface, it would form a layer only about 0.4 centimeter thick. Furthermore, its distribution is not uniform. In the lowest portion of the atmosphere, it represents less than 1 part in 100 million. It is concentrated well above the surface, between 10 to 50 kilometers with a peak near the altitude of 25 kilometers. But even here, where its concentration is greatest, ozone accounts for, at most, only about 1 part in 100 thousand. It is also important to note that the concentration of ozone at any particular level within the atmosphere is not fixed but fluctuates. Such natural variations are related to latitude, season, time of day, and weather patterns. Fluctuations may also be linked to explosive volcanic activity and to changes in solar activity. The greatest variations appear to occur in the Antarctic region during the southern hemisphere spring. At this time of year, the ozone level declines sharply only to rise again. The cause of this so-called hole in the ozone layer is still under investigation.

The formation of ozone in the 10 to 50 kilometer height range is the result of a complicated series of processes that involve the absorption of solar radiation. Molecules of oxygen (O_2) are split into single atoms of oxygen after absorbing shortwave solar energy. Ozone is then formed when an atom of oxygen collides with a molecule of oxygen in the presence of a third, neutral molecule that allows the reaction to occur without being consumed in the process. Ozone is probably concentrated where it is because of a balance between two factors: (1) the availability of shortwave ultraviolet energy from the sun that is necessary to produce atomic oxygen and (2) an atmosphere dense enough to allow the required collisions between molecular oxygen and atomic oxygen.

The presence of the ozone layer in our atmosphere is of vital importance to those of us on earth. The reason lies in the capability of ozone to absorb damaging ultraviolet radiation from the sun. If ozone did not act to filter a great deal of the ultraviolet radiation and if these rays were allowed to reach the surface of the earth, our planet would likely be uninhabitable for most life as we know it. Thus anything that would act to reduce the amount of ozone in the atmosphere could affect the well-being of life on earth. Just such a concern has been raised and this is the subject of the following section.

THE OZONE PROBLEM

Since the early 1970s there has been continuing scientific and public concern about the possibility that a reduction in the amount of ozone in the atmosphere may be occurring as the result of human activities. If true, consequences could be serious. In the preface to a report Harold I. Schiff, Chairman of a National Research Council panel on stratospheric chemistry and transport, summarized the problem this way:

> The dependence of life on the earth's surface on the very small amount of ozone in the stratosphere has been known for some time. Only recently, however, have scientists realized that the quantity of stratospheric ozone is controlled by trace amounts of other substances whose concentrations are thousands of times smaller than that of ozone itself. It is this fact that has led to concern about excessive depletion of the ozone layer, since it is likely that human activities can add these trace substances to the atmosphere, appreciably altering their concentrations and thereby affecting the balance of life-supporting ozone.[1]

It is believed that the greatest human impact on the ozone layer is caused largely by a group of chemicals known as chlorofluorocarbons (CFCs for short).[2] CFCs are used as propellants for aerosol sprays, in the production of certain plastics, and in air-conditioning and refrigeration equipment (Figure 1–3). Because CFCs are practically inert (that is, not chemically active) in the lower atmosphere, a portion of these gases gradually makes its way to the ozone layer, where sunlight separates the chemicals into their constituent atoms. The chlorine atoms released in this way would, through a complicated series of reactions, have the net effect of converting some of the ozone into oxygen.

Because ozone filters out most of the ultraviolet radiation in sunlight, a decrease in its concentration would permit more of these harmful wavelengths to reach

---

[1] *Stratospheric Ozone Depletion by Halocarbons: Chemistry and Transport* (Washington, D.C.: National Academy of Sciences, 1979), p. vii. (*Note*: Most ozone is located in the stratosphere, a region of the atmosphere between 10 and 50 kilometers above the surface.)
[2] CFCs are also called chlorofluoromethanes (CFMs) or halocarbons and go by the trade name Freons.

FIGURE 1–3
Refrigerant leakage from vehicles such as these is one way that chlorofluorocarbons are released into the atmosphere. (Photo by E. J. Tarbuck)

the earth's surface. Two important questions thus arise: (1) What will be the magnitude of the ozone depletion? and (2) what will be the effects of the increased ultraviolet radiation?

A report from the National Academy of Sciences indicates that the release of CFCs could lead to an eventual depletion of between 2 and 4 percent by late in the next century.[3] Such figures, however, should be viewed cautiously because estimates could change as refinements occur in atmospheric models and measurements. Over the 5-year period 1979–1984, for instance, the projected risk to atmospheric ozone was reduced with successive reports from the National Academy of Sciences. A 1979 report indicated that the reduction in stratospheric ozone might approach or possibly exceed 16.5 percent, whereas a 1982 report suggested an ozone depletion of between 5 and 9 percent. Moreover, the value for ozone depletion projected by such reports could be altered by other human-generated materials. It is possible, for example, that the release of nitrous oxide (from combustion, nitrogen fertilizers, and aircraft exhausts) could increase ozone depletion because nitrous oxide, like CFCs, acts as a catalyst in reactions that destroy ozone. On the other hand, continued growth in atmospheric carbon dioxide from the burning of fossil fuels could partially offset ozone depletion. Scientists predict that additional carbon dioxide in the lowest portion of the atmosphere may cause the temperatures in the stratosphere to drop and thereby slow the rates at which certain chemical reactions destroy ozone. When we consider the fact that ozone

---

[3] *Causes and Effects of Changes in Stratospheric Ozone: Update 1983* (Washington, D.C.: National Academy Press, 1984).

concentrations fluctuate naturally, it becomes clear that the magnitude of the ozone depletion problem is very difficult to assess. Natural and human-induced variations are not easily distinguished. Nevertheless, although uncertainties abound in the search for trends, most indicators of ozone abundance appear to be heading down.

Each 1-percent decrease in the concentration of stratospheric ozone increases the amount of ultraviolet radiation that reaches the earth's surface by about 2 percent. Therefore because ultraviolet radiation is known to induce certain types of skin cancer, ozone depletion could seriously affect human health. More specifically, a National Academy of Sciences report states:

> We estimate that there will be a 2 percent to 5 percent increase in basal cell cancer incidence per 1 percent decrease in stratospheric ozone. The increase in squamous cell skin cancer incidence will be double that.[4]

The fact that up to a half million cases of these cancers occur in the United States each year means that ozone depletion could ultimately lead to many thousands of additional cases each year. The effects of additional ultraviolet radiation on animal and plant life may be as important as the direct effects on human health. Still, too little experimental data yet exist for scientists to be able to make specific predictions about possible effects on particular crops or ecosystems. However, analyses by the U.S. Environmental Protection Agency indicate that increased numbers of aquatic organisms would die and that crop yields and quality could be adversely affected in many cases.

What can be done to protect the atmospheric ozone layer? The obvious answer is to stop or greatly reduce the production of CFCs. Is this being done? In December 1978 the United States government banned the manufacture of all nonessential aerosol products using CFC propellants. Although a positive step, this action did not solve the problem because domestic production for other uses, as well as production outside the United States, continued. Realizing that the risks of not curbing CFC emissions were difficult to ignore, 28 countries signed the Vienna Convention for the Protection of the Ozone Layer in 1985. The convention was established under the auspices of the United Nations Environmental Program as a mechanism for framing an international strategy for protecting the ozone layer. In September 1987, following a year of negotiations, 23 countries endorsed a plan to reduce emissions of CFCs. The proposed treaty calls for cutting world consumption of CFCs in half by 1999. Under the agreement, most industrialized countries will have to steadily reduce their usage of a number of CFC products. Some scientists do not believe that this action is adequate. However, treaty now represents the first positive international action on the ozone problem.

The series of National Academy of Sciences reports are certainly not the last word on CFCs and ozone depletion. One observer summarizes the situation this way:

---

[4] *Causes and Effects of Stratospheric Ozone Reduction: An Update* (Washington, D.C.: National Academy Press, 1982), p. 5.

As more sophisticated models are developed and implemented, it should be possible to obtain a much better picture of potential effects. . . . The debate on CFCs . . . has the potential to be carried into the next decade and potentially into the next century.[5]

The following list summarizes the significant aspects of the ozone problem:

1. A portion of the CFCs from aerosol sprays, certain plastics, and refrigeration equipment are believed to eventually reach the ozone layer, where chlorine atoms convert some ozone to oxygen.
2. Reduced ozone concentrations permit more harmful ultraviolet radiation to reach the earth's surface.
3. If the production and release of CFCs continue at projected rates, some estimates indicate an eventual depletion of between 2 and 4 percent by late in the twenty-first century. These numbers, however, are far from firm and should be viewed cautiously.
4. If stratospheric ozone is reduced, there will be a rise in the number of cases of certain types of skin cancer. Crops and ecosystems may also be adversely affected.

ORIGIN OF THE ATMOSPHERE

The earth's atmosphere is unlike that of any other body in the solar system. No other planet is as hospitable or exhibits the same life-sustaining mixture of gases as the earth (Figure 1–4). The atmosphere did not always consist of the same relatively stable mixture of gases that we breathe today. On the contrary, the present mixture of gases that makes up our atmosphere is the result of very gradual change, a slow evolutionary process that began soon after the earth came into being 4.5 to 5 billion years ago.

Degassing

Scientists believe that the earth's earliest atmosphere was swept away by solar winds, vast streams of particles emitted by the sun. As the earth cooled, a solid crust formed and the gases that had been dissolved in the molten rock were gradually released, a process called **degassing.** Thus an atmosphere believed to be made up of gases similar to those released during volcanic eruptions came into being. The principal components of this "new" atmosphere were probably water vapor, carbon dioxide, and nitrogen.

As the earth continued to cool, clouds formed and great rains commenced. At first, the water evaporated before reaching the surface or was quickly boiled away. This step helped to speed the cooling of the earth's surface. When the earth had cooled sufficiently, the torrential rains continued, filling the ocean basins. This event not only diminished the amount of water vapor in the air, but it carried away much carbon dioxide as well.

---

[5] Thomas H. Maugh II, "What Is the Risk from Chlorofluorocarbons?" *Science*, 223, no. 4640 (1984), 1052.

FIGURE 1—4
Like the other planets in our solar system, the atmosphere of Mars is very different from that of Earth. The Martian atmosphere is only 1 percent as dense as that of Earth and is composed primarily of carbon dioxide with very small amounts of water vapor. Although the atmosphere of Mars is very thin, extensive dust storms do occur, sometimes creating dunes that resemble those on Earth. (Courtesy of NASA)

We are now faced with an interesting paradox. If the earth's primitive atmosphere resulted from volcanic degassing, it could not have contained free oxygen because free oxygen is not emitted during this process. How then did our present oxygen-rich atmosphere come into existence? Scientists have proposed two probable sources of the free oxygen in the atmosphere. It is known that water vapor that is carried into the upper atmosphere is dissociated into hydrogen and oxygen by the action of the sun's ultraviolet radiation. Hydrogen, being a very light gas, escapes the atmosphere, whereas the heavier oxygen atoms remain and combine to form molecular oxygen (O_2). Although there is little doubt that some of the atmosphere's free oxygen was created in this manner, this very slow process is not adequate to account for the present percentage of oxygen in our atmosphere.

Photosynthesis

A second more important source of oxygen is believed to have been (and, indeed, continues to be) green plants. By the process of **photosynthesis** plant life uses sunlight to generate oxygen by changing water and atmospheric carbon dioxide into organic matter. This method of oxygen production obviously implies the need for life on earth prior to the time that free oxygen was present in the atmosphere. Scientists believe that the first life forms, probably bacteria, carried out their metabolic processes without oxygen. Even today many of these anaerobic bacteria still exist. Next, primitive green plants evolved, which, in turn, supplied most of the free oxygen to support higher forms of life. Slowly the amount of oxygen in the atmosphere increased. Evidence from the geologic record of this ancient time suggests that the first free oxygen combined with substances dissolved in water, especially iron. Then, once these mineral oxidation needs were met, substantial quantities of free oxygen began to accumulate in the atmosphere. By the beginning of the Paleozoic Era, some 570 million years ago, the fossil record reveals that organisms that require oxygen were abundant in the sea. Two hundred million years later, during the Devonian Period, land plants became widespread. Thus the makeup of the atmosphere is directly related to the life-forms on earth, and its composition evolved through time from an oxygen-free environment to one that contained significant amounts of free oxygen.

In summary, the basic steps in the evolution of our atmosphere are as follows:

1. Because of strong solar winds, the earth's first atmosphere was swept into space.
2. A new atmosphere was created by the gradual release of gases that had been dissolved in molten rock. This atmosphere, however, contained no free oxygen.
3. As the earth cooled, torrential rains filled the ocean basins, thus reducing the amount of water vapor in the air as well as "washing out" much of the carbon dioxide.
4. The free oxygen of the atmosphere was slowly added primarily by green plants carrying on the process of photosynthesis.

EXPLORING THE ATMOSPHERE

For more than 200 years people have attempted to explore that part of the atmosphere that seemed out of reach. In 1752 Benjamin Franklin, using a kite, made his famous discovery that lightning is an electrical charge. Not many years later kites were being used to observe temperatures above the earth's surface. In the late eighteenth century manned balloons were used in an attempt to investigate the composition and properties of the "upper" atmosphere. Although several manned ascents were attempted over the years, they were dangerous undertakings and seldom exceeded heights of 5 to 8 kilometers. Unmanned balloons, on the other hand, could rise to higher altitudes, but there was no assurance that the instruments carried aloft would be recovered. Notwithstanding the difficulties and dangers, considerable data were gathered on the nature of the air above.

Today balloons continue to play a significant role in the systematic investigation of the atmosphere. Giant balloons similar to those pictured in Figure 1–5(a) and (b), for example, are launched regularly, primarily for research purposes. Such balloons can stay aloft for extended periods and represent an important means of carrying monitoring instruments into the region of the atmosphere known as the stratosphere (see Figure 1–13). Since the late 1920s balloons have carried aloft **radiosondes.** These lightweight packages of instruments are fitted with radio transmitters that send back data on temperature, pressure, and relative humidity in the lower portions of the atmosphere (Figure 1–5c). When the radiosonde is tracked by radio-location devices in order to obtain information about winds at various levels, the system is known as **rawinsonde.** Radiosondes are sent aloft twice each day from an extensive network of stations worldwide. The data that they supply are essential for making accurate weather forecasts.

Radiosonde

Rawinsonde

FIGURE 1–5

Exploring the atmosphere using balloons. (a) Giant balloons, such as this pair which are being filled with helium prior to launch, carry instrument packages high into the atmosphere. (b) A giant balloon lifting off from the National Scientific Balloon Facility in Palestine, Texas. Although the balloon looks baggy, the helium will expand at higher altitudes where air pressure is lower. (c) A lightweight package of instruments, the radiosonde, is carried aloft by a small weather balloon. (a and b courtesy of Robert S. Nock, NASA; c courtesy of NOAA)

(a)

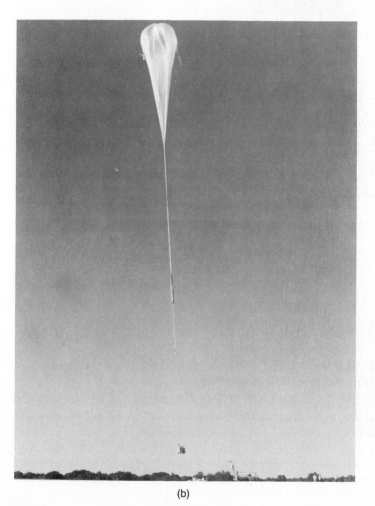

(b)

(c)

FIGURE 1–6
Airborne instruments developed in recent years have been crucial to our understanding of cloud processes. (Courtesy of National Center for Atmospheric Research)

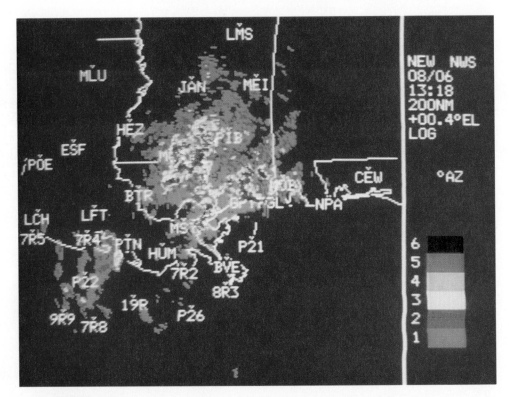

FIGURE 1–7
Color weather radar display. Colors indicate different intensities of precipitation. Level 1 (light green) represents less than 0.1 inch per hour; level 6 (dark red), 5 inches or more per hour. (Photo courtesy of Alden Electronics)

Other important means for exploring the atmosphere include rockets and airplanes. Immediately after World War II the United States acquired rockets, vehicles that revolutionized the study of the upper atmosphere. Prior to this time knowledge of the atmosphere beyond about 30 kilometers came almost exclusively from indirect, ground-based measurements. Airplanes also play a significant role in atmospheric studies. Some high-flying aircraft are capable of reaching portions of the stratosphere, whereas others are designed to measure such relatively small-scale, yet complex phenomena as cloud systems (Figure 1–6).

Among the methods of studying the atmosphere that are best known to the general public and most useful to atmospheric scientists are weather radar and satellites. Today when we watch a television weather report, it is common to see satellite images that show us moving cloud patterns and radar displays that depict the intensity and regional extent of precipitation (Figure 1–7). Recent technological advances greatly enhance the value of weather radar for the purpose of storm detection, warning, and research. Moreover, since 1960 when TIROS I, the first meteorological satellite, was launched, satellites have given us a perspective of the atmosphere that was previously unobtainable and that has proven invaluable (Figure 1–8). Meteorological satellites, for example, provide images that allow us to study the distribution of clouds and the circulation patterns that they reveal

FIGURE 1–8
The Geostationary Operational Environment Satellite (GOES). From a position almost 36,000 kilometers in space over the equator, GOES can monitor a substantial portion of the western hemisphere with good resolution, and obtain and transmit data from any point on the earth within its view. (Courtesy of NOAA)

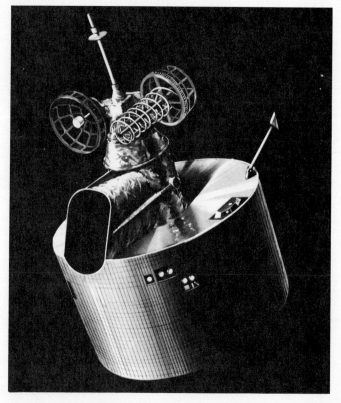

FIGURE 1–9
This satellite image of North America reveals the distribution of clouds associated with a mature storm system. When photographed in visible light, clouds appear bright because they reflect more light than the surface below. (Courtesy of NOAA)

(Figure 1–9). Moreover, they let us see the structure and determine the speed of weather systems over the oceans and other regions where observations are scanty or nonexistent. One obvious benefit is that storms can be detected early and tracked with great precision.

The vessel shown in Figure 1–10 is a ship that is capable of drilling into the floor of the deep ocean basins. Such a ship represents an important, although indirect, means of exploring the atmosphere. As we shall see, the earth's climate is variable and changes on different time scales. Since instrumental records go

FIGURE 1–10
The *JOIDES Resolution*, the drilling ship of the Ocean Drilling Program. The seafloor sediments recovered by this and other research vessels help scientists reconstruct past climates. (Courtesy of the Ocean Drilling Program)

back only a couple of hundred years (at best), scientists must decipher and reconstruct past climates by using indirect evidence. Seafloor sediment is one source of such information because it contains the remains of organisms that once lived near the surface of the ocean. Since the numbers and types of these organisms change when atmospheric conditions fluctuate, seafloor sediments represent an important source of data about our atmosphere and how it has changed over long time spans. Other indirect methods of exploring past atmospheric variations include the study of the growth rings of trees and the chemistry of ice cores taken from glaciers. It is thus possible not only to study and explore the earth's present atmosphere but through indirect means to sample climates of the past as well.

VERTICAL STRUCTURE OF THE ATMOSPHERE

To say that the atmosphere begins at the earth's land-sea surface and extends upward is rather obvious. But where does the atmosphere end and outer space begin? In order to understand the vertical extent of the atmosphere, let us examine the changes in atmospheric pressure with height. The pressure of the atmosphere

TABLE 1–2 Percent of Sea-Level Pressure Encountered at Selected Altitudes

Altitude (km)	Percent of Sea-Level Pressure
0	100
5.6	50
16.2	10
31.2	1
48.1	0.1
65.1	0.01
79.2	0.001
100	0.00003

is simply the weight of the air above. At sea level the average pressure is slightly more than 1000 millibars, a pressure that corresponds to a weight of 1 kilogram of air above every square centimeter of earth. Obviously the pressure at higher altitudes is less. This fact was borne out in a dramatic way by some nineteenth-century balloonists attempting to study the upper atmosphere. Several were known to have passed out on reaching heights in excess of 6 kilometers, and some even perished in the rarefied upper air.

By examining Table 1–2, you can see that one-half of the atmosphere lies below an altitude of 5.6 kilometers. At about 16 kilometers 90 percent of the atmosphere has been traversed, and above 100 kilometers only 0.00003 percent of all the gases making up the atmosphere remains. At this latter altitude the atmosphere is so tenuous that the density of air is less than could be found in the most perfect artificial vacuum at the earth's surface. Nevertheless, the atmosphere continues to even greater heights. The truly rarefied nature of the outer atmosphere is described very well by Richard Craig:

> The earth's outermost atmosphere, the part above a few hundred kilometers, is a region of extremely low density. Near sea level, the number of atoms and molecules in a cubic centimeter of air is about 2×10^{19}; near 600 km it is only about 2×10^7, which is the sea-level value divided by a million million. At sea level, an atom or molecule can be expected, on the average, to move about 7×10^{-6} cm before colliding with another particle; at the 600-km level this distance, called the "mean free path," is about 10 km. Near sea level, an atom or molecule, on the average, undergoes about 7×10^9 such collisions each second; near 600 km, this number is reduced to about 1 each minute.[6]

A graphic portrayal of pressure data (Figure 1–11) shows that the rate of pressure decrease is not constant. Rather, pressure decreases at a decreasing rate with an increase in altitude until, beyond an altitude of about 50 kilometers, the decrease is slight. Put another way, data illustrate that air is highly compressible—

---

[6] *The Edge of Space: Exploring the Upper Atmosphere* (New York, N.Y.: Doubleday & Company, Inc., 1968), p. 130.

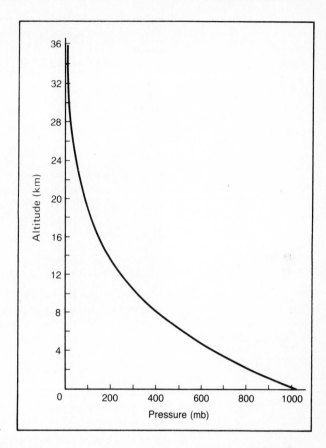

FIGURE 1–11
Pressure variations with altitude.

that is, it expands with decreasing pressure and becomes compressed with increasing pressure. Consquently, traces of our atmosphere extend for thousands of kilometers beyond the earth's surface. Thus to say where the atmosphere ends and outer space begins is arbitrary and, to a large extent, depends on what phenomena one is studying. It is apparent that there is no sharp boundary.

In summary, data on vertical pressure changes reveal that the vast bulk of the gases making up the atmosphere is very near the earth's surface and that the gases gradually merge with the emptiness of space. When compared with the size of the earth, with its radius of about 6400 kilometers, the envelope of air surrounding our planet is indeed very shallow.

Temperature Changes in the Vertical

By the early part of the twentieth century much had been learned about the lower atmosphere, and indirect methods had provided some knowledge of the upper air. Data from balloons and kites had revealed that the air temperature dropped with increasing height above the earth's surface. This fact is familiar to anyone who has climbed a high mountain or who has seen pictures of snow-capped mountaintops rising above snow-free lowlands (Figure 1–12).

FIGURE 1–12
Temperatures drop with an increase in altitude in the troposphere. Therefore it is possible to have snow on a mountaintop and warmer, snow-free lowlands below. (Photo by E. J. Tarbuck)

Although measurements had not been taken above a height of about 10 kilometers, scientists believed that the temperature continued to decline with height to a value of absolute zero ($-273°C$) at the outer edge of the atmosphere. In 1902, however, the French scientist, Leon Philippe Teisserenc de Bort, refuted the notion that temperature decreases continuously with an increase in altitude. In studying the results of more than 200 balloon launchings, Teisserenc de Bort found that the temperature stopped decreasing and leveled off at an altitude between 8 and 12 kilometers. This surprising discovery was at first doubted, but subsequent data gathering confirmed his findings. Later, through the use of radiosondes and rocket-sounding techniques, the temperature structure of the atmosphere up to great heights became clear.

Today the atmosphere is divided vertically into four layers on the basis of temperature (Figure 1–13). The bottom layer, where temperature decreases with an increase in altitude, is known as the **troposphere.** The term was coined in 1908 by Teisserenc de Bort and literally means the region where air "turns over," a reference to the appreciable vertical mixing of air in this lowermost zone. The temperature decrease in the troposphere is called the **environmental lapse rate.** Although its average value is 6.5°C per kilometer, a figure known as the *normal lapse rate*, its value is quite variable. In fact, sometimes shallow layers where temperatures actually increase with height are observed in the troposphere. When such a reversal occurs, a **temperature inversion** is said to exist.

The temperature decrease continues to an average height of approximately

Troposphere

Environmental Lapse Rate

Temperature Inversion

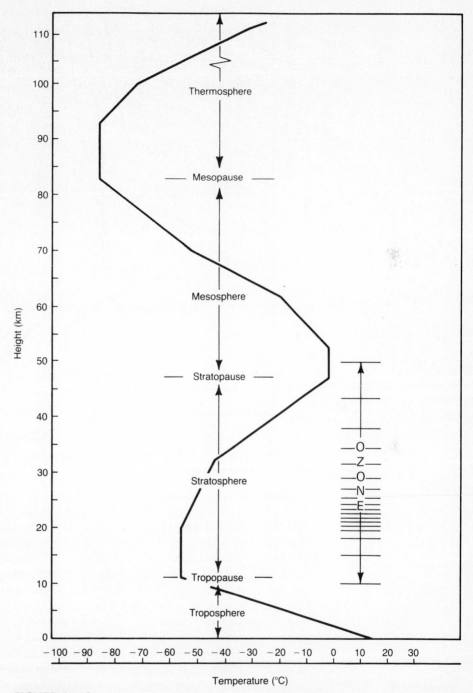

FIGURE 1–13
Thermal structure of the atmosphere to a height of about 110 kilometers.

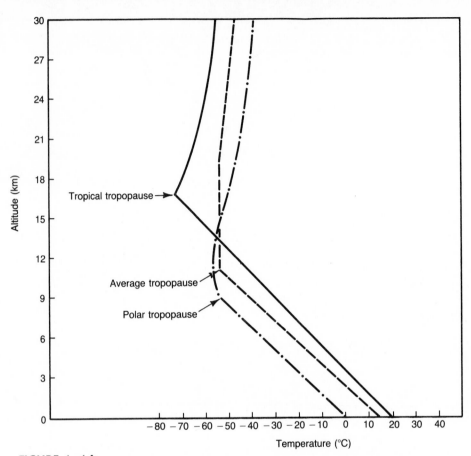

FIGURE 1–14
Differences in the height of the tropopause. (After *Weather for Aircrews*, AFM 51–12)

12 kilometers. Yet the thickness of the troposphere is not everywhere the same. It reaches heights in excess of 16 kilometers in the tropics, but in polar regions it is more subdued, extending to 9 kilometers or less (Figure 1–14). Warm temperatures and highly developed thermal mixing are responsible for the greater vertical extent of the troposphere near the equator. As a result, the lapse rate extends to great heights; and in spite of relatively high temperatures below, the lowest tropospheric temperatures are found aloft in the tropics and not at the poles.

The troposphere is the chief focus of meteorologists, for it is in this layer that essentially all phenomena that we collectively refer to as weather occur. Almost all clouds and certainly all precipitation, as well as all our violent storms, are born in this lowermost layer of the atmosphere. There should be little wonder why the troposphere is often called the "weather sphere."

Stratosphere

Tropopause

Beyond the troposphere lies the **stratosphere;** the boundary between the troposphere and the stratosphere is known as the **tropopause.** Below the tropopause atmospheric properties are readily transferred by large-scale turbulence and mixing, but above it, in the stratosphere, they are not. In the stratosphere

Stratopause

the temperature at first remains nearly constant to a height of about 20 kilometers before it begins a rather sharp increase that continues until the **stratopause** is encountered at a height of about 50 kilometers above the earth's surface. Higher temperatures occur in the stratosphere because it is in this layer that the atmosphere's ozone is concentrated. Recall that ozone absorbs ultraviolet radiation from the sun. Consequently, the stratosphere is heated. Although the maximum ozone concentration exists between 15 and 30 kilometers, the smaller amounts of ozone above this height range absorb enough ultraviolet energy to cause the higher observed temperatures (see Figure 1–13).

Mesosphere

Mesopause

Thermosphere

In the **mesosphere** temperatures again decrease with height until at the **mesopause,** some 80 kilometers above the surface, the temperature approaches $-90°C$. Extending upward from the mesopause and having no well-defined upper limit is the **thermosphere,** a layer that accounts for but a minute fraction of the atmosphere's mass. Temperatures again increase in the extremely rarefied air of this outermost layer as a result of the absorption of very short wavelength solar energy by atoms of oxygen and nitrogen. Although temperatures rise to extremely high values of more than $1000°C$, such temperatures are not strictly comparable with those experienced near the earth's surface. Temperature is defined in terms of the average speed at which molecules are moving. Because the gases of the thermosphere are moving at very high speeds, the temperature is obviously very high. The gases are so sparse, however, that very few of these fast-moving air molecules would collide with a foreign body; therefore only an insignificant quantity of energy would be transferred. Thus the temperature of a satellite orbiting the earth in the thermosphere is determined chiefly by the amount of solar radiation it absorbs and not by the temperature of the surrounding air; if an astronaut inside were to expose his or her hand, it would not feel "hot" outside the space capsule.

Vertical Variations in Composition

In addition to the layers defined by varying temperatures in the vertical, other layers or zones are also recognized in the atmosphere. Based on composition, the atmosphere is often divided into two layers, the homosphere and the heterosphere. From the earth's surface to an altitude of about 80 kilometers, the makeup of the air is uniform in terms of the proportions of its component gases. That is, the composition is the same as that shown earlier in Table 1–1. This lower uniform layer is termed the **homosphere;** the zone of homogeneous composition. In contrast, the rather tenuous atmosphere above 80 kilometers is not uniform. Because it has a heterogeneous composition, the term **heterosphere** is used. Here the gases are arranged into four roughly spherical shells, each with a distinctive composition. The lowermost layer is dominated by molecular nitrogen (N_2); next, a layer of atomic oxygen (O) is encountered, followed by a layer dominated by helium (He) atoms, and finally a region consisting of hydrogen (H) atoms. The stratified nature of the gases making up the heterosphere varies according to their weights. Molecular nitrogen is the heaviest and so it is lowest. The lightest gas, hydrogen, is outermost.

Homosphere

Heterosphere

The Ionosphere

Ionosphere

Located in the altitude range between 80 to 400 kilometers, and thus coinciding with the lower portions of the thermosphere and heterosphere, is an electrically charged layer known as the **ionosphere.** Here molecules of nitrogen and atoms of oxygen are readily ionized as they absorb high-energy, short-wavelength solar energy. In this process each affected molecule or atom loses one or more electrons and becomes a positively charged ion, and the electrons are set free to travel as electric currents. Although ionization occurs at heights as great as 1000 kilometers and extends as low as perhaps 50 kilometers, positively charged ions and negative electrons are most dense in the range 80 to 400 kilometers. The concentration of ions is not great below this zone because much of the short-wavelength radiation needed for ionization has already been depleted. In addition, the atmospheric density at this level results in a large percentage of free electrons being swiftly captured by positively charged ions. Beyond the 400-kilometer upward limit of the ionosphere the concentration of ions is low because of the extremely low density of the air. Because so few molecules and atoms are present, relatively few ions and free electrons can be produced.

The electrical structure of the ionosphere is not uniform. It consists of three layers of varying ion density. From bottom to top, these layers are called the D, E, and F layers, respectively. Because the production of ions requires direct solar radiation, the concentration of charged particles changes from day to night, particularly in the D and E zones. That is, these layers weaken and disappear at night and reappear during the day. The uppermost or F layer, on the other hand, is present both day and night. The density of the atmosphere in this layer is very low, and positive ions and electrons do not meet and recombine as rapidly as they do at lesser heights, where density is higher. Consequently, the concentration of ions and electrons in the F layer does not change rapidly, and the layer, although weak, remains through the night.

As best we can tell, the ionosphere has little impact on our daily weather. The only contact that most people are likely to have with the ionosphere is through its effect on AM radio reception. We have all turned on an AM radio at night and, to our surprise, received a broadcast from a station many hundreds of kilometers away. Moreover, if we leave the radio dial set at the exact spot and try to receive a broadcast from the same distant station the next morning, its signal has disappeared. What has happened?

Recall that the F layer is present both day and night. This is the layer that is largely responsible for reflecting radio signals back to earth. Also remember that the lower layers of the ionosphere exist only during the daytime. In this lower portion, in particular the D layer, radio waves tend to be absorbed rather than reflected. We now have the necessary information to solve the puzzle of nighttime reception of distant AM radio stations. Although the F layer reflects radio waves both day and night, during the sunlit hours most signals are absorbed in the D layer. Those that do penetrate the absorbing layer and are reflected are almost sure to be absorbed as they enter the D layer on their downward journey (Figure 1–15). At night, however, the absorbing D layer all but disappears. As a result, radio waves can reach the F layer easily and be reflected back toward the earth's

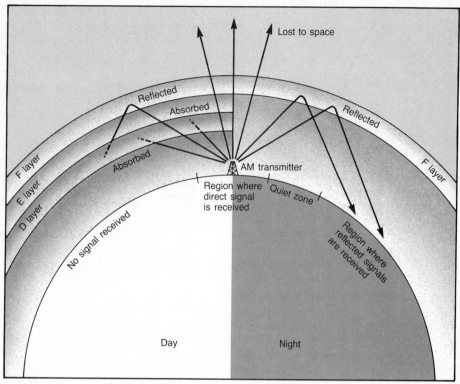

FIGURE 1–15
During the daylight hours the D layer absorbs AM radio waves. At night the D and E layers disappear, allowing the F layer to reflect radio signals back to the earth's surface.

surface. The only waves that are not reflected are those that strike the F layer nearly perpendicularly. These waves will pass through the layer and be lost to space. Consequently, there is usually a "quiet" zone on the ground that is too far from the transmitter to receive direct signals and too close to receive reflected radio waves (Figure 1–15).

Aurora

The auroras, certainly one of nature's most interesting spectacles, take place in the ionosphere (Figure 1–16). The **aurora borealis** (northern lights) and its southern hemisphere counterpart, the **aurora australis** (southern lights), appear in a wide variety of forms. Sometimes the displays consist of vertical streamers in which there can be considerable movement. At other times the auroras appear as a series of luminous expanding arcs or as a quiet glow that has an almost foglike quality.

The occurrence of auroral displays is closely correlated in time with solar flare activity and, in geographic location, with the earth's magnetic poles. Solar flares are massive magnetic storms on the sun that emit enormous amounts of energy and great quantities of fast-moving atomic particles. As the clouds of protons and electrons from the solar storm approach the earth, they are captured by the earth's magnetic field, which, in turn, guides them toward the magnetic poles.

FIGURE 1–16
Aurora borealis (northern lights). (Courtesy of the
National Center for Atmospheric Research/National
Science Foundation. Photo by D. Baumhefner)

Then as the ions impinge on the ionosphere, they energize the atoms of oxygen and molecules of nitrogen and cause them to emit light—the glow of the auroras. Because the occurrence of solar flares is closely correlated with sunspot activity, auroral displays increase conspicuously at times when sunspots are most numerous.

REVIEW

1. List and briefly define the four "spheres" that constitute our environment.

2. Define the terms "weather" and "climate."

3. The following statements refer to either weather or climate. On the basis of the definitions you wrote for question 2, determine which statements refer to weather and which refer to climate. (*Note*: One statement includes aspects of both weather and climate.)
   a. The baseball game was rained out today.
   b. January is Peoria's coldest month.
   c. North Africa is a desert.
   d. The high this afternoon was 25°C.
   e. Last evening a tornado ripped through Canton.
   f. I am moving to southern Arizona because it is warm and sunny.
   g. The highest temperature ever recorded at this station is 43°C.
   h. Thursday's low of −20°C is the coldest temperature ever recorded for that city.
   i. It is partly cloudy.

4. What are the basic elements of weather and climate?

5. What are the major components of clean, dry air?

6. What is responsible for the increasing carbon dioxide content of the air? What is one possible effect of increased carbon dioxide in the atmosphere?

7. Why are water vapor and dust important constituents of our atmosphere?

8. **a.** Why is ozone important to life on earth?
   b. What are CFCs and what is their connection to the ozone problem?
   c. If the ozone layer is reduced, what will be the most important impact on human health?

9. Outline the stages in the formation of our atmosphere.

10. What is a radiosonde? What is a rawinsonde?

11. Recall that average sea-level pressure is about 1000 millibars, which is equivalent to a weight of 1 kilogram per square centimeter. What are the pressure (in millibars) and weight of the atmosphere (in kilograms per square centimeter) above the following heights: 5.6 kilometers, 16 kilometers, 100 kilometers?

12. With an increase in altitude, pressure decreases at a(n) (constant, increasing, decreasing) rate. Underline the correct answer.

13. The atmosphere is divided vertically into four layers on the basis of temperature. List the names of these layers and their boundaries in order (from lowest to highest) and list as many characteristics of each as you can.

14. If the temperature at sea level were 23°C, what would air temperature be at a height of 2 kilometers under average conditions?

15. **a.** On a spring day a middle-latitude city (about 40°N latitude) has a surface (sea-level) temperature of 10°C. If vertical soundings reveal a nearly constant environmental lapse rate of 6.5°C per kilometer and a temperature at the tropopause of −55°C, what is the height of the tropopause?
   b. On the same spring day a station near the equator has a surface temperature of 25°C, 15°C higher than the middle-latitude city mentioned above. Vertical soundings reveal an environmental lapse rate of 6.5°C per kilometer and indicate that the tropopause is encountered at 16 kilometers. What is the air temperature at the tropopause?

16. Why does the temperature increase in the stratosphere?

17. Why are temperatures in the thermosphere not strictly comparable to those experienced near the earth's surface?

18. Distinguish between homosphere and heterosphere.

19. What is the primary significance of the ionosphere?

20. What is the primary cause of auroral displays?

VOCABULARY REVIEW

Review your understanding of important terms in this chapter by defining and explaining the importance of each term listed here. Terms are listed in order of occurrence in the chapter.

lithosphere	climate
hydrosphere	elements (of weather and climate)
atmosphere	air
biosphere	ozone
weather	degassing *(continued)*

photosynthesis
radiosonde
rawinsonde
troposphere
environmental lapse rate
temperature inversion
stratosphere
tropopause
stratopause

mesosphere
mesopause
thermosphere
homosphere
heterosphere
ionosphere
aurora borealis
aurora australis

2

Solar Radiation

From our experiences we know that the sun's rays feel hotter on a clear day than on an overcast day. After taking a barefoot walk on a sunny day, we realize that city pavement gets much hotter than does a grassy boulevard. A picture of a snow-capped mountain leads us to conclude that temperature decreases with altitude. And we know that the fury of winter is always replaced by the newness of spring. You may not know that these occurrences are manifestations of the same phenomenon that causes the blue color of the sky and the red color of a brilliant sunset, however. All such common occurrences are a result of the interaction of solar energy with the earth's atmosphere and its land–sea surface. That is the essence of this chapter.

EARTH–SUN RELATIONSHIPS

The earth intercepts only a minute percentage of the energy given off by the sun—less than one two-billionth. It may seem a rather insignificant amount until we realize that it is several hundred thousand times the electrical generating capacity of the United States. Solar radiation, in fact, represents more than 99.9 percent of the energy that heats the earth. Solar energy is not distributed equally over the earth's land–sea surface, however. It is this unequal heating that drives the oceans and creates the winds, which, in turn, transport heat from the tropics to the poles in an unending attempt to reach an energy balance. If the sun were to be "turned off," the global winds would quickly subside. Yet as long as the sun shines, the winds will blow and the phenomena we know as weather will persist. So in order to have a basic understanding of atmospheric processes, we must understand what causes the temporal and spatial variations in the amount of solar energy reaching the earth. As we shall see, these variations are caused by the motions of the earth relative to the sun and by variations in the earth's surface.

Motions of the Earth

Rotation

The earth has two principal motions—rotation and revolution. **Rotation** is the spinning of the earth about its axis, which is an imaginary line running through the poles. As you know, our planet rotates once every 24 hours, producing the daily cycle of daylight and darkness. In the following chapter we will examine the effects that the daily variation in solar heating has on the atmosphere.

Revolution

The other motion of the earth, **revolution,** refers to the movement of the earth in its orbit around the sun. Hundreds of years ago most people believed that the earth was stationary in space. The reasoning was that if the earth was moving, people would feel the movement of the wind rushing past them. Today we know that the earth is traveling at nearly 113,000 kilometers per hour in an elliptical orbit about the sun. Why don't we feel the air rushing past us? The answer is that the atmosphere, bound by gravity to the earth, is carried along at the same speed as the earth.

The distance between the earth and sun averages about 150 million kilometers. Because the earth's orbit is not perfectly circular, however, the distance varies during the course of a year. Each year, on about January 3, our planet is 147 million kilometers from the sun, closer than at any other time. This position is called **perihelion.** About 6 months later, on July 4, the earth is 152 million kilometers from the sun, farther away than at any other time. This position is called **aphelion.** The variations in the amount of solar radiation received by the earth as the result of its slightly elliptical orbit, however, are slight and play a minor role in producing major seasonal temperature variations. By way of illustration, consider that the earth is closest to the sun during the northern hemisphere winter.

Perihelion

Aphelion

The Seasons

We know that it is colder in winter than in summer, but if variations in solar distance do not cause the seasonal temperature changes, what does? All of us have made adjustments for the continuous change in the length of daylight that occurs throughout the year by planning our outdoor activities accordingly. The gradual but significant change in length of daylight certainly accounts for some of the difference we notice between summer and winter. Furthermore, a gradual change in the **altitude** (angle above the horizon) of the noon sun during the course of a year's time is evident to most people. At midsummer the sun is seen high above the horizon as it makes its daily journey across the sky. But as summer gives way to autumn, the noon sun appears lower in the sky and sunset occurs earlier each evening.

Altitude (of the sun)

The seasonal variation in the altitude of the sun affects the amount of energy received at the earth's surface in two ways. First, when the sun is directly overhead (at a 90-degree angle), the solar rays are most concentrated. The lower the angle, the more spread out and less intense is the solar radiation that reaches the surface. This idea is illustrated in Figure 2–1. You have probably experienced this situation when using a flashlight. If the beam is directed straight at an object, a small intense spot is produced. When the flashlight beam strikes the object at an oblique angle, however, the area illuminated is larger and dimmer. Second, and of lesser importance, the angle of the sun determines the amount of atmosphere that the rays must traverse (Figure 2–2). When the sun is directly overhead, the rays pass through a thickness of only 1 atmosphere, whereas rays entering at a 30-degree angle travel through twice this amount and 5-degree rays travel through a thickness roughly equal to 11 atmospheres (Table 2–1). The longer the path, the greater is the chance for absorption, reflection, and scattering by the atmosphere, which

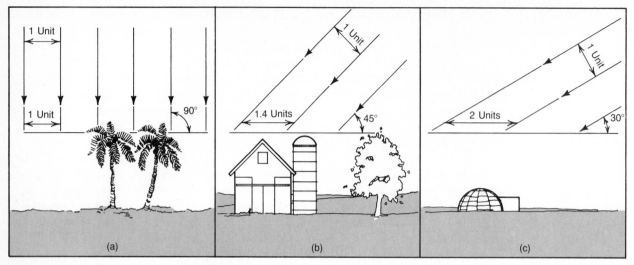

FIGURE 2–1
Changes in the sun angle cause variations in the amount of solar energy reaching the earth's surface. The higher the angle, the more intense the solar radiation.

FIGURE 2–2
Rays striking the earth at a low angle must traverse more of the atmosphere than rays striking at a high angle and thus are subject to greater depletion by reflection and absorption.

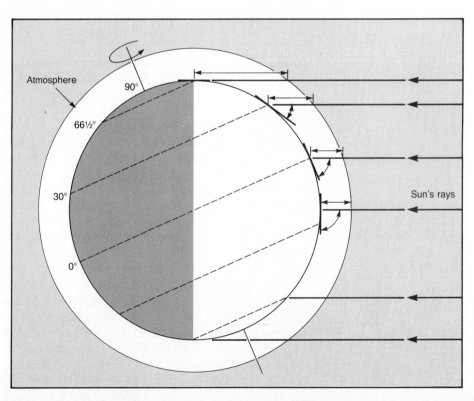

TABLE 2–1 Distance Radiation Must Travel
Through Atmosphere

Altitude of Sun	Number of Atmospheres
90°	1.00
80°	1.02
70°	1.06
60°	1.15
50°	1.31
40°	1.56
30°	2.00
20°	2.92
10°	5.70
5°	10.80
0°	45.00

reduces the intensity at the surface. These same effects account for the fact that the midday sun can be literally blinding and that the setting sun can be a sight to behold.

It is important to remember that the earth has a spherical shape. Hence on any given day only places located by a particular latitude will receive vertical (90-degree) rays from the sun. As we move either north or south of this location, the sun's rays strike at an ever-decreasing angle. Thus the nearer a place is situated to the latitude receiving the vertical rays of the sun, the higher will be its noon sun.

In summary, the most important reasons for the variation in the amount of solar energy reaching the earth are the seasonal changes in the angle at which the sun's rays strike the surface and in the length of day.

What causes the yearly fluctuations in the sun angle and length of day? They occur because the earth's orientation to the sun continually changes. The earth's axis is not perpendicular to the plane of its orbit around the sun; instead it is tilted 23½ degrees from the perpendicular. This is termed the **inclination of the axis** and, as we shall see, if the axis were not inclined, we would have no seasonal changes. In addition, because the axis remains pointed to the same direction (toward the North Star) as the earth journeys around the sun, the orientation of the earth's axis to the sun's rays is always changing (Figure 2–3). On one day each year the axis is such that the northern hemisphere is "leaning" 23½ degrees toward the sun. Six months later, when the earth has moved to the opposite side of its orbit, the northern hemisphere leans 23½ degrees away from the sun. On days between these extremes the earth's axis is leaning at amounts less than 23½ degrees to the rays of the sun. This change in orientation causes the vertical rays of the sun to make a yearly migration from 23½ degrees north of the equator to 23½ degrees south of the equator. In turn, this migration causes the altitude of the noon sun to vary by as much as 47 degrees (23½ plus 23½) for many locations during the course of a year. A midlatitude city like New York, for instance,

Inclination of the Axis

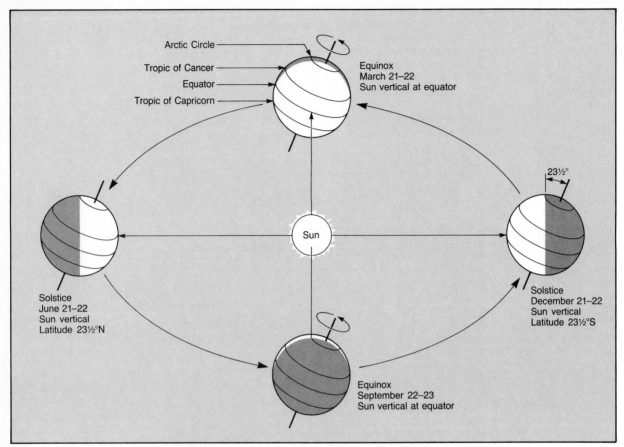

FIGURE 2–3
Earth–sun relationships.

has a maximum noon sun angle of 73½ degrees when the sun's vertical rays have reached their farthest northward location and a minimum noon sun angle of 26½ degrees 6 months later (Figure 2–4).

Historically 4 days a year have been given special significance based on the annual migration of the direct rays of the sun and its importance to the yearly cycle of weather. On June 21 or 22 the earth is in a position where the axis in the northern hemisphere is tilted 23½ degrees toward the sun (Figure 2–3). At this time the vertical rays of the sun are striking 23½°N latitude (23½ degrees north of the equator), a line of latitude known as the **Tropic of Cancer.** For people living in the northern hemisphere, June 21 or 22 is known as the **summer solstice.** Six months later, on about December 21 or 22, the earth is in an opposite position, where the sun's vertical rays are striking at 23½°S latitude. This line is known as the **Tropic of Capricorn.** For those of us in the northern hemisphere, December 21 or 22 is the **winter solstice.** However, at the same time in the southern hemisphere people are experiencing just the opposite—the summer solstice.

Tropic of Cancer

Solstice

Tropic of Capricorn

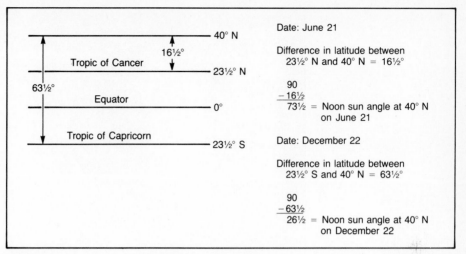

FIGURE 2–4
Calculating the noon sun angle. The noon sun angle at 23½°N latitude on June 21 is 90 degrees. A place located 1 degree away receives an 89-degree angle; a place 2 degrees away, an 88-degree angle, and so forth. To calculate the noon sun angle, simply find the number of degrees of latitude separating the location receiving the vertical rays of the sun and the location in question; then subtract that answer from 90 degrees. In this figure we see the large variation in sun angles experienced at 40°N latitude for the summer and winter solstices (the extremes).

Equinox

The equinoxes occur midway between the solstices. September 22 or 23 is the date of the **autumnal equinox** in the northern hemisphere, and March 21 or 22 is the date of the **vernal** or **spring equinox.** On these dates the vertical rays of the sun are striking at the equator (0 degrees latitude), for the earth is in such a position in its orbit that the axis is tilted neither toward nor away from the sun.

In addition, the length of daylight versus darkness is also determined by the position of the earth in its orbit. The length of daylight on June 21, the summer solstice in the northern hemisphere, is greater than the length of night. This fact can be established by examining Figure 2–5, which illustrates the **circle of illumi-**

Circle of Illumination

nation—that is, the boundary separating the dark half of the earth from the lighted half. The length of daylight is established by comparing the fraction of a given latitude that is on the lighted side of the globe with the fraction on the dark side. Notice that on June 21 all locations in the northern hemisphere experience longer periods of daylight than darkness. The opposite is true for the December solstice; then the length of darkness exceeds the length of daylight at all locations in the northern hemisphere. Again for comparison let us consider New York City. It has 15 hours of daylight on June 21 and only 9 hours on December 21. Also note from Table 2–2 that the farther north you are of the equator on June 21, the longer the period of daylight until the Arctic Circle is reached; here the length of daylight is 24 hours (Figure 2–6). During an equinox (meaning "equal night") the length of daylight is 12 hours everywhere on earth, for the circle of illumination passes directly through the poles, thus dividing the latitudes in half.

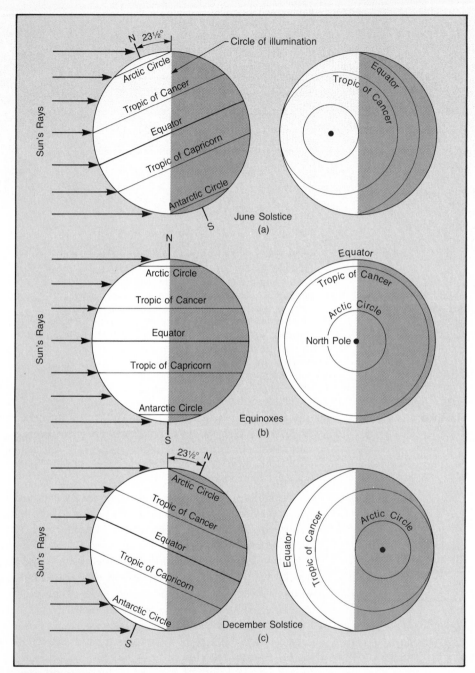

FIGURE 2–5
Characteristics of the solstices and equinoxes.

FIGURE 2–6
The midnight sun seen in late July in northern Greenland (78°N latitude). Eight exposures were taken at 20 minute intervals, four before and four after midnight. (Negative #230863, Courtesy Department of Library Services, American Museum of Natural History)

As a review of the characteristics of the summer solstice for the northern hemisphere, examine Figure 2–5 and Table 2–2 and consider the following facts:

1. The date of occurrence is June 21 or 22.
2. The vertical rays of the sun are striking the Tropic of Cancer (23½°N latitude).
3. Locations in the northern hemisphere are experiencing their longest days and highest sun angles (opposite for the southern hemisphere).
4. The farther north you are of the equator, the longer the period of daylight until the Arctic Circle is reached, where the day is 24 hours long (opposite for the southern hemisphere).

TABLE 2–2 Length of Daylight

Latitude (degrees)	Summer Solstice	Winter Solstice	Equinoxes
0	12 hr	12 hr	12 hr
10	12 hr 35 min	11 hr 25 min	12
20	13 12	10 48	12
30	13 56	10 04	12
40	14 52	9 08	12
50	16 18	7 42	12
60	18 27	5 33	12
70	2 mo	0 00	12
80	4 mo	0 00	12
90	6 mo	0 00	12

The facts about the winter solstice will be just the opposite. It should now be apparent why a midlatitude location is warmest in the summer. It is then when the days are longest and the altitude of the sun is the highest.

In summary, seasonal fluctuations in the amount of solar energy reaching places on the earth's surface are caused by the migrating vertical rays of the sun and the resulting variations in sun angle and length of daylight.

All places situated at the same latitude have identical sun angles and lengths of days. If the earth–sun relationships just described were the only controls of temperature, we would expect these places to have identical temperatures as well. Obviously such is not the case. Although the altitude of the sun is the main control of the temperature, it is not the only control, as we shall see.

What Are the Seasons?

The idea of dividing the year into four seasons clearly originated from the earth–sun relationships just discussed. This astronomical definition of the seasons defines winter (northern hemisphere) as the period from the winter solstice (December 21–22) to the vernal equinox (March 21–22), and so forth. This is also the definition used most widely by the news media. Yet it is not unusual for portions of the United States and Canada to have significant snowfalls weeks before the "official" start of winter.

Because such occurrences are annual events, many meteorologists prefer to divide the year into four 3-month periods based primarily on temperature. Thus winter is defined as December, January, and February, the three coldest months of the year in the northern hemisphere. Summer, on the other hand, is defined as the three warmest months, June, July, and August. Spring and autumn are the transition periods between these two seasons. Inasmuch as these four 3-month periods better reflect the temperatures and weather that we associate with the respective seasons, this definition of the seasons is more useful for meteorological discussions.

SOLAR RADIATION

Radiation

As noted, the sun is the source of energy that drives the weather machine. For this reason, we consider the nature of solar energy in more detail. From our everyday experience we know that the sun emits light and heat as well as the rays that give us a suntan. Although these forms of energy constitute a major portion of the total energy that radiates from the sun, they are only a part of a large array of energy called **radiation** or **electromagnetic radiation.** This array or spectrum of electromagnetic energy is shown in Figure 2–7. All radiation, whether x rays, radio, or heat waves, is capable of transmitting energy through the vacuum of space at 300,000 kilometers per second and only slightly slower through air. Nineteenth-century physicists were so puzzled by the seemingly impossible task of energy traveling without the aid of an intervening medium to transmit

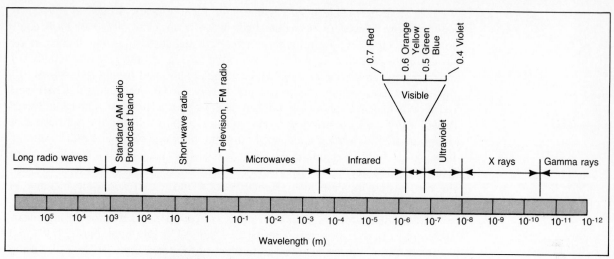

FIGURE 2–7
The electromagnetic spectrum.

it that they assumed that a material, which they named ether, existed between the sun and the earth. This medium was thought to transmit radiant energy in much the same way that air transmits sound waves produced by the vibration of one person's vocal cords to another person's eardrums. Today we know that, like gravity, radiation requires no material to transmit it; yet physicists still do not know why this is possible.

In some respects, the transmission of radiant energy parallels the motion of the gentle swells that one sees in the open ocean. Not unlike ocean swells, these electromagnetic waves, as they are called, come in various sizes. For our purposes, the most important difference between electromagnetic waves is their wavelength or distance from one crest to the next (Figure 2–8). Radio waves have the longest wavelengths, ranging to tens of kilometers. Gamma waves are the shortest, being less than a billionth of a centimeter long. Radiation is often identified by the effect that it produces when it interacts with an object. The retinas of our eyes, **Visible Light** for instance, are sensitive to a range of wavelengths that we call **visible light.**

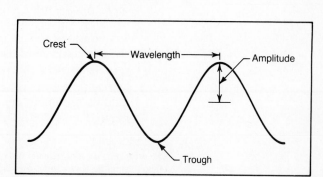

FIGURE 2–8
Characteristics of a wave.

Infrared

Ultraviolet

We often refer to visible light as white light, for it appears "white" in color. It is easy to show, however, that white light is really an array of colors, each color corresponding to a specific wavelength. By using a prism, white light can be divided into the colors of the rainbow, violet having the shortest wavelength, 0.4 micrometer (1 micrometer is 0.0001 centimeter) and red having the longest, 0.7 micrometer. Located adjacent to red, and having a longer wavelength, is **infrared** radiation, which we cannot see but can detect as heat. The closest invisible waves to violet are called **ultraviolet** rays and are responsible for the sunburn that can occur after an intense exposure to the sun. Although we divide radiant energy into groups based on our ability to perceive them, all forms of radiation are basically the same. When any form of radiant energy is absorbed by an object, the result is an increase in molecular motion and a corresponding increase in temperature.

Most radiant energy from the sun is concentrated in the visible and near-visible parts of the spectrum. The narrow band of visible light, between 0.4 and 0.7 micrometer, represents over 43 percent of the total emitted. The bulk of the remainder lies in the near-infrared section (49 percent) and ultraviolet section (7 percent). Less than 1 percent of solar radiation is emitted as x rays, gamma rays, and radio waves.

To obtain a better appreciation of how the sun's radiant energy interacts with the earth's atmosphere and surface, a general understanding of the basic laws governing radiation is necessary. The principles that follow were set forth by physicists during the late 1800s and early 1900s. Although the mathematical implications of these laws are beyond the scope of this book, the concepts themselves are well within your grasp.

1. All objects, at whatever temperature, emit radiant energy. Thus not only do hot objects like the sun continually emit energy, but the earth, including its polar ice caps, does as well.

2. Hotter objects radiate more total energy per unit area than colder objects. The sun, which has a surface temperature of 6000°K, emits hundreds of thousands of times more energy than the earth, which has an average surface temperature of 288°K.

3. The hotter the radiating body, the shorter is the wavelength of maximum radiation. This law can be easily visualized if we examine something from our everyday experiences. For example, a very hot metal rod will emit visible radiation and produce a white glow. On cooling, it will emit more of its energy in longer wavelengths and will glow a reddish color. Eventually no light will be given off, but if you place your hand near the rod, the still longer infrared radiation will be detectable as heat. The sun, which has a surface temperature of 6000°K,[1] radiates maximum energy at 0.5 micrometer, which is in the visible range (Figure 2–9). The maximum radiation for the earth occurs at a

---

[1] Because the sun is a gaseous body, it has no true surface. Most energy radiated from the sun is emitted from the *photosphere*, the visible layer of the sun.

FIGURE 2–9

Radiation spectrums for the emissions from a blackbody at 6000°K (temperature of the sun's surface) and 250°K (approximate temperature of the earth's surface). Note that these two curves are separated into two spectral ranges, greater and less than 4 micrometers. Thus we call the sun's radiation shortwave and the earth's radiation long wave. (From R. G. Fleagle and J. A. Businger, *An Introduction to Atmospheric Physics.* © 1963 by Academic Press; reprinted by permission of the publisher)

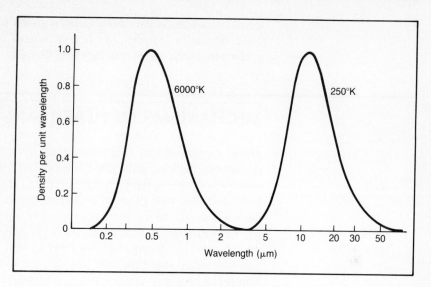

wavelength of 10 micrometers, well within the infrared (heat) range. Because the maximum earth radiation is roughly 20 times longer than the maximum solar radiation, terrestrial radiation is often referred to as long-wave radiation and solar radiation is called shortwave radiation.

Blackbody

4. Objects that are good absorbers of radiation are also good emitters. The perfect absorber (emitter) is a theoretical object called a **blackbody.** The term "black" is misleading, for the object need not be black in color. However, a dull black surface is nearly a blackbody because it absorbs roughly 90 percent of the radiation striking it.

Technically a blackbody is any object that radiates, for every wavelength, the maximum intensity of radiation possible for that temperature. The earth's surface and the sun approach being blackbodies (perfect radiators) because they absorb and radiate with nearly 100 percent efficiency for their respective temperatures. On the other hand, gases are selective absorbers and radiators. Thus the atmosphere, which is nearly transparent to (does not absorb) certain wavelengths of radiation, is nearly opaque to (a good absorber of) others. Our experience tells us that the atmosphere is transparent to visible light; hence it readily reaches the earth's surface. Such is not the case for long-wave terrestrial radiation, as we shall see later. Freshly fallen snow is another example of a selective absorber. It is a poor absorber of visible light (reflects 90 percent), but at the same time it is a very good absorber (and emitter) of infrared heat. Consequently, the air directly above a snow-covered surface is colder than it would otherwise be had the snow surface been barren ground, for much of the incoming radiation is reflected away. A blanket of snow has the opposite effect on the ground below it, however. As the ground radiates heat (infrared) upward, the lower layer of the snow absorbs this energy and reradiates most of it downward. Thus the depth at which a winter's frost can penetrate into the ground is much less when the ground has a snow cover than in an equally cold region without snow. The statement "The ground is blanketed with snow" can be taken literally.

MECHANISMS OF HEAT TRANSFER

Three mechanisms of heat transfer are recognized: radiation, conduction, and convection. Because radiation is the only one that can travel through the relative emptiness of space, most energy coming to and leaving the earth must be in this form. Radiation also plays an important role in transferring heat from the earth's land–sea surface to the atmosphere and vice versa.

Conduction

Conduction is familiar to most of us through our everyday experiences. Anyone who has attempted to pick up a metal spoon that was left in a hot pan is sure to realize that heat was conducted through the spoon. Conduction is the transfer of heat through matter by molecular activity. The energy of molecules is transferred through collisions from one molecule to another, with the heat flowing from the higher temperature to the lower temperature. The ability of substances to conduct heat varies considerably. Metals are good conductors, as those of us who have touched a hot spoon have quickly learned. Air, on the other hand, is a very poor conductor of heat. Consequently, conduction is only important between the earth's surface and the air directly in contact with the surface. As a means of heat transfer for the atmosphere as a whole, conduction is the least significant and can be disregarded when considering most meteorological phenomena.

Convection

Heat gained by the lowest layer of the atmosphere from radiation or conduction is most often transferred by **convection.** Convection is the transfer of heat by the movement of a mass or substance from one place to another. It can only take place in liquids and gases. Convective motions in the atmosphere are responsible for the redistribution of heat from equatorial regions to the poles and from the surface upward. The term **advection** is usually reserved for horizontal convective motions, such as winds; the term "convection" is usually restricted to vertical heat transfer in the atmosphere.

Advection

Figure 2–10 summarizes the various mechanisms of heat transfer. A portion of the radiant energy generated by the campfire is absorbed by the coffeepot and frying pan. This energy is readily transferred through these metal containers by the process of conduction. In the case of the coffeepot, the conduction process also increases the temperature of the water at the bottom. Once warmed, this layer of water moves upward and is replaced by cool water descending from above. Thus convection currents that redistribute the newly acquired energy throughout the pot are established. Meanwhile, the camper is warmed by radiation emitted by the fire and the pan. Furthermore, because metals are good conductors, the camper's hand is likely to be burned if he does not use a potholder. Like this example, the heating of the earth's atmosphere involves the processes of conduction, convection, and radiation, all of which occur simultaneously.

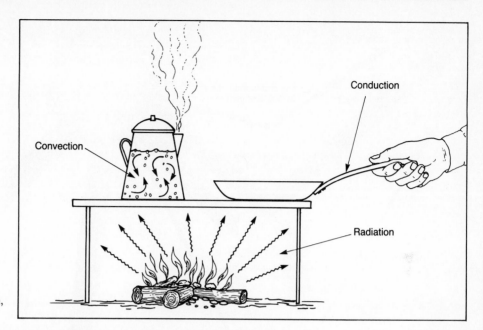

FIGURE 2–10
Conduction, convection, and radiation.

INCOMING SOLAR RADIATION

Although the atmosphere is very transparent to incoming solar radiation, only about 25 percent penetrates directly to the earth's surface without some sort of interference on the part of the atmosphere. The remainder is either absorbed by the atmosphere, scattered about until it reaches the earth's surface or returns to space, or is reflected back to space (Figure 2–11). What determines whether radiation will be absorbed, scattered, or reflected outward? As we shall see, it depends greatly on the wavelength of the energy being transmitted as well as on the size and nature of the intervening material.

Scattering

Although solar radiation travels in a straight line, the gases and dust particles in the atmosphere can redirect this energy. This process, called scattering, is responsible for illuminating a shaded area or a room when direct sunlight is absent. Scattering also produces the so-called **diffused light** that accounts for the brightness of the daytime sky. In contrast, bodies like the moon and Mercury, which are without atmospheres, have dark skies and "pitch black" shadows, even during daylight hours.

   To a large extent, the degree of scattering in the atmosphere is determined by the size of the intervening gas molecules and dust particles. When light is scattered by very small particles, primarily gas molecules, it is distributed in all

Diffused Light

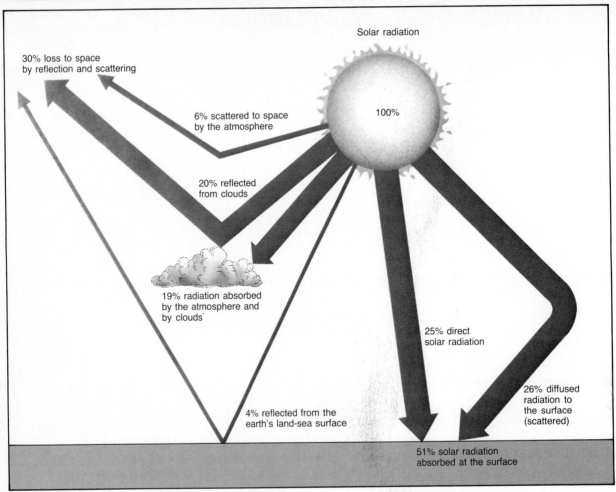

FIGURE 2–11
Distribution of incoming solar radiation by percentage. More solar energy is absorbed by the earth's surface than by the atmosphere. Consequently, the air is not heated directly by the sun but is heated indirectly from the earth's surface.

directions, forward as well as backward. Some light that is backscattered is lost to space, but the remainder continues toward the earth's surface where it interacts with other molecules that further scatter it by changing the direction of the light beam but not its wavelength.

Gas molecules more effectively scatter the shorter wavelengths (blue and violet) of "white" light than the longer wavelengths (red and orange). This fact, in turn, explains the blue color of the sky and the orange and red colors seen at sunrise and sunset (Figure 2–12). Remember, sunlight appears white in color, but it is composed of all colors. When the sun is overhead, an observer can look in any direction away from the direct sun and see predominantly blue light, which was more readily scattered by the atmosphere. On the other hand, the sun appears to have an orangish-to-reddish tint when viewed near the horizon. With the sun in this position, the solar beam must travel through a great deal of

FIGURE 2–12
At sunset the solar disk appears red because most of the blue light has been scattered. (Photo by E. J. Tarbuck)

atmosphere before it reaches your eye (see Table 2–1). Therefore most blues and violets will be scattered out, leaving a beam of light that consists mostly of reds and oranges. This phenomenon is especially obvious on a day when fine dust or smoke particles are present.

The reddish appearance of clouds during sunrise and sunset also results because the clouds are illuminated by light from which the blue color has been subtracted by scattering. The most spectacular sunsets occur when large quantities of fine particles penetrate into the stratosphere. For three years after the great eruption of the Indonesian volcano of Krakatoa in 1883, brilliant sunsets occurred worldwide. The European summer that followed this colossal explosion was cooler than normal, a fact that has been attributed to the added loss of radiation caused by backscattering.

Large particles associated with haze, fog, or smog scatter light more equally in all wavelengths. Because no color is predominant over any other, the sky appears white on days when numerous large particles are abundant. We can conclude, then, that the color of the sky gives an indication of the number of large particles present. The bluer the sky, the cleaner the air.

Reflection

About 30 percent of the solar energy reaching the outer atmosphere is reflected back to space. Included in this figure is the amount sent skyward by backscattering. This energy is lost to the earth and does not play a role in heating the atmosphere.

TABLE 2–3 Albedo (Reflectivity) of Various Surfaces

Surface	Percent
Fresh snow	80–85
Old snow	50–60
Sand	20–30
Grass	20–25
Dry earth	15–25
Wet earth	10
Forest	5–10
Water (sun near horizon)	50–80
Water (sun near zenith)	3–5
Thick cloud	70–80
Thin cloud	25–50
Earth and atmosphere	30

Albedo

The fraction of the total radiation encountered that is reflected by a surface is called its **albedo.** Thus the albedo for the earth as a whole (planetary albedo) is 30 percent. The albedo from place to place as well as from time to time in the same locale, however, varies considerably, depending on the amount of cloud cover and particulate matter in the air, plus the angle of the sun's rays and the nature of the surface. A lower sun angle means that more atmosphere must be penetrated, thereby making the "obstacle course" longer and the loss of solar radiation greater.

Table 2–3 gives the albedo for various surfaces. Note that the angle at which the sun's rays strike a water surface greatly affects its albedo. The amount of light reflected from the earth's land–sea surface represents only about 4 percent of the total planetary albedo of 30 percent. Clouds are responsible for most of the earth's "brightness" as seen from space. The high reflectivity of clouds should not surprise anyone who has tried to drive with bright lights on a foggy night. In comparison, the moon, which is without clouds or an atmosphere, has an average albedo of only 7 percent. Even though a full moon gives us a good bit of light on a clear night, the much brighter earth would provide an explorer of the moon with far more light for an "earth-lit" walk at night.

Absorption within the Atmosphere

As stated earlier, gases are selective absorbers, meaning that they absorb strongly in some wavelengths, moderately in others, and only slightly in still others. When a gas molecule absorbs light waves, this energy is transformed into internal molecular motion, which is detectable as a rise in temperature. Thus it is the gases that are good absorbers of the available radiation that play the primary role in heating the atmosphere as a whole.

Figure 2–13 gives the absorptivity of the principal atmospheric gases in various wavelengths. Nitrogen, the most abundant constituent in the atmosphere, is a

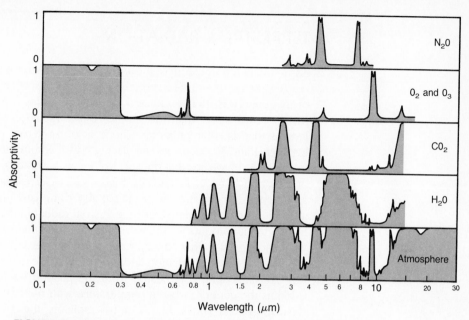

FIGURE 2–13
The absorptivity of selected gases of the atmosphere and the atmosphere as a whole.
(From R. G. Fleagle and J. A. Businger, *An Introduction to Atmospheric Physics.* © 1963
by Academic Press; reprinted by permission of the publisher)

rather poor absorber of incoming solar radiation, the vast bulk of which comes
in wavelengths between 0.2 and 2 micrometers. Oxygen (O_2) and ozone (O_3)
are efficient absorbers of ultraviolet radiation shorter than 0.29 micrometer. Oxygen
removes most of the shorter ultraviolet radiation high in the atmosphere, and
ozone absorbs longer-wavelength ultraviolet rays in the stratosphere between 10
and 50 kilometers. The absorption of ultraviolet radiation in the stratosphere
accounts for the high temperatures experienced there. The importance of the
removal of harmful ultraviolet radiation by ozone was discussed in Chapter 1.
The only other significant absorber of incoming solar radiation is water vapor,
which along with oxygen and ozone accounts for most of the 19 percent of the
total solar radiation that is absorbed within the atmosphere.

From Figure 2–13 we see that, for the atmosphere as a whole, none of the
gases is an effective absorber of radiation that has wavelengths between 0.3 and
0.7 micrometer; so a large "gap" exists. This region of the spectrum corresponds
to the visible range, to which a large fraction of solar radiation belongs. This
explains why most visible radiation reaches the ground and why we say that the
atmosphere is transparent to incoming solar radiation. Thus direct solar energy
is a rather ineffective "heater" of the earth's atmosphere. The fact that the atmosphere
does not acquire the bulk of its energy directly from the sun but is heated by
reradiation from the earth's surface is of the utmost importance to the dynamics
of the weather machine.

TERRESTRIAL RADIATION

Approximately 51 percent of the solar energy that strikes the top of the atmosphere reaches the earth's surface directly or indirectly (scattered) and is absorbed. Most of this energy is then reradiated skyward. Because the earth has a much lower surface temperature than the sun, terrestrial radiation is emitted in longer wavelengths than is solar radiation. The bulk of terrestrial radiation has wavelengths between 1 and 30 micrometers, placing it well within the infrared range. Referring to Figure 2–13, we see that the atmosphere as a whole is a rather efficient absorber of radiation between 1 and 30 micrometers (terrestrial radiation). Water vapor and carbon dioxide are the principal absorbing gases in that range. Water vapor absorbs roughly five times more terrestrial radiation than do all other gases combined, and it accounts for the warm temperatures found in the lower troposphere, where it is most highly concentrated. Because the atmosphere is very transparent to solar (shortwave) radiation and more absorptive to terrestrial (long-wave) radiation, the atmosphere is heated from the ground up instead of vice versa. This explains the general drop in temperature with increased altitude experienced in the troposphere. The farther from the "radiator," the colder it gets. On the average, the temperature drops 6.5°C for each kilometer increase in altitude, a figure known as the normal lapse rate.

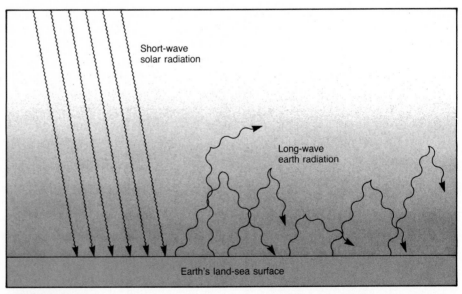

FIGURE 2–14
Much of the short-wavelength radiation from the sun passes through the atmosphere and is absorbed by the earth's land–sea surface. This energy is then reradiated as longer-wavelength terrestrial radiation, most of which is absorbed by certain gases in the atmosphere. Some of the energy absorbed by the atmosphere will be reradiated earthward. This so-called greenhouse effect is responsible for keeping the earth's surface much warmer than it would otherwise be.

When gases in the atmosphere absorb terrestrial radiation, they warm, but eventually they radiate this energy away. Some travels upward, where it may be reabsorbed by other gas molecules, a possibility that becomes much less likely with increasing height because the concentration of water vapor decreases with altitude. The remainder travels downward and is again absorbed by the earth. Thus the earth's surface is being continually supplied with heat from the atmosphere as well as from the sun. This energy will again be emitted by the earth's surface, and some will be returned to the atmosphere, which will, in turn, radiate some earthward and so forth. This complicated game of "pass the hot potato" that is played between the earth's surface and selected gases in the atmosphere keeps the earth's average temperature 35°C warmer than it would otherwise be (Figure 2–14). Without these absorptive gases in our atmosphere, the earth would not provide a suitable habitat for humans and numerous other life-forms.

Greenhouse Effect

This extremely important phenomenon has been termed the **greenhouse effect** because it was once thought that greenhouses were heated in a similar manner. The glass in a greenhouse allows shortwave radiation to enter and be absorbed by the objects inside. These objects, in turn, reradiate the heat but at longer wavelengths, to which glass is nearly opaque. The heat, therefore, is trapped in the greenhouse. It has been demonstrated, however, that greenhouses attain higher temperatures than the outside air primarily because the glass restricts the vertical motion that usually occurs when air is heated. Nevertheless, the term "greenhouse effect" remains.[2]

The importance of water vapor and carbon dioxide in keeping the atmosphere warm is well known to those living in mountainous regions. More radiant energy is received on mountaintops than in the valleys below because there is less atmosphere to hinder its arrival. The less dense mountain air, however, also allows much of the heat to escape these lofty peaks. This factor more than compensates for the extra radiation received. As a result, the valleys remain warmer than adjacent mountains even though they receive less solar radiation.

Clouds, like water vapor and carbon dioxide, are good absorbers of infrared (terrestrial) radiation and play a key role in keeping the earth's surface warm, especially at night. A thick cloud cover will absorb most terrestrial radiation and return it to the surface. This explains why on clear, dry nights the surface cools considerably more than on cloudy or humid evenings.

HEAT BUDGET

Except for a small amount of energy stored as fossil fuels, a balance exists between the amount of incoming solar radiation and the amount of terrestrial radiation returned to space; otherwise the earth would be getting progressively colder or progressively warmer. The balance of incoming and outgoing radiation has been

---

[2] Because the term greenhouse effect is misleading, Fleagle and Businger (*An Introduction to Atmospheric Physics* [Orlando, Fla.: Academic Press, 1963]) suggest that this trapping of radiation in the lower atmosphere be called the *atmosphere effect*.

Heat Budget

termed the earth's **heat budget.** An examination of the earth's heat budget provides a good review of the processes just discussed. Figure 2–15 quantitatively illustrates this balance by using 100 units to represent the solar radiation intercepted at the outer edge of the atmosphere.

Of the total radiation intercepted by the earth, roughly 30 units are reflected back to space. The remaining 70 units are absorbed, 19 units within the atmosphere and 51 units by the earth's land–sea surface. If all the energy absorbed by the earth were reradiated directly back to space, the earth's heat budget would be a

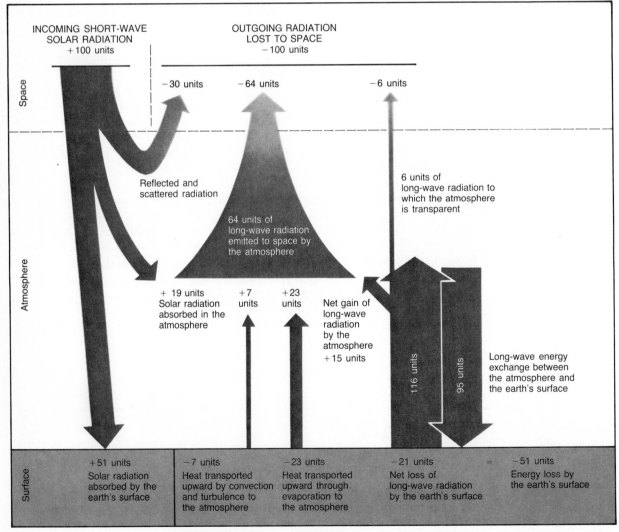

FIGURE 2–15
Heat budget of the earth and atmosphere. (Data from *Understanding Climate Change*, U.S. National Academy of Science, Washington, D.C., 1975)

simple matter. But as previously stated, certain gases in the atmosphere act to delay the loss of terrestrial radiation by absorbing a large portion of it and reradiating much of this energy earthward. As a result of this process, the earth's surface receives what might appear to be an unusually large amount of long-wave radiation from the atmosphere (95 units).[3] The earth's surface, in turn, radiates 116 units of long-wave energy back to the atmosphere. The right-hand side of Figure 2–15 illustrates the exchange of long-wave radiation between the atmosphere and the earth's surface. Notice that during this exchange the atmosphere receives a net gain of 15 units, whereas the earth's surface has a net loss of 21 units. The remaining 6 units pass directly through the atmosphere and are lost to space. Radiation between 8 and 11 micrometers in length escapes the troposphere most readily because water vapor and carbon dioxide do not absorb these wavelengths (see

Atmospheric Window

Figure 2–13). This zone is called the **atmospheric window,** for it is transparent to terrestrial radiation of those wavelengths.

We have accounted for 21 of the 51 units of shortwave radiation absorbed by the earth's surface. But what about the remaining 30 units? This energy is transferred from the earth's land–sea surface to the atmosphere by water molecules during the process of evaporation (23 units) and by conduction and convection (7 units). An overall balance is achieved because the atmosphere emits 64 units of energy to space as long-wave radiation. A careful examination of Figure 2–15 confirms that incoming radiation is balanced by outgoing radiation.

LATITUDINAL HEAT BALANCE

Earlier it was pointed out that the amount of incoming solar radiation is roughly equal to the amount of outgoing terrestrial radiation for the earth as a whole. Thus the average temperature of the earth remains rather constant. The balance of incoming and outgoing radiation that holds for the entire earth is not maintained at each latitude, however. At latitudes below about 36 degrees, more solar radiation is received than is lost to space by the earth. The opposite is true for higher latitudes, where more heat is lost through radiation than is received. The conclusion that might be drawn is that the tropics are getting hotter and the poles are getting colder. But we know that it is not happening; instead the atmosphere and, to a lesser extent, the oceans act as giant thermal engines that transfer heat from the tropics poleward. In effect, it is the imbalance of heat that drives the winds and the ocean currents.

It should be of interest to those who live in the midlatitudes (30 to 50 degrees) that most heat transfer takes place across this region. Consequently, much of the stormy weather we associate with the midlatitudes can be attributed to this unending transfer of heat from the tropics toward the poles. These processes are discussed in more detail in later chapters.

---

[3] The atmosphere actually emits more energy than the amount of solar energy absorbed by the earth. It does so because of the greenhouse effect, which retains energy in the lower atmosphere.

REVIEW

1. Can distance variations between the earth and sun adequately explain seasonal temperature changes? Explain.

2. Why does the amount of solar energy received at the earth's surface change when the altitude of the sun changes?

3. After referring to Figure 2–4, calculate the noon sun angle on June 21 and December 21 at 50°N latitude, 0 degrees latitude (the equator), and 20°S latitude. Which of these latitudes has the greatest variation in noon sun angle between summer and winter?

4. For the latitudes listed in question 3, determine the length of daylight and darkness on June 21 and December 21 (refer to Table 2–2). Which of these latitudes has the largest seasonal variation in length of daylight? Which latitude has the smallest variation?

5. How would our seasons be affected if the earth's axis were not inclined 23½ degrees to the plane of its orbit but were instead perpendicular?

6. Most solar radiation is concentrated in what portion of the electromagnetic spectrum?

7. Describe the relationship between the temperature of a radiating body and the wavelengths it emits.

8. Describe the three basic mechanisms of heat transfer. Which mechanism is least important meteorologically?

9. What is the difference between convection and advection?

10. Why does the sky usually appear blue?

11. Although the sky is usually blue during the day, it may appear to have a red or orange tint near sunrise or sunset. Explain.

12. What factors might influence albedo from time to time and from place to place?

13. Explain why the atmosphere is heated chiefly by reradiation from the earth's surface.

14. Which gases are the primary heat absorbers in the lower atmosphere? Which one of these gases is most important?

VOCABULARY REVIEW

rotation	radiation or electromagnetic radiation
revolution	visible light
perihelion	infrared
aphelion	ultraviolet
altitude (of the sun)	blackbody
inclination of the axis	conduction
Tropic of Cancer	convection
summer solstice	advection
Tropic of Capricorn	diffused light
winter solstice	albedo
autumnal equinox	greenhouse effect
vernal or spring equinox	heat budget
circle of illumination	atmospheric window

3
Temperature

Temperature is one of the basic elements of weather and climate. In the preceding chapter, we learned how air is heated and examined the role of earth–sun relationships in causing seasonal temperature variations. In Chapter 3 we will focus on several other aspects of this very important atmospheric property, including how it is measured and expressed, and what factors cause temperature patterns to change from place to place and from time to time. We will also see that temperature data can be of very practical value to people. Applications include calculations that are useful in evaluating energy consumption, crop maturity, and degrees of human discomfort.

HEAT AND TEMPERATURE

Heat

Temperature

The concepts of heat and temperature are sometimes confused. The phrase "in the heat of the day" is but one common expression in which the word "heat" is used to describe the concept of temperature. Essentially **heat** is a form of energy. When heat is added to a substance, the temperature of that substance rises. Therefore the term **temperature** refers to intensity—that is, the degree of "hotness." The temperature of a cup of boiling water, for example, is the same as the temperature of a pail of boiling water. The pail of boiling water has a greater quantity of energy than the cup of boiling water, however. Moreover, a cup of boiling water obviously has a higher temperature than a tub of lukewarm water, but the cup does not contain as much energy. Far more ice would be melted in the tub of lukewarm water than in the cup of boiling water. The water temperature in the cup is higher, but the amount of energy is smaller. Thus the quantity of heat depends on the mass of material considered; temperature does not. It is now easier to understand why the extremely rarefied air of the thermosphere (discussed in Chapter 1), although having a high temperature, contains little energy.

Although the ideas of temperature and heat are distinct, they are nevertheless related. Certainly the addition or subtraction of heat causes temperatures to increase or decrease. In addition, differences in temperature determine the direction of heat flow. When two bodies having different temperatures are in contact, heat moves from the higher-temperature body to the lower-temperature body. In fact, a common working definition of temperature is: the property that determines whether heat will flow out of or into an object when in contact with another object.

TEMPERATURE MEASUREMENT

Thermometer

Liquid-in-Glass Thermometer

Thermometers are devices designed to measure temperature. Most substances expand when heated and contract when cooled, and the designs of most commonly used thermometers are based on this property. More precisely, the design of these instruments relies on the fact that different substances react to temperature changes differently. The **liquid-in-glass thermometer** shown in Figure 3–1 is a simple instrument that provides relatively accurate readings over a wide range of temperatures. Its design has remained essentially unchanged since it was developed in the late seventeenth century. When temperatures rise, the expansion of the fluid in the bulb is much greater than the corresponding change in the enclosing glass. As a consequence, a thin thread of fluid moves up the capillary tube. Con-

FIGURE 3–1
The main components of a liquid-in-glass thermometer.

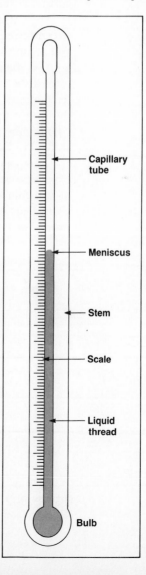

Capillary tube

Meniscus

Stem

Scale

Liquid thread

Bulb

versely, when temperatures fall, the liquid contracts and the thread of fluid moves back down the tube toward the bulb. The movement of the end of this thread (known as the *meniscus*) can be calibrated against an established scale to indicate the temperature.

The highest and lowest temperatures that occur each day are of considerable importance and are often obtained by using specially designed liquid-in-glass thermometers. Mercury is the liquid used in the **maximum thermometer,** which has a narrowed passage called a constriction in the bore of the glass tube just above the bulb (Figure 3–2). As the temperature rises, the mercury expands and is forced through the narrow opening. When the temperature falls, the constriction prevents a return of mercury to the bulb. As a result, the top of the mercury column remains at the highest point. The instrument is reset by shaking or by whirling it around a mounting. Once the thermometer is set, it indicates the current air temperature.

Maximum Thermometer

In contrast to a maximum thermometer that contains mercury, a **minimum thermometer** contains a liquid of low density, such as alcohol. Within the alcohol, and resting at the top of the column, is a small dumbbell-shaped index (Figure 3–2). As the air temperature drops, the column shortens and the index is pulled toward the bulb by the effect of surface tension with the meniscus. When the temperature subsequently rises, the alcohol flows past the index, leaving it at the lowest temperature reached. To return the index to the top of the alcohol column, the thermometer is simply tilted. Because the index is free to move, it will fall to the bottom if the minimum thermometer is not mounted horizontally.

Minimum Thermometer

Bimetal Strip

Another commonly used thermometer that works on the principle of differential expansion is the **bimetal strip.** As the name indicates, this thermometer consists of two thin strips of metal that are welded together and have widely different coefficients of thermal expansion. When the temperature changes, the two metals expand or contract unequally and cause changes in the curvature of the element.

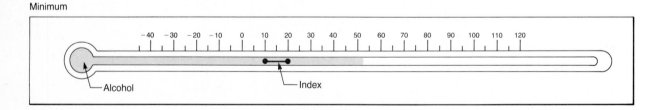

FIGURE 3–2
Maximum and minimum thermometers.

FIGURE 3–3
A common use of the bimetal strip is in the construction of a thermograph, an instrument that continuously records temperature. (Courtesy of WeatherMeasure Corporation)

These changes are then translated into temperature values on a calibrated dial. The primary meteorological use of the bimetal strip is in the construction of a **thermograph,** an instrument that continuously records temperature. The changes in the curvature of the strip can be used to move a pen arm that records the temperature on a calibrated chart that is attached to a clock-driven, rotating drum (Figure 3–3). Although very convenient, thermograph records are generally less accurate than readings obtained from a mercury-in-glass thermometer. To obtain the most reliable values, it is necessary to check and correct the thermograph periodically by comparing it with an accurate, similarly exposed thermometer.

Thermograph

Thermometers that do not rely on differential expansion but instead measure temperatures electrically have also been devised. A common meteorological use of electrical thermometers is in gathering upper-air data. One of two principles is commonly employed in such instruments. **Thermocouples** operate on the principle that differences in temperature between the junction of two unlike metal wires in a circuit will induce a current to flow. **Thermistors** consist of a conductor whose resistance to the flow of current is temperature dependent. Temperature is therefore indicated as a function of current. This instrument is commonly used in radiosondes.

Thermocouple

Thermistor

The accurate determination of air temperature depends not only on the care with which the thermometer is constructed but on its proper exposure as well. To obtain a meaningful temperature reading, thermometers must be shaded from direct sunlight and be shielded from radiating surfaces, such as buildings and the ground. Radiation should be prevented from reaching the instruments because thermometers are much more efficient absorbers than the air. It is the air tempera-

FIGURE 3–4
A standard instrument shelter. A shelter protects instruments from direct sunlight and allows for the free flow of air. (Courtesy of Robert E. White Instruments)

ture that is desired, not the temperature of the thermometer after absorbing radiation. In addition, good ventilation is essential. Thermometers sheltered from freely moving air will not indicate the true air temperature. Therefore to avoid obtaining inaccurate and misleading data, thermometers are mounted in an instrument shelter (Figure 3–4). The shelter is a white box that has louvered sides that permit the free movement of air through it while shielding the instruments from sunshine and precipitation. Furthermore, the shelter is placed over grass whenever possible and is mounted about 1 meter above the ground and as far from buildings as circumstances permit.

TEMPERATURE SCALES

In order to make quantitative measurements of temperature possible, it was necessary to establish scales or units. Such temperature scales are based on the use of reference points, sometimes called **fixed points.** In 1714 Gabriel Daniel **Fahrenheit,** a German physicist who spent much of his life in Holland, devised the temperature scale that bears his name. He constructed a mercury-in-glass thermometer in which the zero point was the lowest temperature he could attain with a mixture of ice, water, and common salt. For his second fixed point he chose human body temperature, which he arbitrarily set at 96°. On this scale he determined that the melting point of ice (the **ice point**) was 32° and the boiling point of

Fixed Points

Fahrenheit Scale

Ice Point

Steam Point

Celsius Scale

Kelvin Scale

Absolute Zero

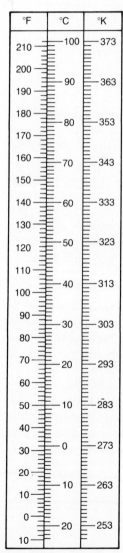

FIGURE 3–5
Temperature scales.

water (the **steam point**) was 212°. Because Fahrenheit's original reference points were difficult to reproduce accurately, his scale is now defined by using the ice point and the steam point. As thermometers improved, human body temperature was later shown to be 98.6°F instead of 96°F.

In 1742, 28 years after Fahrenheit invented his scale, Anders Celsius, a Swedish astronomer, devised a decimal scale on which the melting point of ice was set at 0° and the boiling point of water at 100°.[1] For many years it was called the *centigrade scale*, but it is now known as the **Celsius scale,** after its inventor.

Because the interval between the melting point of ice and the boiling point of water is 100° on the Celsius scale and 180° on the Fahrenheit scale, a Celsius degree (°C) is larger than a Fahrenheit degree (°F) by a factor of 180/100 or 1.8. So to convert from one system to the other, allowance must be made for the difference in the size of units as well as for the fact that the ice point on the Celsius scale is at 0° rather than at 32° as in the Fahrenheit system. This relationship is shown graphically in Figure 3–5 and by the following formulas:

$$°F = (1.8 \times °C) + 32 \qquad \text{or} \qquad °C = \frac{°F - 32}{1.8}$$

The Fahrenheit scale is best known in English-speaking countries, where, along with the British system of weights and measures, its official use is declining. In other parts of the world as well as in the scientific community, where the metric system is used, the Celsius temperature scale is also used.

For some scientific purposes, a third temperature scale is used, the **Kelvin** or **absolute scale.** It is similar to the Celsius scale because its divisions are exactly the same; there are 100° separating the melting point of ice and the boiling point of water. However, on the Kelvin scale the ice point is set at 273° and the steam point at 373° (see Figure 3–5). Moreover, the zero point (called **absolute zero**) represents the temperature at which all molecular motion is presumed to cease. Thus unlike the Celsius and Fahrenheit scales, it is not possible to have a negative value when using the Kelvin scale, for there is no temperature lower than absolute zero. The relationship between Kelvin and Celsius scales is easily written as follows:

$$°C = °K - 273 \qquad \text{or} \qquad °K = °C + 273$$

AIR TEMPERATURE DATA

Temperatures collected daily at thousands of stations throughout the world serve as the basis for much of the temperature data compiled by meteorologists and climatologists. Hourly values are either recorded by an observer or, more fre-

[1] The boiling point referred to in the Celsius and Fahrenheit scales pertains to pure water at standard sea-level pressure. It is necessary to remember this fact, for the boiling point of water gradually decreases with a decrease in air pressure.

Daily Mean

quently, are obtained from a thermograph record. At many locations only the maximum and minimum temperatures are obtained. The **daily mean** for a given day may therefore be determined by averaging the 24 hourly readings or, more commonly, by adding the maximum and minimum temperatures for a 24-hour

Daily Range

period and dividing by 2. From the maximum and minimum the **daily range** is also computed by finding the difference between these figures. In addition to the daily mean and range, other data involving longer time periods are also compiled:

Monthly Mean

1. The **monthly mean** is calculated by adding together the daily means for each day of the month and dividing by the number of days in the month.

Annual Mean

2. The **annual mean** is an average of the 12 monthly means.

Annual Temperature Range

3. The **annual temperature range** is computed by finding the differences between the warmest and coldest monthly means.

Mean temperatures are especially useful for making comparisons, whether on a daily, monthly, or annual basis. It is common to hear a weather reporter state that "Last month was the warmest February on record" or "Today Omaha was 10 degrees warmer than Chicago." Temperature ranges are also useful statistics because they give an indication of extremes, a necessary part of understanding the weather and climate of a place or an area.

Isotherm

In order to examine the distribution of air temperatures over large areas, isotherms are commonly used. An **isotherm** is a line that connects places that have the same air temperature. Therefore all points through which an isotherm passes have identical mean temperatures for the time period indicated, either an hour, day, month, or year. Generally isotherms representing 5° or 10° temperature differences are used, but any interval may be chosen. Figure 3–6 illustrates how isotherms are drawn on a map. Notice that most isotherms do not pass directly through the observing stations, because the station readings do not coincide with the values chosen for the isotherms. Since only an occasional station temperature will be exactly the same as the value of the isotherm, it is usually necessary to draw the lines by estimating the proper position between stations.

Isothermal maps are valuable tools because they clearly make the important aspects of temperature distribution visible at a glance. Areas of low and high temperatures and the degree of temperature change per unit of distance (the *temperature gradient*) are easily seen. On the other hand, a map covered with numbers representing temperatures at tens or hundreds of places would make such patterns difficult to grasp.

APPLICATIONS OF TEMPERATURE DATA

In order to make weather data more useful to people, many different applications have been developed over the years. In this section we examine several commonly

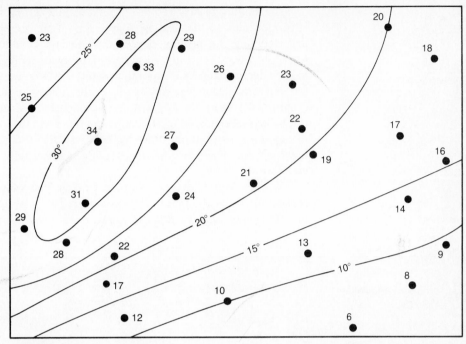

FIGURE 3–6
Isotherms are lines that connect points of equal temperature. Showing temperature distribution in this way makes patterns easier to see.

used practical applications. First, we look at three indices that all have the words *degree days* as part of their name. Two of these applications are relative measures that allow us to evaluate the weather-produced needs and costs of heating and cooling. The third is a simple index used by farmers to estimate the maturity of crops. The second group of applications explores human comfort and discomfort. Here we shall see that several factors influence the sensation of temperature that the human body feels.

Degree Days

Heating Degree Day

Developed by heating engineers early in this century, **heating degree days** represent a practical method for evaluating energy demand and consumption. This easy-to-use application of temperature data is based on the assumption that heating is not required in a building when the daily mean is 65°F (18.3°C) or higher.[2] Simply, each degree of temperature below 65°F is counted as one heating degree day. Therefore heating degree days are determined each day by subtracting the daily mean below 65°F from 65°F. Thus a day with a mean temperature of 50°F has 15 heating degree days (65 − 50 = 15) and one with an average temperature of 65°F or higher has none.

---

[2] Because the National Weather Service and the media in the United States still compute and report degree-day information in Fahrenheit degrees, we will use °F throughout this discussion.

The amount of heat required to maintain a certain temperature in a building is proportional to the heating degree days total. This linear relationship means that doubling the heating degree days usually doubles the fuel consumption. Consequently, a fuel bill will generally be twice as high for a month with 1000 heating degree days as for a month with just 500. When seasonal totals are compared for different places, we can estimate differences in seasonal fuel consumption (Figure 3–7). More than four times as much fuel is required to heat a building in Chicago (about 6200 total heating degree days) than to heat a similar building in New Orleans (1400 heating degree days), for instance. This statement is true, however, only if we assume that building construction and living habits in these areas are the same.

Heating degree day totals are a familiar part of many newspaper weather reports. Each day the previous day's accumulation is reported as well as the total thus far in the season. For the purpose of reporting heating degree days, the

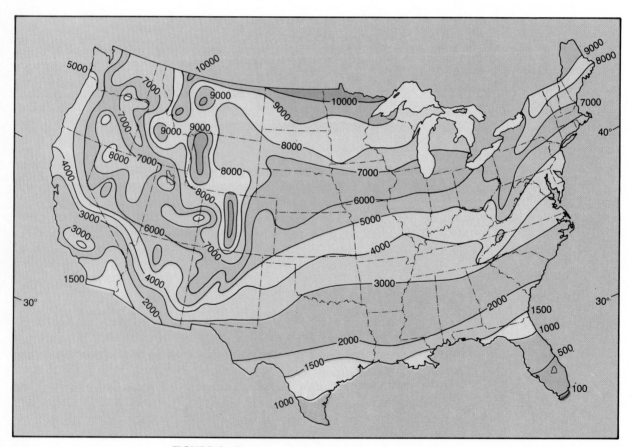

FIGURE 3–7
Mean annual total heating degree days. Caution should be used when interpolating on this generalized map, particularly in mountainous regions. (Source: Environmental Data Service, NOAA)

heating season is defined as the period from July 1 through June 30. These reports often include a comparison with the total up to this date last year or with the long-term average for this date or both, and so it is a relatively simple matter to judge whether the season so far is above, below, or near normal.

Cooling Degree Days

Just as fuel needs for heating can be estimated and compared by using heating degree days, the amount of power required to cool a building can be estimated by using a similar index called the **cooling degree day.** Because the 65°F base temperature is also used in calculating this index, cooling degree days are determined each day by subtracting 65°F from the daily mean. Thus if the mean temperature for a given day is 80°F, 15 cooling degree days would be accumulated. Mean annual totals of cooling degree days for selected cities are shown in Table 3–1. By comparing the totals for St. Louis and San Antonio, we can see that the fuel requirements for cooling a building in San Antonio are about twice as great as for a similar building in St. Louis. The "cooling season" is conventionally measured from January 1 through December 31. Therefore when cooling degree day totals are reported, the number represents the accumulation since January 1 of that year.

Although indices that are more sophisticated than heating and cooling degree days have been proposed to take into account the effects of wind speed, solar radiation, and humidity, degree days continue to be widely used.

Growing Degree Days

Another practical application of temperature data is used in agriculture to determine the approximate date when crops will be ready for harvest. This simple index is called the **growing degree day.** The number of growing degree days for a particular crop on any day is the difference between the daily mean temperature and the base temperature of the crop—that is, the minimum temperature required for growth of that crop. The base temperature for sweet corn is 50°F and for peas it is 40°F, for example. Thus on a day when the mean temperature is 75°F, the number of growing degree days for sweet corn is 25 and the number for

TABLE 3–1 Average Annual Cooling Degree Days for Selected Cities

City	Cooling Degree Days
Miami, FL	4000
San Antonio, TX	3000
Tucson, AZ	2800
St. Louis, MO	1500
Washington, DC	1400
Los Angeles, CA	1200
Chicago, IL	1000
Boston, MA	700
Detroit, MI	700
Great Falls, MT	350
International Falls, MN	200
Seattle, WA	200

SOURCE: National Oceanic and Atmospheric Administration

peas is 35. Starting with the onset of the growth season, the daily growing degree day values are added; then if 2000 growing degree days are needed for that crop to mature, it should be ready to harvest when the accumulation reaches 2000. Although many factors important to plant growth, such as moisture conditions and sunlight, are not included in the growing degree day index, this system nevertheless serves as a simple and widely used tool in determining approximate dates of crop maturity.

Indices of Human Discomfort

The sensation of temperature that the human body feels is often quite different from the actual temperature of the air as recorded by a thermometer. The human body is a heat engine that is continually releasing energy, and anything that influences the rate of heat loss from the body also influences the sensation of temperature that the body feels, thereby affecting human comfort. Several factors play a part in controlling the thermal comfort of the human body, and certainly air temperature is a major factor. Other environmental conditions, such as relative humidity, wind, and solar radiation, are also significant, however.

Temperature-Humidity Index (THI)

Because evaporation is a cooling process, the evaporation of perspiration from skin is a natural means of regulating body temperature. When the air is very humid, however, heat loss by evaporation is reduced. As a result, a hot and humid day feels warmer and more uncomfortable than a hot and dry day. The **temperature-humidity index (THI)** is a well-known and often used summertime guide to human comfort or discomfort based on the conditions of temperature and humidity. In order to calculate the THI by using information commonly given in weather reports, the following formula is used:[3]

$$\text{THI} = T - (0.55 - 0.55\ \text{RH})(T - 58)$$

Here T is the temperature of the air in degrees Fahrenheit, and RH is the relative humidity expressed as a decimal fraction. Results can be interpreted by checking Table 3–2.

Wind is another significant factor affecting the sensation of temperature. Because wind increases the rate of evaporation, a cold and windy winter day may feel much colder than the air temperature would seem to indicate. Not only is cooling by evaporation heightened in this situation, but the wind also acts to carry heat away from the body by constantly replacing warmer air with colder

---

[3] Because the U.S. National Weather Service and the media still report temperatures in Fahrenheit degrees, the formula that uses this scale is given emphasis here. However, if it is the Celsius temperature that is available, the following formula is used:

$$\text{THI} = T - 0.55(1 - \text{RH})(T - 14)$$

Here T is the temperature of the air (in degrees Celsius) and RH is the relative humidity expressed as a decimal fraction. Values in excess of 25 indicate that most people will feel uncomfortable, whereas a THI between 15 and 20 is accepted by most as comfortable.

TABLE 3–2 Temperature-Humidity Index

Temperature (°F)	Relative Humidity (%)									
	10	20	30	40	50	60	70	80	90	100
70	64	64	65	66	66	67	68	68	69	70
75	66	67	68	69	70	71	72	73	74	75
80	69	70	71	72	73	75	76	77	78	80
85	71	73	74	76	77	79	80	82	83	85
90	74	75	77	79	81	82	84	86	88	90
95	76	78	80	82	84	86	88	90	92	95
100	79	81	83	86	88	90	93	95	97	100
105	82	84	87	89	92	95	97	100	102	105

Legend:
- Increasing chill
- No discomfort
- Slight discomfort
- Increased discomfort
- Great discomfort (caution, heatstroke)
- Maximum discomfort (danger, heatstroke)

Wind Chill

air. On a day when the temperature is −8°C and the wind speed is 30 kilometers per hour, for instance, the sensation of temperature would be approximately −25°C, or 17°C less than the actual air temperature. The **wind-chill** chart (Table 3–3) illustrates the effects of wind and temperature on the cooling rate of the human body by translating the cooling power of the atmosphere with wind to a temperature under calm conditions. By examining Table 3–3, it is clear that the cooling power of the wind rises as wind speed increases and as temperature decreases. As a result, wind chill is mainly important in the winter. In contrast to the cold and windy day, a calm and sunny day in winter often feels warmer than the thermometer reading. In this situation, the warm feeling is caused by the absorption of direct solar radiation by the body.

Weather-Stress Index

Early in 1983 the National Weather Service began using a new tool for gauging human discomfort. Developed jointly by the National Oceanographic and Atmospheric Administration (NOAA) and the University of Delaware, the **weather-stress index** is intended to estimate relative human discomfort and stress attributed to the weather for both summer and winter seasons.[4] The index is based on

[4] This discussion is based largely on National Weather Service Technical Procedures Bulletin No. 324, *Weather Stress Index*, by Laurence S. Kalkstein, December 1982; and "An Evaluation of Winter Weather Severity in the United States Using the Weather Stress Index," by Laurence S. Kalkstein and Kathleen M. Valimont, in *Bulletin of the American Meteorological Society*, 68, no. 12 (December, 1987), pp. 1535–40.

TABLE 3–3 Wind-Chill Equivalent Temperature as a Function of Wind Speed and Air Temperature

Actual Temperature (°C)	Wind Speed (km per hr)										
	6	10	20	30	40	50	60	70	80	90	100
20	20	18	16	14	13	13	12	12	12	12	12
16	16	14	11	9	7	7	6	6	5	5	5
12	12	9	5	3	1	0	−0	−1	−1	−1	−1
8	8	5	0	−3	−5	−6	−7	−7	−8	−8	−8
4	4	0	−5	−8	−11	−12	−13	−14	−14	−14	−14
0	0	−4	−10	−14	−17	−18	−19	−20	−21	−21	−21
−4	−4	−8	−15	−20	−23	−25	−26	−27	−27	−27	−27
−8	−8	−13	−21	−25	−29	−31	−32	−33	−34	−34	−34
−12	−12	−17	−26	−31	−35	−37	−39	−40	−40	−40	−40
−16	−16	−22	−31	−37	−41	−43	−45	−46	−47	−47	−47
−20	−20	−26	−36	−43	−47	−49	−51	−52	−53	−53	−53
−24	−24	−31	−42	−48	−53	−56	−58	−59	−60	−60	−60
−28	−28	−35	−47	−54	−59	−62	−64	−65	−66	−66	−66
−32	−32	−40	−52	−60	−65	−68	−70	−72	−73	−73	−73
−36	−36	−44	−57	−65	−71	−74	−77	−78	−79	−79	−79
−40	−40	−49	−63	−71	−77	−80	−83	−85	−86	−86	−86
−44	−44	−53	−68	−77	−83	−87	−89	−91	−92	−92	−92
−48	−48	−58	−73	−82	−89	−93	−96	−98	−99	−99	−99
−52	−52	−62	−78	−88	−95	−99	−102	−104	−105	−105	−105
−56	−56	−67	−84	−94	−101	−105	−109	−111	−112	−112	−112
−60	−60	−71	−89	−99	−107	−112	−115	−117	−118	−118	−118

Source: NOAA, National Weather Service.

Apparent Temperature

studies conducted by R. G. Steadman, who evaluated human physiological responses to various weather situations. The result of Steadman's research is the **apparent temperature,** which is defined as the perceived air temperature for an individual.

The weather-stress index is simple to understand and contains features not found in the commonly used temperature-humidity index or wind-chill index. The same units are used to express the index in both summer and winter, for example. Moreover, the weather-stress index makes use of temperature, relative humidity, and wind, whereas the other indices evaluate only two of these parameters.

The weather-stress index assumes that the degree of human discomfort depends on variation from the mean condition and that people are relatively comfortable when average conditions prevail in their locale. To illustrate, if it were 33°C, with a relative humidity of 40 percent and a wind speed of 5 kilometers per hour at Duluth, Minnesota, on a particular summer day, this would represent a most uncomfortable situation and the local residents would suffer accordingly. Yet a similar situation in New Orleans would be relatively common to most inhabitants because they expect and are prepared for such weather.

The weather-stress index is calculated by evaluating how the apparent temperature for a particular day varies from the mean apparent temperature for that day at a particular place. Based on 40 years of data, mean apparent temperatures for each day of the year for about 100 evenly spaced stations around the United States have been determined. It is assumed that people are adjusted to the average condition in their region and that they undergo increasing stress when apparent temperatures drop below normal in winter and rise above normal in summer. Thus the weather-stress index is defined as a proportion of days with weather more comfortable (or less stressful) than the day under review. The index has a range of 0 to 100 percent, with 100 percent representing the most uncomfortable (or stressful) situation. For that day of the year, virtually all the days would be expected to have more comfortable weather. During the summer a 100 percent index value indicates that this day's apparent temperature is unusually high, and in the winter the same value represents an unusually low apparent temperature. An index value of 50 percent represents a "normal" or "average" apparent temperature for a given day, and it is assumed that an individual is not under stress and hence that daily activities are not hampered. Of course, a 50 percent value in Bismarck, North Dakota, in January corresponds to an apparent temperature approaching −18°C, whereas the same value during January in Tampa, Florida, exceeds 15°C.

The index is used to generate weather-stress maps of the United States, such as those in Figure 3–8. Besides being of interest to the general public, the maps are also useful for industry, agriculture, recreation, transportation, and medicine.

CONTROLS OF TEMPERATURE

Controls of Temperature

The **controls of temperature** are those factors that cause variations in temperature from place to place. Chapter 2 examined the single greatest cause for temperature variations—differences in the receipt of solar radiation. Because variations in sun angle and length of daylight are a function of latitude, they are responsible for warm temperatures in the tropics and colder temperatures at more poleward locations. Still, latitude is not the only control of temperature; if it were, we would expect all places along the same parallel to have identical temperatures. Such is clearly not the case. Eureka, California, and New York City, for instance, are both coastal cities at about the same latitude, and both places have an average annual mean temperature of 11°C. Yet New York City is 9.4°C warmer than Eureka in July and 9.4°C colder than Eureka in January. Two cities in Ecuador, Quito and Guayaquil, are relatively close to one another, but the mean annual temperatures at these two cities differ by 12.2°C. To explain these situations and countless others, we must realize that factors other than variations in solar radiation also exert a strong influence on temperatures. In the following pages we examine these other controls, which include

1. Differential heating of land and water
2. Ocean currents
3. Altitude
4. Geographic position

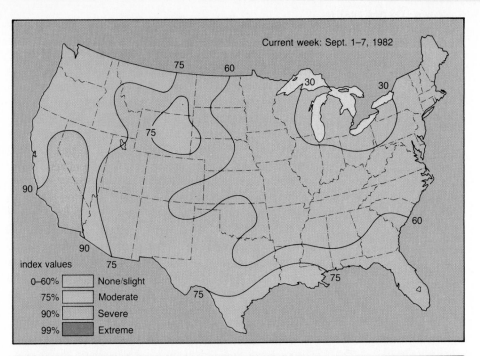

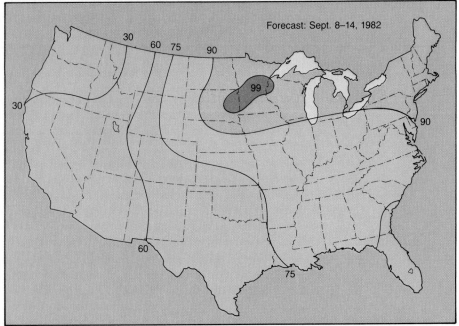

FIGURE 3–8

The weather-stress index estimates the reaction of people to variable weather conditions. It assumes that human reactions are dependent upon variation from the "normal" condition, and that people are relatively comfortable during normal weather at their locale. The index is expressed as a percentage of days with weather less stressful to the local populace than today. Thus a value of 95% represents severely stressful conditions because 95 days out of 100 would have less stressful weather. (Courtesy of NOAA)

Land and Water

The heating of the earth's surface controls the heating of the air above. Therefore in order to understand variations in air temperatures, we must examine the heating properties of various surfaces. Different land surfaces reflect and absorb varying amounts of incoming solar energy, which, in turn, causes variations in the temperature of the air above. The greatest contrast, however, is not between different land surfaces but between land and water. Land heats more rapidly and to higher temperatures than water, and it cools more rapidly and to lower temperatures than water. Variations in air temperatures, therefore, are much greater over land than over water.

An important reason for surface water temperatures rising and falling much more slowly than surface temperatures on land is the fact that water is highly mobile. As water is heated, turbulence distributes the heat through a considerably larger mass. On a daily basis temperature changes occur to depths of 6 meters or more below the surface, and on a yearly basis oceans and deep lakes are subject to temperature variations through a layer between 200 and 600 meters thick.

In contrast, heat does not penetrate deeply into soil or rock; it remains in a thin surface layer. Obviously no turbulent mixing occurs on land; heat must be transferred by the slow process of conduction. Consequently, daily temperature changes are small below a depth of 10 centimeters, although some change can occur to a depth of perhaps 1 meter. Annual temperature variations usually reach depths of 15 meters or less. Thus as a result of the mobility of water and the lack of it in the solid earth, a relatively thick layer of water is heated to moderate temperatures during the summer. On land, only a thin layer is heated but to much higher temperatures. During winter the shallow layer of rock and soil that was heated in summer cools rapidly. Water bodies, on the other hand, cool slowly as they draw on the reserves of heat stored within. As the surface of the water cools, vertical motions are established. The chilled surface water sinks and is replaced by warmer water from below. Consequently, a larger mass of water must be cooled before the temperature at the surface will drop appreciably.

Other factors also contribute to the differential heating of land and water.

Specific Heat

1. Because land surfaces are opaque, heat is absorbed only at the surface. Water, being more transparent, allows some solar radiation to penetrate to a depth of several meters.
2. The **specific heat** (the amount of heat needed to raise 1 gram of a substance 1°C) is almost three times greater for water than for land. Thus water requires considerably more heat to raise its temperature the same amount as an equal quantity of land.
3. Evaporation (a cooling process) from water bodies is greater than from land surfaces.

All these factors collectively cause water to warm more slowly, store greater quantities of heat energy, and cool more slowly than land.

TABLE 3–4 Monthly Temperature Data: Marine and Continental Stations

Vancouver, British Columbia (Marine)

	J	F	M	A	M	J	J	A	S	O	N	D	Annual
Temperature (°C)	2	4	6	9	13	15	18	17	14	10	6	4	10

Winnipeg, Manitoba (Continental)

	J	F	M	A	M	J	J	A	S	O	N	D	Annual
Temperature (°C)	−18	−15	−8	3	11	17	20	19	13	6	−5	−13	2.5

Monthly temperature data for two cities will demonstrate the moderating influence of a large water body and the extremes associated with land (Table 3–4). Vancouver, British Columbia, is located along a windward coast, whereas Winnipeg, Manitoba, is in a continental position far from the influence of water. Both cities are at about the same latitude and thus experience similar sun angles and lengths of daylight. Winnipeg, however, has a mean January temperature that is 20°C lower than Vancouver's. Conversely, Winnipeg's July mean is 2°C higher than Vancouver's. Although their latitudes are nearly the same, Winnipeg, which has no water influence, experiences much greater temperature extremes than Vancouver, which does.

On a different scale, the moderating influence of water may also be demonstrated when temperature variations in the northern and southern hemispheres are compared. The views of earth in Figure 3–9 show the uneven distribution of land and water over the globe. Water covers 61 percent of the northern hemisphere; land represents the remaining 39 percent. However, the figures for the southern hemisphere (81 percent water, 19 percent land) reveal why it is correctly called the *water hemisphere*. Between 45°N and 70°N latitude there is actually more land than water, whereas between 40°S and 65°S latitude there is almost no land to interrupt the oceanic and atmospheric circulation. Table 3–5 portrays the considerably smaller annual temperature variations in the water-dominated southern hemisphere compared with the northern hemisphere.

Ocean Currents

The effects of ocean currents on temperatures of adjacent land areas are variable. The moderating effect of poleward-moving warm ocean currents is well known. The North Atlantic Drift, an extension of the warm Gulf Stream (Figure 3–10), keeps wintertime temperatures in Great Britain and much of western Europe warmer than would be expected for their latitudes. Because of the prevailing westerly winds, the moderating effects are carried far inland. Berlin (52°N latitude), for example, has a mean January temperature similar to that experienced at New York City, which lies 12° latitude farther south, whereas the January mean at London (51°N latitude) is 4.5°C higher than at New York City.

In contrast to warm ocean currents whose effects are felt most during the

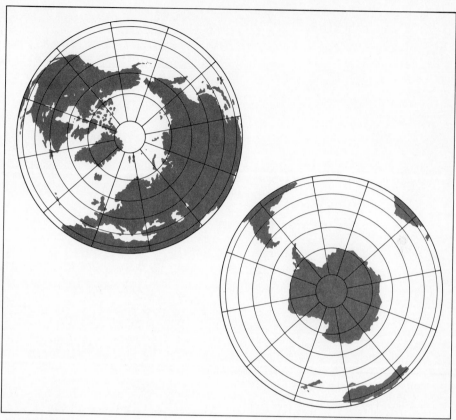

FIGURE 3–9
These views of the earth show the uneven distribution of land and water between the northern and southern hemispheres. Almost 81 percent of the southern hemisphere is covered by the oceans—20 percent more than the northern hemisphere.

TABLE 3–5 Variation in Mean Annual Temperature Range with Latitude (°C)

Latitude	Northern Hemisphere	Southern Hemisphere
0	0	0
15	3	4
30	13	7
45	23	6
60	30	11
75	32	26
90	40	31

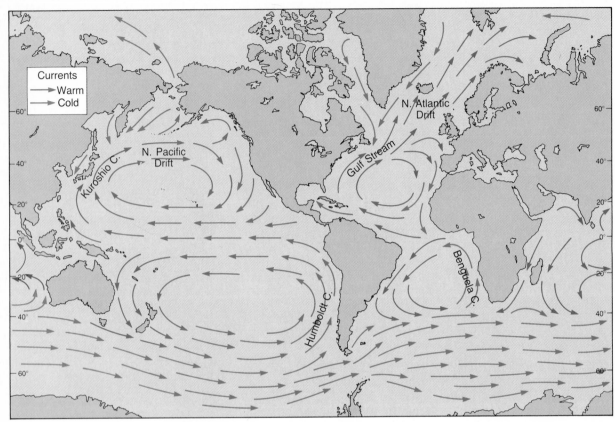

FIGURE 3–10
Major ocean currents. Poleward-moving currents are warm, and equatorward-moving currents are cold.

winter, the influence of cold currents is most pronounced in the tropics or during the summer months in the middle latitudes. Cool currents, such as the Benguela Current off the western coast of southern Africa, moderate the tropical heat. Walvis Bay (23°S latitude), a town adjacent to the Benguela Current, is 5°C cooler in summer than Durban, which is 6° latitude farther poleward but on the eastern side of South Africa away from the influence of the current. Because of the cold California Current, summer temperatures in subtropical coastal southern California are lower by 6°C or more compared to East Coast stations in the subtropical United States.

Altitude

The two cities in Ecuador mentioned earlier, Quito and Guayaquil, demonstrate the influence of altitude on mean temperature. Both cities are near the equator and relatively close to one another, but the annual mean at Guayaquil is 25.5°C compared with Quito's mean of 13.3°C. The difference may be understood when the cities' elevations are noted. Guayaquil is only 12 meters above sea level, whereas Quito is high in the Andes Mountains at 2800 meters. Recall that temperatures drop an average of 6.5°C per kilometer in the troposphere; thus cooler temperatures are to be expected at greater heights. Yet the magnitude of the difference is not totally explained by the normal lapse rate. If this figure were used, we would expect Quito to be about 18.2°C cooler than Guayaquil, but the difference is only 12.2°C. The fact that high-altitude places, such as Quito, are warmer than the value calculated using the normal lapse rate results from the absorption and reradiation of solar energy by the ground surface.

In addition to the effect of altitude on mean temperatures, the daily temperature range also changes with variations in height. Not only do temperatures drop with an increase in altitude, but atmospheric pressure and density also diminish. Because of the reduced density at high altitudes, the overlying atmosphere absorbs and reflects a smaller portion of the incoming solar radiation. Consequently, with an increase in altitude, the intensity of insolation increases, resulting in rapid and intense daytime heating. Conversely, rapid nighttime cooling is also the rule in high mountain locations. Therefore stations located high in the mountains generally have a greater daily temperature range than stations at lower elevations.

Geographic Position

The geographic setting may greatly influence the temperatures experienced at a specific location. A windward coastal location—that is, a place that is subject to prevailing onshore winds—experiences considerably different temperatures than a coastal location where the prevailing winds are directed from the land toward the ocean. In the first situation, the place will experience the full moderating influence of the ocean—cool summers and mild winters compared to an inland station at the same latitude. A leeward coastal situation, however, will have a more continental temperature regime because the winds do not carry the ocean's influence onshore. Eureka, California, and New York City, the two cities mentioned earlier, illustrate this aspect of geographic position (Table 3–6). The annual temperature range at New York City is 19°C higher than Eureka's.

Seattle and Spokane, both in the state of Washington, illustrate a second aspect of geographic position: mountains acting as barriers. Although Spokane is only

---

TABLE 3–6 Monthly Temperature Data: Windward and Leeward Coastal Stations

Eureka, California (Windward)

	J	F	M	A	M	J	J	A	S	O	N	D	ANNUAL
Temp. (°C)	9	9	9	10	12	13	14	14	14	12	11	9	11

New York City (Leeward)

	J	F	M	A	M	J	J	A	S	O	N	D	ANNUAL
Temp. (°C)	−1	−1	3	9	16	21	23	23	21	15	7	2	11

TABLE 3–7 Monthly Temperature Data: Mountains Acting as Barriers

Seattle, Washington

	J	F	M	A	M	J	J	A	S	O	N	D	ANNUAL
Temp. (°C)	4	5	7	9	12	15	17	17	14	11	8	6	11

Spokane, Washington

	J	F	M	A	M	J	J	A	S	O	N	D	ANNUAL
Temp. (°C)	−3	−1	3	9	13	16	21	20	16	9	2	−1	9

about 360 kilometers east of Seattle, the towering Cascade Range separates the cities. Consequently, while Seattle's temperatures show a marked marine influence, Spokane's are more typically continental (Table 3–7). Spokane is 7°C cooler than Seattle in January and 4°C warmer than Seattle in July. The annual range at Spokane is 11°C greater than at Seattle. The Cascade Range effectively cuts Spokane off from the moderating influence of the Pacific Ocean.

WORLD DISTRIBUTION OF TEMPERATURES

By examining isothermal maps for January and July (Figures 3–11 and 3–12), global temperature patterns can be studied and the effects of the controls of temperature, especially latitude, the distribution of land and water, and ocean currents, can be seen. January and July represent the seasonal extremes of temperature for most places on earth; for this reason, these months are usually selected for analysis. As is the case for most isothermal maps of large regions, all temperatures on these world maps have been reduced to sea level in order to eliminate the complications caused by differences in altitude.

On both maps the isotherms generally trend east and west and show a decrease in temperatures poleward from the tropics. They illustrate one of the most fundamental and best-known aspects of the world distribution of temperature: the fact that the effectiveness of incoming solar radiation in heating the earth's surface and the atmosphere above is largely a function of latitude. Moreover, there is a latitudinal shifting of temperatures caused by the seasonal migration of the sun's vertical rays.

If the effect of latitudinal variations in the receipt of solar energy were the only control of temperature distribution, our analysis could end at this point, but such, of course, is not the case. The added effect of the differential heating of land and water is also reflected on the January and July temperature maps. The warmest and coldest temperatures are found over land. Consequently, because temperatures do not fluctuate as much over water as over land, the north-south migration of isotherms is greater over the continents than over the oceans. In addition, it is clear that the isotherms in the southern hemisphere, where there

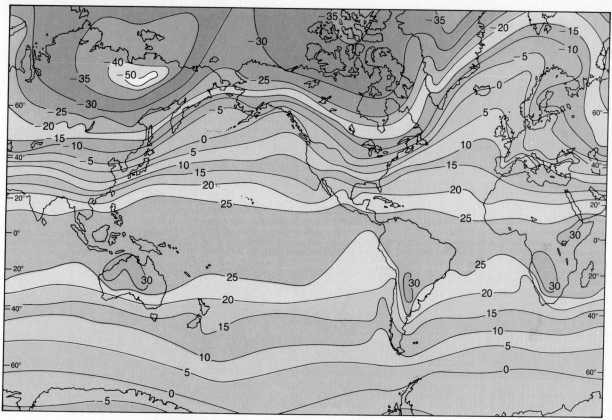

FIGURE 3–11
World mean sea-level temperatures in January in °C.

is little land and where the oceans predominate, are much more regular than in the northern hemisphere, where they bend sharply northward in July and southward in January over the continents.

Isotherms also reveal the presence of ocean currents. Warm currents cause isotherms to be deflected toward the poles, whereas cold currents cause an equatorward bending. The horizontal transport of water poleward warms the overlying air and results in air temperatures that are higher than would otherwise be expected for the latitude. Conversely, currents moving toward the equator produce cooler than expected air temperatures.

Because Figures 3–11 and 3–12 show the seasonal extremes of temperature, it is possible to evaluate variations in the annual range of temperature from place to place. A comparison of the two maps shows that a station near the equator will record a very small annual range because it experiences little variation in the length of daylight and it always has a relatively high sun angle. A station in the middle latitudes, however, experiences much wider variations in sun angle and length of daylight and hence large variations in temperature. Therefore we can state that the annual temperature range increases with an increase in latitude.

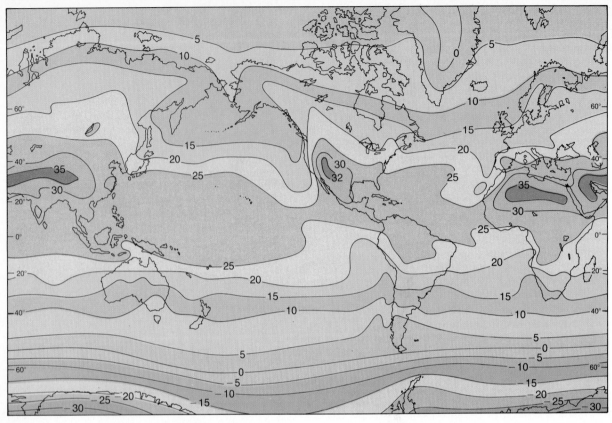

FIGURE 3–12
World mean sea-level temperatures in July in °C.

Moreover, land and water also affect seasonal temperature variations, especially outside the tropics. A continental location must endure hotter summers and colder winters than a coastal location. Consequently, the annual range will increase with an increase in continentality.

A classic example of the effect of latitude and continentality on annual temperature range is Yatkusk, USSR. This city is located in Siberia at approximately 60°N latitude and far from the influence of water. As a result, Yatkusk has an average annual temperature range of 62.2°C, among the highest in the world.

CYCLES OF AIR TEMPERATURE

Daily March of Temperature

A thermograph record such as Figure 3–13 or your own experience reveals that there is a rhythmic rise and fall of air temperatures during the day, a phenomenon commonly called the **daily march of temperature.** After reaching a minimum about sunrise, the temperature curve climbs steadily until midafternoon to late afternoon. Attaining a maximum between 2 P.M. and 5 P.M., the temperature then declines until sunrise the following day. The primary control of this daily cycle of air temperature is probably as obvious as the cycle itself; it is the sun. As the

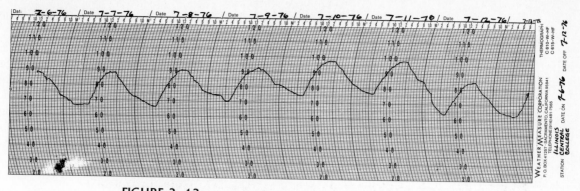

FIGURE 3–13
This thermograph record for a July week at Peoria, Illinois, illustrates the daily rhythm of temperatures typical of the middle latitudes.

sun angle increases throughout the morning, the intensity of sunlight also rises, reaching a peak at noon and gradually diminishing in the afternoon. During the night the atmosphere and the surface of the earth cool as they radiate heat away that is not replaced by incoming solar energy. The minimum temperature therefore occurs about the time of sunrise, after which the sun again heats the ground, which, in turn, heats the air. It is apparent to most of us, however, that the time of highest temperature does not generally coincide with the time of maximum radiation. The delay in the occurrence of the maximum until midafternoon to late afternoon is termed the *lag of the maximum.*

The lag of the daily maximum is mainly a result of the process by which the atmosphere is heated. Recall that air is a poor absorber of most solar radiation; consequently, it is heated primarily by energy acquired from the earth's surface. The rate at which the earth supplies heat to the atmosphere through radiation, conduction, and other means, however, is not in balance with the rate at which the atmosphere radiates heat away. Generally for a few hours after the period of maximum solar radiation, more heat is supplied to the atmosphere by the earth's surface than is emitted by the atmosphere to space. Consequently, most locations experience an increase in air temperature during the afternoon hours. In dry regions, particularly on cloud-free days, the amount of radiation absorbed by the earth's surface will generally be high. As a result, the time of the maximum temperature at these locales will often occur quite late in the afternoon. Humid locations, on the other hand, will frequently experience a shorter time lag in the occurrence of their temperature maximum.

The magnitude of daily temperature changes is variable and may be influenced by locational factors or local weather conditions or both. Three common examples will illustrate this point. The first two relate to location and the third pertains to the influence of clouds.

1. Variations in sun angle are rather great during the course of a day in the middle and low latitudes; however, points near the poles experience a low sun angle all day. Consequently, the amount of temperature

change experienced during the course of a day in the high latitudes is small.

2. A location on a windward coast is likely to experience only modest variations in the daily cycle. During a typical 24-hour period the ocean warms less than 1°C. As a result, the air above shows a correspondingly slight change in temperature. For example, Eureka, California, a windward coastal station, consistently has a lower daily temperature range than Des Moines, Iowa, an inland city at about the same latitude. On an annual basis the daily range at Des Moines averages 10.9°C compared with 6.1°C at Eureka, a difference of 4.8°C.

3. By comparing the temperature curve for July 25 (a clear day) to that of July 27 (an overcast day), the effect of cloud cover on the daily march of temperature may be seen in Figure 3–14. It is apparent that an overcast day is responsible for a flattened daily temperature curve. By day clouds block incoming solar radiation and so reduce daytime heating. At night the clouds retard the loss of radiation by the ground and air and reradiate heat earthward. Therefore nighttime temperatures are not as low as they otherwise would have been.

Although the rise and fall of daily temperatures usually reflect the general rise and fall of solar radiation, such is not always the case. If thermograph records for a station were examined for a period of several weeks, nonperiodic variations that are obviously not sun controlled would be seen. Such irregularities are caused primarily by the passage of atmospheric disturbances that are often accompanied by variable cloudiness and winds that bring air having contrasting temperatures. Under these circumstances the maximum and minimum temperatures may occur at any time of the day or night.

FIGURE 3–14
The daily march of temperature at Peoria, Illinois, on a clear day and on an overcast day. As is typical, the maximum temperature on the clear day was higher and the minimum temperature was lower than for the cloudy day.

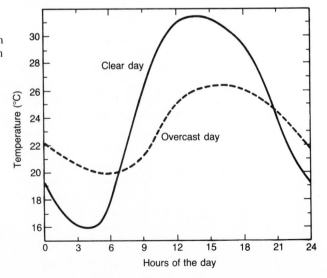

Annual March of Temperature

Before leaving the topic of temperature cycles, the **annual march of temperature** should also be mentioned. On a yearly basis the dates of the high and low temperatures do not coincide with the times of maximum and minimum solar radiation. Although the greatest intensity of solar radiation occurs at the time of the summer solstice, the months of July and August are generally the warmest of the year in the northern hemisphere. Conversely, a minimum of solar energy is received in December at the time of the winter solstice, but January and February are usually colder.

The fact that the occurrence of annual maximum and minimum radiation does not coincide with the times of temperature maximums and minimums attests to the fact that the amount of solar radiation received is not the only factor that determines the temperature at a particular location. Recall that places located equatorward of about 36° receive more solar radiation than is lost to space and that the opposite is true of more poleward regions. Based on the imbalance between incoming and outgoing radiation, any location in the southern United States, for example, should continue to get warmer late into autumn. This situation does not occur because more poleward locations begin experiencing a negative radiation balance shortly after the summer solstice. As the temperature contrasts become greater, the atmosphere and ocean currents "work harder" to transport heat from lower latitudes poleward. Shortly after the solar maximum, the temperature contrasts that persist across latitudes become much more pronounced over the continents than over the oceans. Consequently, the lag of the warmest month is greater in areas that have a strong oceanic influence. Whereas July is the warmest month at practically all continental stations, windward coastal cities, such as Los Angeles and San Diego, demonstrate even greater lags of the maximum. In both instances, the months of August and September are warmer than July.

REVIEW

1. Distinguish between the concepts of heat and temperature. Include an example to illustrate the difference.

2. Describe how each of the following thermometers works: liquid-in-glass, maximum, minimum, bimetal strip, thermocouple, thermistor.

3. What is a thermograph? Which one of the thermometers listed in question 2 is commonly used in the construction of a thermograph?

4. In addition to having an accurate thermometer, which other factors must be considered in order to obtain a meaningful air temperature reading?

5. **a.** What is meant by the terms steam point and ice point?
   b. What values are given these points on each of the three temperature scales presented in this chapter?

6. Why is it impossible to have a negative value when using the Kelvin temperature scale?

7. How are the following temperature data calculated: daily mean, daily range, monthly mean, annual mean, annual range?

8. What are isotherms and what is their purpose?

9. The mean temperature is 55°F on a particular day. The following day the mean drops to 45°F. Calculate the number of heating degree days for each day. How much more fuel would be needed to heat a building on the second day compared with the first day?

10. When heating- and cooling-degree day totals for different places are examined for the purpose of comparing fuel consumption, what important assumption is made?

11. How are growing degree days calculated? For what purpose is this index used?

12. If the air temperature is 85°F and the relative humidity is 80 percent, what is the THI? If the relative humidity were 30 percent, what would the THI be?

13. Using Table 3–3, determine equivalent temperatures under the following circumstances:
    a. Temperature = −12°C, wind speed = 20 km/hr.
    b. Temperature = −12°C, wind speed = 50 km/hr.

14. Briefly explain why a calm and sunny winter day will feel warmer than the thermometer reading indicates.

15. If Atlanta, Georgia, and Chicago, Illinois, experienced nearly similar weather conditions on a particular day, would both cities have the same weather-stress index? Explain.

16. What does it mean when the weather-stress index is 90 percent?

17. a. State the relationship between the heating and cooling of land versus water.
    b. List and explain the factors that cause the difference between the heating and cooling of land and water.
    c. Because we are interested in the atmosphere, why are we concerned with the heating characteristics at the earth's surface?

18. How does the annual temperature range near the equator compare with the annual temperature ranges experienced in the middle to high latitudes? Explain.

19. Three cities are located at the same latitude (about 45°N latitude). One city is located along a windward coast, another in the center of the continent, and the third along a leeward coast. Compare the annual temperature ranges of these cities.

20. Answer the following questions about world temperature distribution (you may wish to refer to the January and July isotherm maps).
    a. Isotherms generally trend east-west. Why?
    b. Isotherms bend (poleward, equatorward) over continents in summer. Underline the correct answer and explain.
    c. Isotherms shift north and south from season to season. Why?
    d. Where do isotherms shift most, over land or water? Explain.
    e. How do isotherms show ocean currents? How can you tell if the current is warm or cold?
    f. Why are the isotherms more irregular in the northern hemisphere than in the southern hemisphere?

21. By referring to the world maps of temperature distribution for January and July (Figures 3–11 and 3–12), determine the approximate January mean, July mean, and annual temperature range for a place located at 60°N latitude, 80°E longitude and a place located at 60°S latitude, 80°E longitude.

22. Although the intensity of incoming solar radiation is greatest at noon, the warmest part of the day is most often midafternoon. Why?

23. How does the daily march of temperature on a completely overcast day compare with that on a cloudless and sunny day? Explain your answer.

VOCABULARY REVIEW

heat

temperature

thermometer

liquid-in-glass thermometer

maximum thermometer

minimum thermometer

bimetal strip

thermograph

thermocouple

thermistor

fixed points

Fahrenheit

ice point

steam point

Celsius scale

Kelvin scale

absolute zero

daily mean

daily range

monthly mean

annual mean

annual temperature range

isotherm

heating degree day

cooling degree day

growing degree day

temperature–humidity index (THI)

wind chill

weather-stress index

apparent temperature

controls of temperature

specific heat

daily march of temperature

annual march of temperature

4

Humidity, Condensation, and Atmospheric Stability

ater vapor constitutes only a small fraction of the atmosphere, varying from almost 0 to 4 percent by volume. The importance of water in the air is far greater than this small percentage would indicate. Indeed, scientists agree that when it comes to understanding atmospheric processes, water vapor is the most important gas in the atmosphere. As you observe day-to-day weather changes, many questions may come to mind concerning the role of moisture in the atmosphere. What is relative humidity and how is it measured? Why do clouds form on some occasions but not on others? Why do some clouds look thin and harmless whereas others appear as gray and ominous towers? In the following pages we investigate these and other questions involving water in the air.

Some of the answers to the questions just posed are related to a concept known as atmospheric stability. The stability of air plays a significant part in controlling many aspects of daily weather and is closely related to a major environmental concern—air pollution. We will see that air quality is not just a function of the quantity and types of pollutants emitted into the air, but it is also closely linked to the atmosphere's ability to disperse these noxious substances. Dispersal, in turn, is related to the stability of the atmosphere. Therefore, in addition to dealing with the concepts of humidity and cloud formation, Chapter 4 will examine the concept of atmospheric stability and relate it to daily weather and to air pollution problems.

THE HYDROLOGIC CYCLE

An adequate supply of water is vital to life on earth. The increasing demands on this finite resource have led scientists to pay a great deal of attention to the continuous exchanges of water between the oceans, the atmosphere, and the continents. This unending circulation of the earth's water supply has come to be called the **hydrologic cycle.** It is a gigantic system powered by energy from the sun in which the atmosphere provides the vital link between the oceans and continents. Water from the oceans, and to a much lesser extent from the continents, is constantly evaporating into the atmosphere. Winds transport the moisture-laden air, often great distances, until the complex processes of cloud formation are set in motion that eventually result in precipitation. The precipitation that falls into the ocean has ended its cycle and is ready to begin another. The water that falls on the continents, however, must still make its way back to the oceans.

Hydrologic Cycle

Once precipitation has fallen on land, a portion of the water soaks into the ground, some of it moving downward, then laterally, and finally seeping into lakes and streams or directly into the ocean. When the rate of rainfall is greater than the earth's ability to absorb it, the additional water flows over the surface into streams and lakes. Much of the water that soaks in or runs off eventually finds its way back to the atmosphere. In addition to evaporation from the soil, lakes, and streams, some water that infiltrates the ground surface is absorbed by plants, which then release it into the atmosphere, a process called **transpiration.**

Transpiration

A diagram of the earth's water balance, a quantitative view of the hydrologic cycle, is shown in Figure 4–1. Although the amount of water vapor in the air at any one time is but a minute fraction of the earth's total water supply, the absolute quantities that are cycled through the atmosphere in one year are immense, some 380,000 cubic kilometers—enough to cover the earth's surface to a depth of about 100 centimeters. Estimates show that over North America almost six times more water is carried within the moving currents of air than is transported by all the continent's rivers. Because the total amount of water vapor in the atmosphere remains about the same, the average annual precipitation over the earth must be equal to the quantity of water evaporated. However, for all the continents taken together, precipitation exceeds evaporation. Conversely, over the oceans evaporation exceeds precipitation. Because the level of the world ocean is not dropping, runoff from land areas must balance the deficit of precipitation over the oceans.

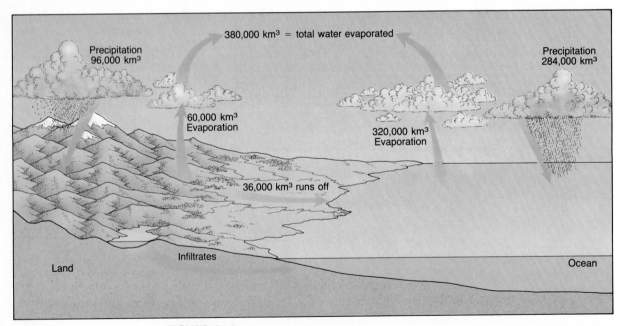

FIGURE 4–1
The earth's water balance. Huge quantities of water are cycled through the atmosphere each year. Notice that on the continents precipitation exceeds evaporation and that the reverse is true for the oceans. Since the level of the world ocean is not dropping, runoff from the continents must balance the deficit of the oceans.

In summary, the hydrologic cycle represents the continuous movement of water from the oceans to the atmosphere, from the atmosphere to the land, and from the land back to the sea. The movement of water through the cycle holds the key to the distribution of moisture over the surface of our planet and is intricately related to all atmospheric phenomena.

CHANGES OF STATE

Water vapor is an odorless, colorless gas that mixes freely with the other gases of the atmosphere. Unlike oxygen and nitrogen—the two most abundant components of the atmosphere—water vapor can change from one state of matter (solid, liquid, or gas) to another at the temperatures and pressures experienced near the surface of the earth. It is because of this ability, which allows water to leave the oceans as a gas and return again as a liquid, that the vital hydrologic cycle exists. The processes that involve a change of state require that heat be absorbed or released, as shown in Figure 4–2. The heat energy involved is often measured in calories. One **calorie** is the amount of heat required to raise the temperature of 1 gram of water 1°C. Thus when 10 calories of heat is added to 1 gram of water, a 10°C temperature rise occurs.

Calorie

Under certain conditions heat may be added to a substance without an accompanying temperature change. This situation occurs during a change in state. When heat is supplied to a glass of ice water (0°C), for example, the temperature remains constant until all the ice has melted. Where has the heat gone? In this case, the energy was used to disrupt the internal crystalline structure of the ice cubes and cause them to melt. Because this heat energy is not associated with a temperature change, it is generally referred to as **latent** (meaning hidden) **heat.** This energy is not available as heat until the liquid returns to the solid state. The importance of latent heat in atmospheric processes is crucial and is considered later.

Latent Heat

FIGURE 4–2
Changes of state.

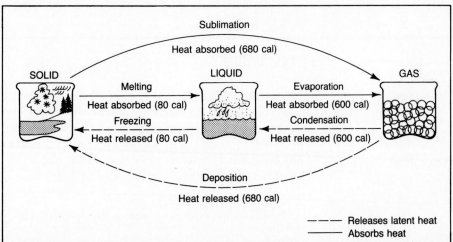

Evaporation

The process of converting a liquid to a gas is termed **evaporation.** It takes approximately 600 calories of energy to convert 1 gram of water to water vapor. The energy absorbed by the water molecules during evaporation is used solely to give them the motion needed to escape the surface of the liquid and become a gas. This energy is subsequently released as heat when the vapor changes back to a liquid and is referred to as *latent heat of vaporization.*[1] During the process of evaporation it is the higher-temperature (faster-moving) molecules that escape the surface. As a result, the average molecular motion (temperature) of the remaining water is reduced—hence the common expression "evaporation is a cooling process." You have undoubtedly experienced this cooling effect on stepping dripping wet from a swimming pool or bathtub.

Condensation

Condensation denotes the process whereby water vapor changes to the liquid state. In order for condensation to occur, the water molecules must release energy (*latent heat of condensation*) equivalent to what was absorbed during evaporation. This energy plays an important role in producing violent weather and can act to transfer great quantities of heat energy from tropical oceans to more poleward locations. When condensation occurs in the atmosphere, it results in the formation of such phenomena as fog and clouds.

Melting

Melting is the process by which a solid is changed to a liquid. It requires absorption of approximately 80 calories of energy per gram of water. **Freezing,** the reverse process, releases these 80 calories per gram as *latent heat of fusion.*

Freezing

Sublimation

The last of the processes illustrated in Figure 4–2 are **sublimation** and **deposition.** Sublimation is the term used to describe the conversion of a solid directly to a gas without passing through the liquid state. You may have observed this change as you watched the sublimation of dry ice (frozen carbon dioxide). The term deposition is used to denote the reverse process, the conversion of a vapor to a solid. This change occurs, for example, during the formation of frost. A household example of the process of deposition is the "frost" that accumulates in a freezer compartment. As shown in Figure 4–2, sublimation and deposition involve an amount of energy equal to the total of the other two processes.

Deposition

HUMIDITY

Humidity

Humidity is the general term used to describe the amount of water vapor in the air. Several methods are used to express humidity quantitatively. Among them are (1) absolute humidity, (2) mixing ratio, and (3) relative humidity.

Saturation

Before considering each of these humidity measures individually, it is important to understand the concept of **saturation.** To do so, imagine a closed container half full of pure water and overlain with dry air as shown in Figure 4–3. As the

---

[1] The latent heat of vaporization depends upon temperature and varies from 569 cal/g to 629 cal/g as the temperature goes from 50°C to −50°C. For most purposes, a value of 600 cal/g is a good approximation.

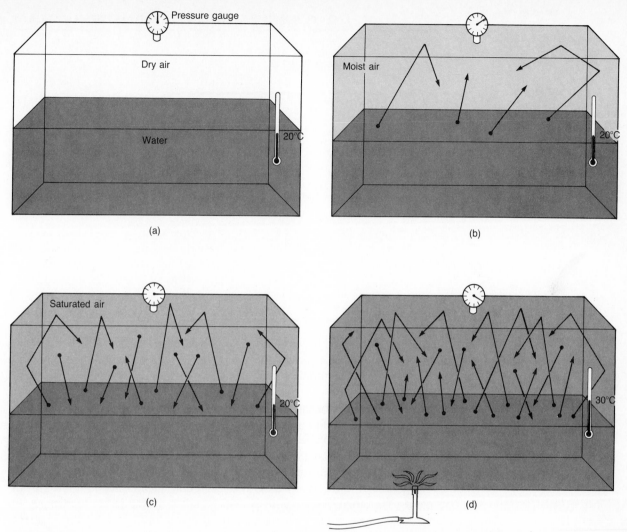

FIGURE 4–3

Schematic illustration of vapor pressure and saturation.

water begins to evaporate from the water surface, a small increase in pressure can be detected in the air above. This increase in pressure is the result of the motion of the water vapor molecules that were added to the air through evaporation. In the open atmosphere this pressure is termed *vapor pressure* and is defined as that part of the total atmospheric pressure attributable to its water vapor content. As more and more molecules escape from the water surface in the closed container, the steadily increasing vapor pressure in the air above forces more and more of these molecules to return to the liquid. Eventually the number of vapor molecules returning to the surface will balance the number leaving. At that point the air is said to be saturated. When air is saturated, the pressure exerted by the water vapor is called **saturation vapor pressure.** If we increased the temperature of the water in the closed container, however, more water would evaporate before a balance was reached. Consequently, the saturation vapor pressure is temperature

Saturation Vapor Pressure

TABLE 4–1 Saturation Mixing Ratio
(at Sea-Level Pressure)

Temperature (°C)	g/kg
−40	0.1
−30	0.3
−20	0.75
−10	2
0	3.5
5	5
10	7
15	10
20	14
25	20
30	26.5
35	35
40	47

dependent and rises with an increase in temperature. Stated more simply, at higher temperatures more water vapor is required for saturation to occur. The amount of water vapor required for saturation at various temperatures is shown in Table 4–1.

Now that we have become familiar with the concept of saturation, we can better appreciate the differences between the various methods of measuring the moisture content of air. Of the methods used to express humidity, absolute humidity and mixing ratio are similar in that both specify the amount of water vapor contained in a unit of air. **Absolute humidity** is expressed as the mass of water vapor in a given volume of air (usually as grams per cubic meter). As air moves from one place to another, variations in pressure and temperature cause changes in volume. When such volume changes occur, the absolute humidity also changes even if no water vapor is added or removed. Consequently, it is difficult to monitor the water vapor content of a moving mass of air if absolute humidity is the index being used. Therefore meteorologists generally use mixing ratio to express the water vapor content of air.[2] The **mixing ratio** is expressed as the mass of water vapor in a unit mass of dry air. Because it is measured in units of mass (usually grams per kilogram), the mixing ratio is not affected by changes in pressure or temperature. Neither the absolute humidity nor the mixing ratio, however, can be easily determined via direct sampling. Fortunately, they can be calculated from a more easily established measure, the relative humidity.

The most familiar and perhaps the most misunderstood term used to describe the moisture content of air is relative humidity. By definition, **relative humidity** is the ratio of the actual mixing ratio to the saturation mixing ratio. Stated another way, relative humidity is a ratio of the air's actual water vapor content compared

Absolute Humidity

Mixing Ratio

Relative Humidity

---

[2] Another commonly used expression is *specific humidity*, which is the mass of water vapor in a unit mass of air, including the water vapor. Because the amount of water vapor in the air rarely exceeds a few percent of the total mass of the air, the specific humidity of air is equivalent to its mixing ratio for all practical purposes.

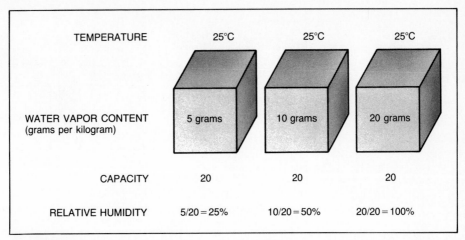

| TEMPERATURE | 25°C | 25°C | 25°C |

FIGURE 4–4
At a constant temperature the relative humidity will increase as water vapor is added to the air. Here, the capacity remains constant at 20 grams per kilogram and the relative humidity rises from 25 percent to 100 percent as the water vapor content increases.

with the amount of water vapor required for saturation at that temperature. Thus relative humidity indicates how near the air is to saturation rather than indicating the actual quantity of water vapor in the air. To illustrate, we see from Table 4–1 that at 25°C the saturation mixing ratio of the air is 20 grams per kilogram. If the air contains 10 grams per kilogram on a 25°C day, the relative humidity is expressed as 10/20 or 50 percent. When air is saturated, the relative humidity is 100 percent.

Because relative humidity is based on the air's water vapor content and the amount of moisture required for saturation is temperature dependent, relative humidity can be changed in either of two ways. First, if moisture is added to or subtracted from air, its relative humidity will change. Notice in Figure 4–4 that when water vapor is added to a parcel of air, its relative humidity increases until saturation occurs (100 percent relative humidity). What if even more moisture is added to this parcel of saturated air? Does the relative humidity exceed 100 percent? Normally this situation does not occur. Instead the excess water vapor condenses to form liquid water. You may have experienced such a situation while taking a hot shower. The water leaving the shower is composed of very energetic (hot) molecules, which means that the rate of evaporation is high. As long as you run the shower, the process of evaporation continually adds water vapor to the unsaturated air in the bathroom. Therefore if you take a long enough shower, the air eventually becomes saturated and the excess water vapor condenses on the mirror or window in the room. In nature moisture is added to the air mainly via evaporation from the oceans, but plants, soil, and smaller bodies of water do make substantial contributions. Unlike your shower, however, the rate of evaporation is generally not great enough to cause saturation to occur directly.

The second condition that affects relative humidity is air temperature (Figure 4–5). We can generalize the effect of temperature on relative humidity as follows. If the water vapor content remains at a constant level, a decrease in temperature

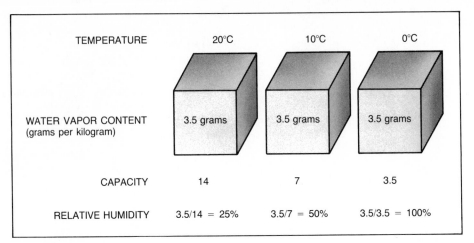

FIGURE 4–5
When the water vapor content (mixing ratio) remains constant, the relative humidity may be changed by increasing or decreasing the air temperature. In this example the mixing ratio remains at 3.5 grams per kilogram. The reduction in temperature from 20°C to 0°C causes a decrease in capacity and thus an increase in the relative humidity.

results in an increase in relative humidity and an increase in temperature causes a decrease in relative humidity. People living in the central United States experience this effect in their homes. In the summer warm, moist air enters their houses and some circulates into the relatively cool basements. As a result, the temperature of this air drops and relative humidity increases. In response, the homeowner often installs a dehumidifier to alleviate the problem of a "damp" basement. During winter months, however, cold air enters these houses and is heated. This process, in turn, causes relative humidity to drop, often to uncomfortably low levels. Consequently, the homeowner may install a humidifier to increase relative humidity to a comfortable level.

    In nature changes in relative humidity caused by temperature variations typically occur in one of three ways:

1. It changes as daily temperatures change.
2. It changes as air moves from one location to another.
3. It changes when air moves vertically in the atmosphere.

The importance of the last two processes will be discussed later. The effect of the typical daily temperature cycle on relative humidity is shown in Figure 4–6. Notice that during the warmer midday period relative humidity reaches its lowest level, whereas the cooler evening hours are associated with higher relative humidities. In this example the actual water vapor content (mixing ratio) of the air remains unchanged; only the relative humidity varies. Now we can better understand why a high relative humidity does not necessarily indicate a high water vapor content. To illustrate this point, let us compare a typical situation at Winnipeg, Manitoba, and Phoenix, Arizona, on a January day. On this hypothetical day the temperature in Winnipeg is a frigid −10°C and the relative humidity is 100 percent.

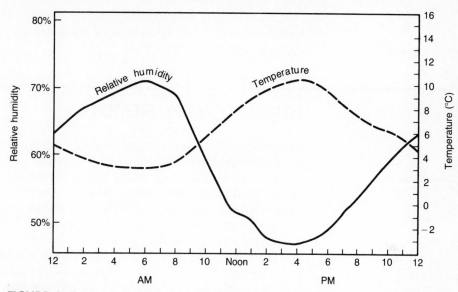

FIGURE 4–6
Typical daily variations in temperature and relative humidity during a spring day at Washington, D.C.

Using Table 4–1, we note that at saturation −10°C air has a mixing ratio of 2 grams per kilogram. In contrast, the desert air at Phoenix is a warm 25°C and the relative humidity is just 20 percent. Note from Table 4–1 that 25°C air has a saturation mixing ratio of 20 grams per kilogram. Therefore with a relative humidity of 20 percent, the air at Phoenix has a mixing ratio of 4 grams per kilogram (20 grams per kilogram × 20 percent). Consequently, the air at Phoenix actually has a higher water vapor content than the air at Winnipeg.

Despite the previous example, we still describe air having a low relative humidity as being "dry" and vice versa. The use of the word "dry" in this context indicates that the air is far from being saturated. Thus the rate of evaporation on a dry day is generally higher than on a humid day. In summary, relative humidity indicates how near the air is to being saturated, whereas the air's mixing ratio denotes the actual quantity of water vapor contained in that air.

Dew Point

Another important idea related to relative humidity is the dew-point temperature. **Dew point** is the temperature to which a parcel of air would need to be cooled in order to reach saturation. Note that in Figure 4–5 unsaturated air at 20°C must be cooled to 0°C before saturation occurs. Therefore 0°C would be the dew-point temperature for this air. If the same air were cooled further, the air's saturation mixing ratio would be exceeded and the excess water vapor would condense, typically as dew, fog, or clouds. The term dew point stems from the fact that during evening hours objects (particularly metallic objects) located near the earth's surface often cool below the dew-point temperature.[3] Air in contact

---

[3] Normally we associate dew with grass. Because of transpiration by the blades of grass, however, the relative humidity on a calm night is much higher near the grass than a few inches above the surface. Consequently, dew forms on grass before it does on most other objects.

with these surfaces also cools by conduction until it becomes saturated and dew begins to form. When the dew-point temperature is below freezing, the water vapor is deposited as white frost.

HUMIDITY MEASUREMENT

Hygrometer

Psychrometer

As noted, absolute humidity and mixing ratio are difficult to measure directly, but if the relative humidity is known, they can be readily computed by consulting an appropriate table or graph. A variety of instruments, called **hygrometers,** can be used to measure relative humidity. One of the simplest hygrometers, a **psychrometer,** consists of two identical thermometers mounted side by side (Figure 4–7). One thermometer, called the wet bulb, has a thin muslin wick tied around the end. To use the psychrometer, the cloth wick is saturated with water and a continuous current of air is passed over the wick, either by swinging the instrument freely in the air or by fanning air past it. As a result, water evaporates from the wick and the temperature of the wet bulb drops. The heat that was required to evaporate water from the wet bulb causes its temperature to drop. The amount of cooling that takes place is directly proportional to the dryness of the air. The drier the air, the greater the cooling. Therefore the larger the difference between the wet- and dry-bulb temperatures, the lower the relative humidity; the smaller the difference, the higher the relative humidity. If the air is saturated, no evaporation will occur and the two thermometers will have identical readings.

Tables have been devised to obtain both the relative humidity and the temperature of the dew point (Tables 4–2 and 4–3). All that is required is to record the air (dry-bulb) temperature and calculate the difference between the wet- and dry-bulb readings. The difference is known as the *depression of the wet bulb.* Assume, for instance, that the dry-bulb temperature is 20°C and that the wet-bulb reading after swinging or fanning is 15°C. To determine the relative humidity, find the dry-bulb temperature on the left-hand column of Table 4–2 and the depression of the wet bulb across the top. The relative humidity is found where the two meet. In this example the relative humidity is 58 percent. The dew point can be determined the same way, using Table 4–3. In this case, it would be 12°C.

Another commonly used instrument for measuring relative humidity, the *hair hygrometer*, can be read directly without using tables. The hair hygrometer operates on the principle that hair changes length in proportion to changes in relative humidity; the hair lengthens as relative humidity increases and it shrinks as relative humidity drops. People with naturally curly hair experience this phenomenon, for in humid weather their hair lengthens and hence becomes curlier. The hair

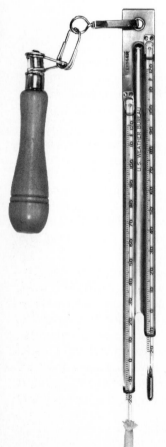

FIGURE 4–7

Sling psychrometer. This instrument is used to determine relative humidity and dew point. The dry-bulb thermometer gives the current air temperature. The thermometers are spun until the temperature of the wet bulb thermometer stops declining. Then the thermometers are read and the data are used in conjunction with Tables 4–2 and 4–3. (Courtesy of Belfort Instrument Company)

TABLE 4–2 Relative Humidity in Percent (1000 Millibars)

Dry-Bulb Temperature (°C)	Wet-Bulb Depression ($T_d - T_w$)																			
	1	2	3	4	5	6	7	8	9	10	11	12	13	14	15	16	17	18	19	20
−20	28																			
−18	40																			
−16	48	0																		
−14	55	11																		
−12	61	23																		
−10	66	33	0																	
−8	71	41	13																	
−6	73	48	20	0																
−4	77	54	32	11																
−2	79	58	37	20	1															
0	81	63	45	28	11															
2	83	67	51	36	20	6														
4	85	70	56	42	27	14														
6	86	72	59	46	35	22	10	0												
8	87	74	62	51	39	28	17	6												
10	88	76	65	54	43	33	24	13	4											
12	88	78	67	57	48	38	28	19	10	2										
14	89	79	69	60	50	41	33	25	16	8	1									
16	90	80	71	62	54	45	37	29	21	14	7	1								
18	91	81	72	64	56	48	40	33	26	19	12	6	0							
20	91	82	74	66	58	51	44	36	30	23	17	11	5	0						
22	92	83	75	68	60	53	46	40	33	27	21	15	10	4	0					
24	92	84	76	69	62	55	49	42	36	30	25	20	14	9	4	0				
26	92	85	77	70	64	57	51	45	39	34	28	23	18	13	9	5				
28	93	86	78	71	65	59	53	47	42	36	31	26	21	17	12	8	4			
30	93	86	79	72	66	61	55	49	44	39	34	29	25	20	16	12	8	4		
32	93	86	80	73	68	62	56	55	46	41	36	32	27	22	19	14	11	8	4	
34	93	86	81	74	69	63	58	52	48	43	38	34	30	26	22	18	14	11	8	5
36	94	87	81	75	69	64	59	54	50	44	40	36	32	28	24	21	17	13	10	7
38	94	87	82	76	70	66	60	55	51	46	42	38	34	30	26	23	20	16	13	10
40	94	89	82	76	71	67	61	57	52	48	44	40	36	33	29	25	22	19	16	13

hygrometer uses a bundle of hairs linked mechanically to an indicator that is calibrated between 0 and 100 percent. Thus we need only glance at the dial to determine the relative humidity. Unfortunately, the hair hygrometer is less accurate than the psychrometer. Furthermore, it requires frequent calibration and is slow in responding to changes in humidity, especially at low temperatures.

A different type of hygrometer is used in remote-sensing instrument packages, such as radiosondes, that transmit upper-air observations back to ground stations. The electric hygrometer contains an electrical conductor coated with a moisture-absorbing chemical. It works on the principle that the passage of current varies as the relative humidity varies.

TABLE 4–3 Dew-Point Temperature (1000 Millibars)

Dry-Bulb Temperature (°C)	Wet-Bulb Depression ($T_d - T_w$)																			
	1	2	3	4	5	6	7	8	9	10	11	12	13	14	15	16	17	18	19	20
−20	−33																			
−18	−28																			
−16	−24																			
−14	−21	−36																		
−12	−18	−28																		
−10	−14	−22																		
−8	−12	−18	−29																	
−6	−10	−14	−22																	
−4	−7	−11	−17	−29																
−2	−5	−8	−13	−20																
0	−3	−6	−9	−15	−24															
2	−1	−3	−6	−11	−17															
4	1	−1	−4	−7	−11	−19														
6	4	1	−1	−4	−7	−13	−21													
8	6	3	1	−2	−5	−9	−14													
10	8	6	4	1	−2	−5	−9	−14	−28											
12	10	8	6	4	1	−2	−5	−9	−16											
14	12	11	9	6	4	1	−2	−5	−10	−17										
16	14	13	11	9	7	4	1	−1	−6	−10	−17									
18	16	15	13	11	9	7	4	2	−2	−5	−10	−19								
20	19	17	15	14	12	10	7	4	2	−2	−5	−10	−19							
22	21	19	17	16	14	12	10	8	5	3	−1	−5	−10	−19						
24	23	21	20	18	16	14	12	10	8	6	2	−1	−5	−10	−18					
26	25	23	22	20	18	17	15	13	11	9	6	3	0	−4	−9	−18				
28	27	25	24	22	21	19	17	16	14	11	9	7	4	1	−3	−9	−16			
30	29	27	26	24	23	21	19	18	16	14	12	10	8	5	1	−2	−8	−15		
32	31	29	28	27	25	24	22	21	19	17	15	13	11	8	5	2	−2	−7	−14	
34	33	31	30	29	27	26	24	23	21	20	18	16	14	12	9	6	3	−1	−5	−12
36	35	33	32	31	29	28	27	25	24	22	20	19	17	15	13	10	7	4	0	−4
38	37	35	34	33	32	30	29	28	26	25	23	21	19	17	15	13	11	8	5	1
40	39	37	36	35	34	32	31	30	28	27	25	24	22	20	18	16	14	12	9	6

CONDENSATION ALOFT AND ADIABATIC TEMPERATURE CHANGES

Up to this point we considered some basic properties of water vapor as well as how its variability is measured. We are now ready to examine one of the important roles that water vapor plays in weather processes. Recall that condensation occurs when water vapor changes to a liquid. The result of condensation may be the formation of dew, fog, or clouds. Although each type of condensation is different, all require saturated air in order to form. As indicated earlier, saturation occurs either when water vapor is added to the air or, more commonly, when the air is cooled to its dew point. Heat near the earth's surface is readily exchanged between the ground and the air above. Thus radiation cooling of the earth's surface during evening hours accounts for the formation of dew and some types of fog. Yet clouds often form during the warmest part of the day. Consequently, some other mechanism must operate during cloud formation.

The process that is responsible for most cloud formation is easily visualized if you have ever pumped up a bicycle tire and noticed that the pump barrel became very warm. The heat energy that you felt was the consequence of the work that you did on the air in order to compress it. When energy is used to compress air, the motion of the gas molecules increases and therefore the temperature of the air rises. Conversely, when air is allowed to escape from a bicycle tire, it expands and cools. Here the expanding air pushes (does work on) the surrounding air and must cool by an amount equivalent to the energy expended. The temperature changes just described, in which heat was neither added nor subtracted, are called **adiabatic temperature changes** and result when air is compressed or allowed to expand. In summary, when air is allowed to expand, it cools; when air is compressed, it warms.

Adiabatic Temperature Change

Anytime a parcel of air moves upward, it passes through regions of successively lower pressure. As a result, ascending air expands and cools adiabatically. Unsaturated air cools at the rather constant rate of 1°C for every 100 meters of ascent (10°C per kilometer). Conversely, descending air comes under increasingly higher pressures and is compressed and heated 1°C for every 100 meters of descent. This rate of cooling or heating applies only to vertically moving unsaturated air and is known as the **dry adiabatic rate.**

Dry Adiabatic Rate

Lifting Condensation Level

If a parcel of air rises high enough, it will cool sufficiently to cause condensation. From this point along its ascent, called the **lifting condensation level,** the latent heat that was stored when water evaporated is liberated. Although the parcel will continue to cool adiabatically, the release of latent heat acts to suppress the rate of cooling. In other words, when a parcel of air ascends above the lifting condensation level, the rate of cooling is reduced because of the release of latent heat. This slower rate of cooling caused by the addition of latent heat is called the **wet adiabatic rate** of cooling. Because the amount of latent heat released depends on the quantity of moisture present in the air, the wet adiabatic rate varies from 0.5°C per 100 meters for air with a high moisture content to 0.9°C per 100 meters for air with a low moisture content. Figure 4–8 illustrates the role of adiabatic cooling in the formation of clouds. Note that from the surface up to the lifting condensation level the air cools at the dry adiabatic rate. The wet adiabatic rate commences at the point of condensation.

Wet Adiabatic Rate

STABILITY

It was pointed out that if air rises, it will cool and eventually produce clouds. Why does air rise on some occasions but not on others? Why do the size of clouds and the amount of precipitation vary so much when air does rise? The

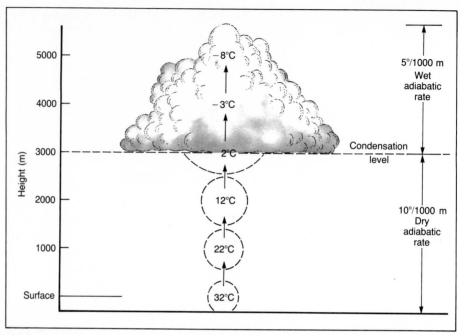

FIGURE 4–8
Rising air cools at the dry adiabatic rate of 10°C per 1000 meters until the air reaches the dew point and condensation (cloud formation) begins. As air continues to rise, the latent heat released by condensation lowers the rate of cooling. The wet adiabatic rate is therefore always less than the dry adiabatic rate.

answers are closely related to the stability of the air. Imagine a large bubble of air with a thin flexible cover that allows it to expand but prevents it from mixing with the surrounding air. If the imaginary bubble were forced to rise, its temperature would decrease because of expansion. By comparing the bubble's temperature to that of the surrounding air, we could determine its stability. If the bubble's temperature were lower than that of its environment, it would be more dense; and if allowed to do so, it would sink to its original position. Air of this type, **Stable Air**   termed **stable,** resists vertical displacement.

  If, however, our imaginary bubble were warmer and hence less dense than the surrounding air, it would continue to rise until it reached an altitude having the same temperature, much as a hot-air balloon would rise as long as it was **Unstable Air**   lighter than the surrounding air. This type of air is classified as **unstable.**

Determination of Stability

Actually, the stability of air is determined by examining the temperature of the atmosphere at various heights. Recall that this measure is called the environmental lapse rate. Do not confuse the environmental lapse rate, which is the temperature of the atmosphere as determined from observations made by balloons and airplanes, with adiabatic temperature changes. The latter measure indicates the change in

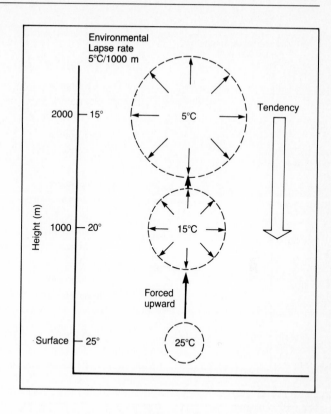

FIGURE 4–9
Schematic representation of a stable atmosphere. Note that the air near the surface is potentially cooler than the air aloft and therefore resists upward motion.

temperature that a parcel of air would experience as it moved vertically through the atmosphere.

To illustrate, consider a situation in which the prevailing environmental lapse rate is 5°C per 1000 meters (Figure 4–9). Under this condition, when the air at the surface has a temperature of 25°C, the air at 1000 meters will be 5°C cooler or 20°C, whereas the air at 2000 meters will have a temperature of 15°C and so forth. At first glance it appears that the air at the surface is lighter than the air at 1000 meters, for it is 5°C warmer. However, if the air near the surface were unsaturated and were to rise to 1000 meters, it would expand and cool at the dry adiabatic rate of 1°C per 100 meters. Therefore on reaching 1000 meters, its temperature would have dropped a total of 10°C to 15°C. Being 5°C cooler than its environment, it would be heavier and would tend to sink to its original position. So we say that the air near the surface is potentially cooler than the air aloft and therefore it will not rise. By similar reasoning, if the air at 1000 meters subsided, adiabatic heating would increase its temperature 10°C by the time it reached the surface, making it warmer than the surrounding air; thus its buoyancy would cause it to return. The air just described is stable and resists vertical movement.

Absolute Stability

Stated quantitatively, **absolute stability** prevails when the environmental lapse rate is less than the wet adiabatic rate. Figure 4–10 depicts this situation by using an environmental lapse rate of 5°C per 1000 meters and a wet adiabatic rate of 6°C per 1000 meters. Note that at 1000 meters the temperature of the

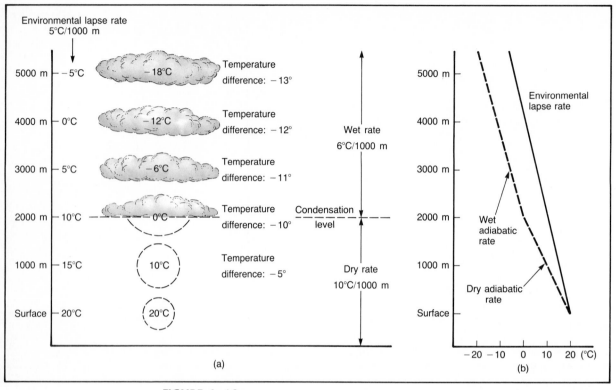

FIGURE 4–10
Absolute stability prevails when the environmental lapse rate is less than the wet adiabatic rate. (a) The rising parcel of air is always cooler and heavier than the surrounding air. (b) Graphic representation of the conditions shown in part (a).

surrounding air is 15°C and that the rising parcel of air has cooled to 10°C and so is the heavier air. Even if this stable air were forced above the condensation level, it would remain cooler and heavier than its environment and would have a tendency to return to the surface.

Absolute Instability   At the other extreme, air is said to exhibit **absolute instability** when the environmental lapse rate is greater than the dry adiabatic rate. As shown in Figure 4–11, the ascending parcel of air is always warmer than its environment and will continue to rise because of its own buoyancy.

Although absolute instability can occur on very warm days, this condition is generally confined to the first few kilometers of the atmosphere. A more common **Conditional Instability**   type of atmospheric instability is called **conditional instability.** This condition prevails when moist air has an environmental lapse rate between the dry and wet adiabatic rates (between 0.5 and 1°C per 100 meters). Notice in Figure 4–12 that the rising parcel of air is cooler than the surrounding air for the first 4000 meters and thus is considered stable. With the addition of latent heat above the lifting condensation level, the parcel eventually becomes warmer than the surrounding air. From this point along its ascent the parcel will continue to rise without an outside force and so is considered unstable. Conditionally unstable air can be described as air that begins its ascent as stable air but at some point above the

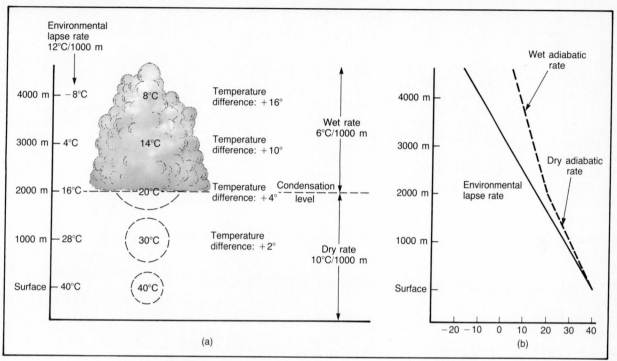

FIGURE 4–11
Absolute instability illustrated by using an environmental lapse rate of 12°C per 1000 meters.
(a) The rising air is always warmer and therefore lighter than the surrounding air. (b)
Graphic representation of the conditions shown in part (a).

lifting condensation level becomes unstable. The word "conditional" is used because the air must be mechanically forced upward, such as over mountainous terrain, before it becomes unstable and rises because of its own buoyancy.

Stability and Daily Weather

From the previous discussion we can conclude that stable air resists vertical movement and that unstable air ascends freely because of its own buoyancy. But how do these facts manifest themselves in our daily weather?

Because stable air resists upward movement, we might conclude that clouds would not form when stable conditions prevail in the atmosphere. Although this premise seems reasonable, processes do exist that force air aloft. They are discussed in the following section. When stable air is forced aloft, the clouds that form are widespread and have little vertical thickness compared with their horizontal dimension, and precipitation, if any, is light to moderate. In contrast, clouds associated with unstable air are towering and are usually accompanied by heavy precipitation. So we can conclude that on a dreary overcast day with light drizzle, stable air was forced aloft. On the other hand, on a day when cauliflower-shaped clouds appear to be growing as if bubbles of hot air were surging upward, we can be relatively certain that the ascending air is unstable.

Instability occurs often on hot summer afternoons when solar heating is intense.

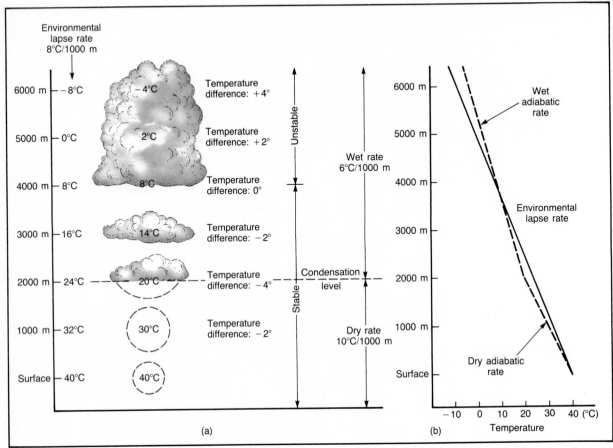

FIGURE 4–12
Conditional instability illustrated by using an environmental lapse rate of 8°C per 1000 meters that lies between the dry adiabatic rate and the wet adiabatic rate. (a) The rising parcel of air is cooler than the surrounding air below 4000 meters and warmer above 4000 meters. (b) Graphic representation of the conditions shown in part (a).

Surface irregularities cause pockets of air to be warmed more than the surrounding air. Consequently, these warmer parcels of air will buoy upward. If they rise above the lifting condensation level, clouds form, which on occasion produce midafternoon rain showers. The height of clouds produced in this fashion is somewhat limited, for instability caused solely by surface heating is confined to, at most, the first few kilometers of the atmosphere. Also, the accompanying rains are of short duration because the precipitation readily cools the surface.

The most stable conditions occur during a temperature inversion when temperature increases with height. In this situation, the air near the surface is cooler and heavier than the air aloft and therefore little vertical mixing occurs between the layers. Because pollutants are generally added to the air from below, a temperature inversion confines them to the lowermost layer where their concentration will continue to increase until the temperature inversion dissipates. Widespread

fog is another sign of stability. If the layer containing fog were mixing freely with the "dry" layer above, evaporation would quickly eliminate the foggy condition.

In summary, the role of stability in determining our daily weather cannot be overemphasized. The air's stability, or lack of it, determines whether the soot from a smokestack is dispersed widely or is confined to a narrow layer near the earth's surface. Of possibly greater importance, it determines to a large degree whether clouds develop and produce precipitation and whether that precipitation will come as a gentle shower or a violent downpour.

Changes in Stability

Most processes that alter stability occur as a result of the movement of air, although daily temperature changes do play an important role. In general, any factor that causes an increase in the environmental lapse rate renders the air more unstable, whereas a reduction in the environmental lapse rate increases the air's stability. Recall that the larger the environmental lapse rate, the more rapidly the temperature drops with increased altitude. Therefore any factor that causes the air near the surface to become warmed in relation to the air aloft increases instability. The opposite is also true; any factor that causes the surface air to be chilled results in the air becoming more stable. So as stated earlier, on a clear day when there is abundant surface heating, the lower atmosphere often becomes unstable and causes parcels of air to rise. After the sun sets, surface cooling generally renders the air stable again.

Similar changes in stability occur as air moves horizontally, traversing a surface having markedly different temperatures. In the winter warm air from the Gulf of Mexico moves northward over the cold land surface of the Midwest. Because the air is cooled from below, it becomes more stable, often producing widespread fog. The opposite occurs when wintertime polar air moves southward over the open waters of the Great Lakes. The moisture and heat added to the frigid polar air from the water below are enough to make it unstable and generate the clouds that produce heavy snowfalls on the downwind shores of these lakes.

Subsidence

Vertical movements of air also influence stability. When there is a general downward airflow, called **subsidence,** the upper portion of the subsiding layer is heated by compression, more so than the lower portion. Usually the air near the surface is not involved in the subsidence and so its temperature remains unchanged. The net effect is to stabilize the air, for the air aloft is warmed in relation to the surface air. The warming effect of a few hundred meters of subsidence is enough to evaporate the clouds found in any layer of the atmosphere. Thus one sign of subsiding air is a cloudless sky. Subsidence can also produce a temperature inversion aloft. The most intense and prolonged temperature inversions and associated air pollution episodes are caused by subsidence, a topic discussed more fully at the end of the chapter.

Upward movement of air generally enhances instability, particularly when the lower portion of the rising layer has a higher moisture content than the upper portion. As the air moves upward, the lower portion becomes saturated first and cools at the lesser wet adiabatic rate. The net effect is to increase the

FIGURE 4–13
On warm days airflow off the ocean onto the peninsula of Florida is partly responsible for the cloudy conditions over the land as shown in this Gemini V photograph. (Courtesy of NASA)

lapse rate within the rising layer. This process is especially important in producing the instability associated with thunderstorms. In addition, recall that conditionally unstable air can become unstable if lifted sufficiently.

Convergence

Whenever air flows together **(convergence),** it results in general upward movement, for as air converges, it occupies a smaller and smaller area, thus necessitating that the height of the air column increase. Consequently, air within the column must move upward, thereby enhancing instability. The Florida peninsula provides an excellent example of the role that convergence plays in initiating instability. On warm days the airflow is off the ocean along both coasts of Florida and causes general convergence over the peninsula. This convergence and associated uplift, aided by intense solar heating, cause more frequent midafternoon thunderstorms in this part of the United States than in any other (Figure 4–13).

On a smaller scale, the loss of heat by radiation from cloud tops during evening hours adds to their instability and growth. Unlike air, which is a poor radiator of heat, cloud droplets emit energy to space nearly as well as does the earth's surface. Towering clouds that owe their growth to surface heating lose their source of energy at sunset. After sunset, however, radiation cooling at their tops steepens the lapse rate and can initiate additional upward flow of warm parcels from below. This process is believed responsible for producing nocturnal thunderstorms from clouds whose growth prematurely ceased at sunset.

A list of factors that modify air's stability provides a summary. Instability is enhanced by

1. Intense solar heating that warms the air from below
2. The heating of an air mass from below as it traverses a warm surface

3. Forceful lifting of air, such as over an elevated land surface
4. Upward movement of air associated with general convergence
5. Radiation cooling from cloud tops

Stability is enhanced by

1. Radiation cooling of the earth's surface after sunset
2. The cooling of an air mass from below as it traverses a cold surface
3. Subsidence of an air column

FORCEFUL LIFTING

Earlier we demonstrated that stable air and conditionally unstable air will not rise on their own; they require some mechanism to force the vertical movement. Three such mechanisms are convergence, orographic lifting, and frontal wedging. The role of convergence in vertical lifting has already been discussed.

Orographic Lifting

Orographic lifting occurs when sloping terrain, such as mountains, act as barriers to the flow of air and force the air to ascend (Figure 4–14). Many of the rainiest places in the world are located on windward mountain slopes. A station at Mt. Waialeale, Hawaii, for example, records the highest average annual rainfall in the world, some 1168 centimeters. The station is located on the windward (northeast) coast of the island of Kauai at an elevation of 1523 meters.[4]

In addition to providing the lift to render air unstable, mountains further remove more than their share of moisture in other ways. By slowing the horizontal flow of air, they cause convergence and retard the passage of storm systems. Also, the irregular topography of mountains enhances differential heating and surface instability. These combined effects account for the generally higher precipitation associated with mountainous regions compared with surrounding lowlands.

By the time air reaches the leeward side of a mountain much of the moisture has been lost; and if the air descends, it warms, making condensation and precipita-

FIGURE 4–14
Orographic lifting.

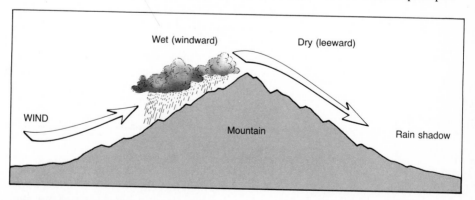

---

[4] For other precipitation extremes, see Appendix F.

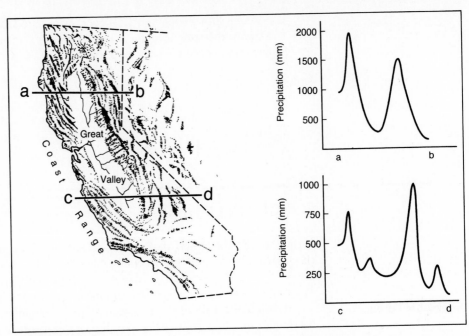

FIGURE 4–15
Relationship between topography and average annual rainfall from the California coast to the dry plateau of Nevada's Great Basin. As shown in these profiles, precipitation maximums occur in the mountainous regions, but precipitation is much lighter in the valleys. (Precipitation data from the U.S. Department of Agriculture)

Rain Shadow Desert

tion even less likely. As shown in Figure 4–14, the result often is a **rain shadow desert.** The Great Basin desert of the western United States lies only a few hundred miles from the Pacific Ocean, but it is effectively cut off by the imposing Sierra Nevada (Figure 4–15). The Gobi desert of Mongolia, the Takla Makan of China, and the Patagonia desert of Argentina are other examples of deserts found on the leeward sides of mountains.

Frontal Wedging

    Frontal wedging occurs when cool air acts as a barrier over which warmer, less dense air rises. This phenomenon is common throughout the continental United States and is responsible for the bulk of the precipitation in many areas, as we shall see later. Figure 4–16 illustrates frontal wedging of stable and unstable air. As this figure shows, forceful lifting is important in producing clouds. The stability of the air, however, determines to a great extent the type of clouds formed and the amount of precipitation that may be expected.

AIR POLLUTION

Air pollution and meteorology are linked in two ways. One concerns the influence that weather conditions have on the dilution of air pollutants. The second connection is the reverse and deals with the effect that air pollution has on weather and

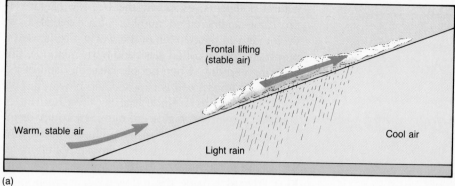

(a)

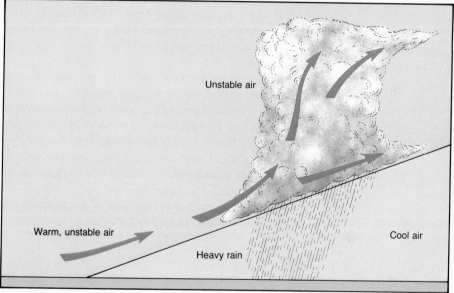

(b)

FIGURE 4–16

(a) When stable air is lifted, layered clouds usually result. (b) When warm, unstable air is forced to rise over cooler air, "towering" clouds develop.

climate. The first of these associations is examined in the following pages. The second, and equally important, relationship is discussed in detail in Chapter 13.

Air pollution is a continuing threat to our health and welfare. An average adult male requires about 13.5 kilograms of air each day compared with about 1.2 kilograms of food and 2 kilograms of water. The cleanliness of air, therefore, should certainly be as important to us as the cleanliness of our food and water.

Air is never perfectly clean. Many natural sources of air pollution have always existed. Ash from volcanic eruptions, salt particles from breaking waves, pollen and spores released by plants, smoke from forest and brush fires, and windblown dust are all examples of "natural air pollution." Ever since people have been on earth, however, they have added to the frequency and intensity of some of these natural pollutants, especially the last two (Figure 4–17). With the discovery of fire came an increased number of accidental as well as intentional burnings. Even today, in many parts of the world, fire is used to clear land for agricultural purposes (the so-called slash-and-burn method), filling the air with smoke and reducing visibility. When people clear the land of its natural vegetative cover for whatever purpose, soil is exposed and blown into the air. Yet when we consider the air in a modern-day industrial city, these human-accentuated forms of pollution, although significant, may seem minor by comparison.

Although some types of air pollution are relatively recent creations, others have been around for centuries. Smoke pollution, for example, plagued London for centuries. Because of the odor and smoke produced by the burning of coal, King Edward I made the following proclamation in 1300: "Be it known to all within the sound of my voice, whosoever shall be found guilty of burning coal shall suffer the loss of his head." Unfortunately, one Londoner did not heed the king's warning and paid the extreme price for his misdeed. As far as is known, however, this is the only case of capital punishment resulting from an air pollution violation!

The ban on burning coal led to the use of an alternative fuel—wood. Then extensive wood burning soon dramatically reduced English forests and coal consumption again increased in spite of royal disapproval. Thus in 1661 when John Evelyn wrote *Fumifugium, or the Inconvenience of Aer and Smoak of London Dissipated, together with some Remidies Humbly Proposed*, the problem of foul air still plagued Londoners. In his book Evelyn noted that a traveler, although many miles from London, "sooner smells than sees the city to which he repairs." In fact, London continued to have severe air pollution problems well into the twentieth century. It was only after a devastating smog disaster in 1952 that truly decisive action was taken to clean the air.

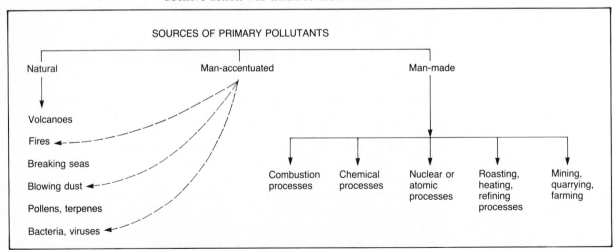

FIGURE 4–17
Sources of primary pollutants. (After Reid A. Bryson and John E. Kutzbach, *Air Pollution*, Commission on College Geography Resource Paper No. 6, Figure 2, p. 8. Copyright by the Association of American Geographers)

London, however, has not monopolized the air pollution scene. With the coming of the Industrial Revolution, many cities began to experience "big-time" air pollution. Instead of just simply accelerating natural sources, people found many new ways to pollute the air (Figure 4–17) and many new things with which to pollute it. This rapid rise in urban air pollution, however, was not always viewed with great alarm. Rather, chimneys belching forth smoke and soot were a symbol of growth and prosperity (Figure 4–18). The following quotation from an 1880 speech by the well-known lawyer and orator Robert Ingersoll, for example, is reported to have elicited great cheering and cries of "Good! Good!" from the audience: "I want the sky to be filled with the smoke of American industry and upon that cloud of smoke will rest forever the bow of perpetual promise. That is what I am for." With the rapid growth of the world's population and accelerated industrialization, the quantities of atmospheric pollutants increased drastically.

The first major episode of air pollution disasters to be studied in depth began in the Meuse valley in Belgium. Here for 5 days in December 1930 a blanket of smog hung in the valley, killing 63 people and causing 6000 to become ill. Since the 1930s many air pollution episodes have demonstrated the devastating effect that dirty air can have on life and property. In October 1948 Donora, Pennsylvania, had such an experience. The grime that settled from the air coated houses, streets, and sidewalks, so that pedestrians and autos actually left distinct footprints and tire tracks. Almost 6000 of the town's 14,000 inhabitants became ill, and 20 died. The most tragic air pollution episode ever occurred in London in December 1952. More than 4000 people died as a result of this 5-day ordeal. The people

FIGURE 4–18
Stacks belching smoke and soot such as these were once a sign of economic prosperity.
(EPA-Documerica, Marc St. Gil)

who suffered most were those with respiratory and heart problems, primarily the elderly. Extreme air pollution darkened London again in 1953 and 1962 and affected New York City in 1953, 1963, and 1966. Although regulations and controls have reduced the frequency and severity of such episodes, health authorities are equally concerned with the slow and subtle effects on our lungs and other organs by air pollution levels that are much lower but that are present every day year after year.

Sources and Types of Air Pollution

Primary Pollutant

Pollutants may be grouped into two categories: primary and secondary. **Primary pollutants** are emitted directly from identifiable sources. Table 4–4 lists the major sources as well as the types and amounts of primary pollutants that characterize each. The significance of the transportation category is obvious. It accounts for more than half of our air pollution (by weight). In addition to highway vehicles,

TABLE 4—4 National Emissions Estimates for 1985 (millions of metric tons)

Source Category	Particulates	Sulfur Oxides	Nitrogen Oxides	Volatile Organics	Carbon Monoxide	Lead*
Transportation						
Highway vehicles	1.1	0.5	7.1	6.0	40.7	14.5
Aircraft	0.1	0.0	0.1	0.2	1.1	—
Railroads	0.0	0.1	0.5	0.1	0.2	—
Vessels	0.0	0.2	0.2	0.4	1.4	—
Other off-highway	0.1	0.1	1.0	0.4	4.1	0.9
Transportation total	1.3	0.9	8.9	7.1	47.5	15.4
Stationary source fuel combustion						
Electric utilities	0.6	14.2	6.8	0.0	0.3	0.1
Industrial	0.3	2.2	2.9	0.1	0.6	0.4
Commercial institutional	0.0	0.4	0.2	0.0	0.0	0.0
Residential	1.2	0.2	0.4	2.4	7.1	0.0
Fuel combustion total	2.1	17.0	10.3	2.5	8.0	0.5
Industrial processes	2.7	2.9	0.6	8.6	4.6	2.3
Solid waste disposal						
Incineration	0.1	0.0	0.0	0.3	1.1	—
Open burning	0.2	0.0	0.1	0.3	0.9	—
Solid waste total	0.3	0.0	0.1	0.6	2.0	2.8
Miscellaneous						
Forest fire	0.7	0.0	0.1	0.6	4.7	—
Other burning	0.1	0.0	0.0	0.1	0.6	—
Misc. organic solvent	0.0	0.0	0.0	1.6	0.0	—
Misc. total	0.8	0.0	0.1	2.3	5.3	—
Total of all sources	7.2	20.8	20.0	21.1	67.4	21.0

Source: National Air Pollutant Emission Estimates, 1940–1985, U.S. Environmental Protection Agency Publication No. EPA-450/4–86–018, 1987.

* Thousands of metric tons.

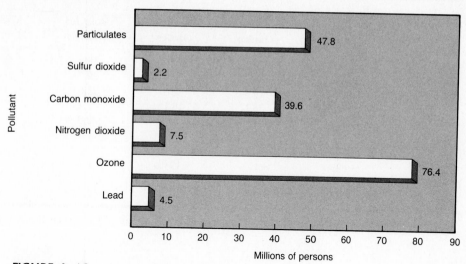

FIGURE 4–19

Number of persons living in counties with air quality levels above the primary National Ambient Air Quality Standards in 1985. (Source: U.S. Environmental Protection Agency, Office of Air Quality Planning and Standards)

this category includes trains, ships, and airplanes. Still, the tens of millions of cars and trucks on U.S. roads are, without a doubt, the greatest contributors in this category.

Although considerable progress has been made controlling air pollution, the quality of the air we breathe still remains a serious public health problem. With the goal of protecting the public health and welfare, the U.S. Environmental Protection Agency has developed the National Ambient Air Quality Standards. Primary standards are established to protect the public health, whereas secondary standards protect the public welfare, as measured by the effects of air pollution on vegetation, materials, and visibility. In 1985 millions of people were still breathing air that was in violation of the standards (Figure 4–19). Pollution levels, however, continue to improve.

An examination of Table 4–5 indicates that estimated emissions of several types of primary pollutants dropped during the 1940–1985 period. Many of these improvements resulted from the implementation of air quality standards and the development and use of pollution control technology to curb emissions. Emissions of carbon monoxide decreased by more than 30 percent from 1970 to 1985, for instance, despite increases in highway vehicle travel. Pollutants that show an increase between 1940 and 1985 would have risen even more had controls not been instituted. Yet not all reductions in emissions resulted entirely from the effects of new laws and improved pollution controls. The decline in the emission of particulates between 1940 and 1960, for example, was largely a result of the phasing out of coal-burning railroad locomotives.

It should be noted that data in Tables 4–4 and 4–5 are only estimates and do not represent the results of any program for the measurement of actual emissions. Therefore their accuracy and usefulness are limited. Moreover, they are nationwide estimates and do not necessarily represent local trends in emissions or air quality.

TABLE 4–5 Summary of National Emission Estimates (millions of metric tons)

Year	Particulates	Sulfur Oxides	Nitrogen Oxides	Volatile Organics	Carbon Monoxide	Lead (thousands of metric tons)
1940	22.8	17.5	6.8	18.4	81.6	NA
1950	24.5	19.8	9.3	20.7	86.3	NA
1960	21.1	19.5	12.8	23.6	88.4	NA
1970	18.1	28.1	18.1	27.2	98.7	203.8
1980	8.4	23.2	20.3	22.8	76.0	70.6
1985	7.2	20.8	20.0	21.1	67.4	21.0
Change 1940–1985	−68%	+19%	+194%	+16%	−17%	NA
Change 1970–1985	−60%	−26%	+10%	−22%	−32%	−90%

Source: National Air Pollutant Emissions Estimates, 1940–1985, U.S. Environmental Protection Agency Publication No. EPA-450/4–86–018, 1987.

Instead these data simply form a basis for estimating national progress in the control of primary pollutants.

Secondary Pollutant

Secondary pollutants are produced in the atmosphere when certain chemical reactions take place among the primary pollutants. Sulfuric acid (H_2SO_4) is one example of a secondary pollutant. It is produced when sulfur dioxide combines with oxygen, yielding sulfur trioxide, which then combines with water to form this irritating and corrosive acid. A later section on acid precipitation discusses the effects in some detail.

Photochemical Reaction

Many reactions that produce secondary pollutants are triggered by strong sunlight and so are termed **photochemical reactions.** One common example occurs when nitrogen oxides absorb solar radiation, initiating a chain of complex reactions. When certain organic compounds called hydrocarbons or volatile organics are present, the result is the formation of a number of undesirable secondary products that are very reactive, irritating, and toxic. Among these products is ozone. Recall from Chapter 1 that ozone is also formed by natural processes in the stratosphere, where it plays a vital role because of its ability to absorb damaging ultraviolet radiation. However, it is considered a pollutant when produced near the earth's surface because it is both unhealthy to breathe and can have a toxic effect on plants.

Smog

Air pollution in urban and industrial areas is often termed **smog.** The word was coined in 1905 by Harold A. Des Veaux, a London physician, and was created by combining the words "smoke" and "fog." Des Veaux's term was indeed an apt description of London's principal air pollution threat, which was associated with the products of coal burning coupled with periods of high humidity. Today, however, the term "smog" is used as a synonym for general air pollution and does not necessarily imply the smoke-fog combination. Therefore when greater clarity is desired, we often find the term "smog" preceded by such modifiers as "London-type," "classical," "Los Angeles–type," or "photochemical." The first two refer to the original meaning of the term, and the last two to air quality problems created by secondary pollutants.

METEOROLOGICAL FACTORS AFFECTING AIR POLLUTION

Certainly the most obvious factor influencing air pollution is the quantity of contaminants emitted into the atmosphere. Still, experience shows that even when emissions remain relatively steady for extended periods, we often find wide variations in air quality from one day to the next. Indeed, when air pollution episodes occur, they are not generally the result of a drastic increase in the output of pollutants; instead they occur because of changes in certain atmospheric conditions.

Perhaps you have heard the well-known phrase, "The solution to pollution is dilution." To a significant degree, it is true. If the air into which the pollution is released is not dispersed, the air will become more toxic. Two of the most important atmospheric conditions affecting the dispersion of pollutants are (1) the strength of the wind and (2) the stability of the air. These factors are critical because they determine how rapidly pollutants are diluted by mixing with the surrounding air after leaving the source.

The manner in which wind speed influences the concentration of pollutants is shown in Figure 4–20. Assume that a burst of pollution leaves the stack every second. If the wind speed were 8 meters per second, as in Figure 4–20(a), the distance between each pollution "cloud" will be 8 meters. If the wind is reduced to 4 meters per second (Figure 4–20b), the distance between "clouds" will be 4 meters. Consequently, because of the direct effect of wind speed, the concentration of pollutants is twice as great with the 4 meters per second wind as with the 8 meters per second wind. It is easy to understand why air pollution problems seldom occur when winds are strong but rather are associated with periods when winds are weak or calm.

A second aspect of wind speed influences air quality. The stronger the wind, the more turbulent the air. Thus strong winds mix polluted air more rapidly with the surrounding air, thereby causing the pollution to be more dilute. Conversely, when winds are light, there is little turbulence and the concentration of pollutants remains high.

Mixing Depth

Whereas wind speed governs the amount of air into which pollutants are initially mixed, atmospheric stability determines the extent to which vertical motions will mix the pollution with cleaner air above. The vertical distance between the earth's surface and the height to which convectional movements extend is termed the **mixing depth.** Generally, the greater the mixing depth, the better the air quality. When the mixing depth is several kilometers, pollutants are mixed through a large volume of cleaner air and dilute rapidly. When the mixing depth is shallow, pollutants are confined to a much smaller volume of air and concentrations can reach unhealthy levels. When air is stable, convectional motions are suppressed and mixing depths are small. Conversely, an unstable atmosphere promotes vertical air movements and greater mixing depths. Since heating of the earth's surface by the sun enhances convectional movements, mixing depths are usually greater during the afternoon hours. For the same reason, mixing depths during the summer months are typically greater than during the winter months.

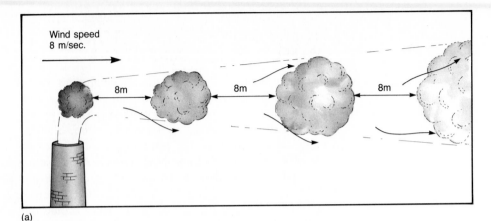

(a)

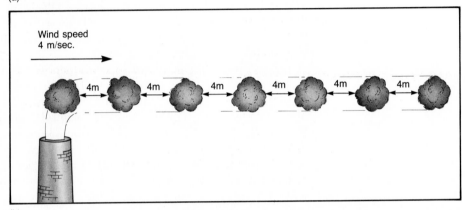

(b)

FIGURE 4–20
The effect of wind speed on the dispersion of pollutants. The stronger wind speed in (a) causes the puffs of pollution to be spread farther than in (b). In addition, since faster winds are more turbulent, the pollutants in (a) are also more diluted by mixing than those in (b).

Temperature Inversion

Temperature inversion represents a situation in which the atmosphere is very stable and the mixing depth is significantly restricted. Warm air overlying cooler air acts as a lid and prevents upward movement, leaving the pollutants trapped in a relatively narrow zone near the ground. This effect is dramatically illustrated by the photograph in Figure 4–21. Most of the air pollution episodes cited earlier were linked to the occurrence of temperature inversions.

Solar heating can result in high surface temperatures during the late morning and afternoon that steepen the lapse rate and render the lower air unstable. During nighttime hours, however, just the opposite situation may occur; temperature inversions, which result in very stable atmospheric conditions, can develop close to the ground. These surface inversions form because the ground is a more effective radiator than the air above. Such being the case, radiation from the ground to a clear night sky causes more rapid cooling at the surface than higher in the atmosphere. Consequently, the coldest air is found next to the ground,

FIGURE 4–21
Temperature inversions such as this one near El Cajon, California, act as lids to trap pollutants below. (Photo by James E. Patterson)

yielding a vertical temperature profile resembling the one shown in Figure 4–22(a). Once the sun rises, the ground is heated and the inversion disappears. Although usually rather shallow, surface inversions may be very deep in regions where the land surface is uneven. Because cold air is denser than warm air, the chilled air near the surface gradually drains from the uplands and slopes into adjacent lowlands and valleys. As might be expected, this deeper surface inversion will not dissipate as quickly after sunrise. Thus although valleys are often preferred sites for manufacturing because they afford easy access to water transportation, they are also more likely to experience relatively thick surface inversions that, in turn, will have a negative effect on air quality.

Many extensive and long-lived air pollution episodes are linked to temperature inversions that develop in association with the sinking air that characterizes slow-moving centers of high air pressure (anticyclones). As the air sinks to lower altitudes, it is compressed and so its temperature rises. Because turbulence is almost always present near the ground, this lowermost portion of the atmosphere is generally prevented from participating in the general subsidence. Thus an inversion develops aloft between the lower turbulent zone and the subsiding warmer layers above (Figure 4–22b). Moreover, the clear skies typically associated with high-pressure zones mean that it is not unusual for a surface inversion to form during the night and early morning hours.

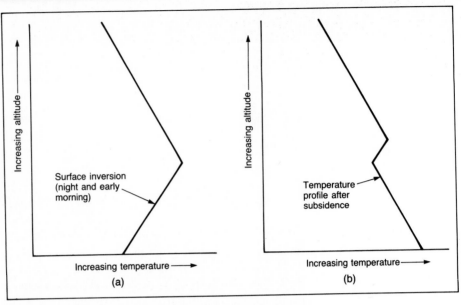

FIGURE 4–22
Types of temperature inversions. (a) The vertical temperature profile as it might appear early in the morning. (b) Subsidence often creates an inversion aloft.

In summary, we saw that when the wind is strong and an unstable environmental lapse rate prevails, the diffusion of pollutants is rapid and high pollution concentrations will not occur except perhaps near a major source. On the other hand, when an inversion exists and winds are light, diffusion is inhibited and high pollution concentrations are to be expected in areas where there are sources.

ACID PRECIPITATION

Newspapers, magazines, and television news reports sometimes carry features with such titles as "April Showers Could Kill May Flowers" and "Acid Rain—No One Really Knows Yet How Bad It Really Is." The stories warn of a significant environmental problem. The opening statement of a report by a panel of scientists of the National Research Council briefly summarizes the problem this way:

> During the past 25 years in Europe and the past 10 years in North America, scientific evidence has accumulated suggesting that air pollution resulting from emissions of oxides of sulfur and nitrogen may have significant adverse effects on ecosystems even when the pollutants or their reaction products are deposited from the air in locations remote from the major sources of the pollution.[5]

---

[5] *Acid Deposition: Atmospheric Processes in Eastern North America* (Washington, D.C.: National Academy Press, 1983), p. 1.

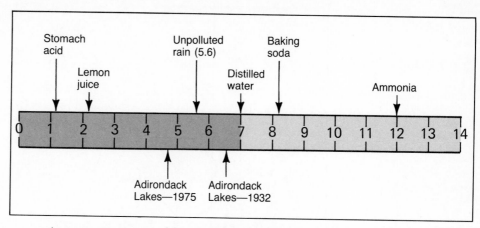

FIGURE 4–23
The pH scale.

As a consequence of burning large quantities of fossil fuels like coal and petroleum products, tens of millions of tons of sulfur and nitrogen oxides are released into the atmosphere each year. The major sources of these emissions include power-generating plants, industrial processes, such as ore smelting and petroleum refining, and vehicles of all kinds (see Table 4–4). Through a series of complex chemical reactions, some of these pollutants are converted into acids that then fall to the earth's surface as rain or snow. Another portion (estimated to be between 20 and 50 percent of the total) is deposited in dry form and subsequently converted into acid after coming in contact with precipitation, dew, or fog. Although a substantial portion of the pollutants is believed to be deposited in dry form, the phenomenon is nevertheless commonly referred to as acid rain or acid precipitation.

pH Scale

Before we can define acid precipitation in a precise way, we must become acquainted with the **pH scale.** The pH is a common measure of the degree of acidity or alkalinity of a solution. As Figure 4–23 illustrates, the scale ranges from 0 to 14, with a value of 7 denoting a solution that is neutral. Values below 7 indicate greater acidity, whereas numbers above 7 indicate greater alkalinity. It is important to note that the pH scale is logarithmic; that is, each whole number increment indicates a tenfold difference. Thus pH 4 is 10 times more acidic than pH 5 and 100 times more acidic than pH 6. Rain is naturally somewhat acidic. Normal, unpolluted precipitation is assumed to have a pH not lower than 5.6.

Acid Precipitation

Acid precipitation, therefore, is defined as rain or snow with pH values of less than 5.6.

Acid precipitation is not a new phenomenon. It was identified more than a century ago by an English chemist, and Swedish scientists have been studying the problem since the 1950s. But only since the early 1970s has acid precipitation become the subject of intensive research in North America. As the map in Figure 4–24 illustrates, the northeastern United States and eastern Canada appear to have been affected most seriously.

In addition to local pollution sources, a portion of the acidity found in the northeastern United States and eastern Canada originates hundreds of kilometers away in industrialized regions to the south and southwest. This situation occurs because many pollutants remain in the atmosphere for periods as long as 5 days,

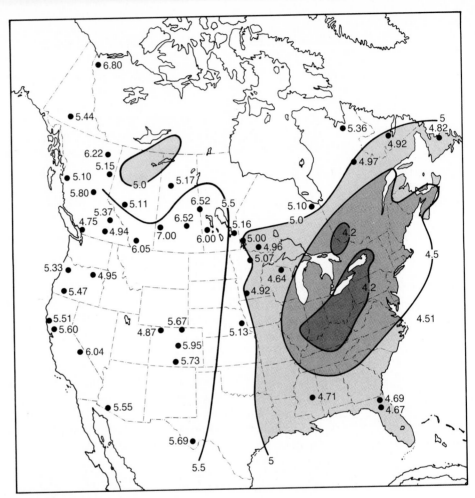

FIGURE 4–24

Annual mean value of pH in precipitation, weighted by the amount of precipitation, in North America for 1980. According to a National Research Council report, available data indicate elevated levels of pollutants in the air and acidic substances in precipitation over much of eastern North America. Concentrations are much higher than can be accounted for by emissions from natural sources. (From U.S./Canada Work Group No. 2, Final Report, Washington, D.C.: U.S. Environmental Protection Agency, 1982)

during which time they may be transported great distances. One contributing factor is, of all things, some of the pollution control technologies that are used to reduce pollution in the immediate vicinity of a source. Taller chimney stacks, for example, reduce local air quality problems by releasing pollutants into the stronger and more persistent winds that exist at greater heights. Although such stacks enhance dilution and dispersion, they also promote the long-distance transport of these unwanted emissions. In this way, individual stack plumes with pollution concentrations considered too dilute to be a direct health or environmental threat locally contribute to interregional pollution problems. Unfortunately, because atmospheric processes in eastern North America lead to a thorough mixing of pollutants, it is not yet possible to distinguish clearly between the relative impact of distant sources compared with local sources.

Unlike most atmospheric pollutants, acid precipitation has not yet been shown to have any direct adverse effects on human health. Still, its damaging effects on the environment are believed to be considerable in some areas and imminent in others. The best-known effect of acid precipitation is the lowering of pH in thousands of lakes in Scandinavia and eastern North America. Accompanying this condition have been substantial increases in dissolved aluminum that is leached from the soil by the acidic water and that, in turn, is toxic to fish. Consequently, some lakes are virtually devoid of fish, whereas others are approaching this condition. Furthermore, ecosystems are characterized by many interactions at many levels of organization, which means that evaluating the effects of acid precipitation on these complex systems is difficult and expensive, and far from complete. It should also be pointed out that, even within small areas, the effects of acid precipitation can vary significantly from one lake to another. Much of this variation is related to the nature of the soil and rock materials in the area surrounding the lake. Because deep mineral soils can neutralize acid solutions, lakes surrounded by such soils are less likely to become acidic. On the other hand, lakes that lack this buffering material can be severely affected. Over a period of time, however, the cumulative effects of acid precipitation could cause the pH of lakes that have not yet been acidified to drop as the buffering material in the surrounding soil is gradually depleted.

In addition to the thousands of lakes in which fish populations have been adversely affected by acidification, some research indicates that acid precipitation may also impair the productivity of forests. Over 200,000 acres of coniferous woodlands in West Germany are believed to be in serious decline at least in part because of acid rain. Forests in the northern Appalachian Mountains may also be similarly affected. It is important to point out, however, that there is still not sufficient data to clearly link acid precipitation to forest decline. In a 1986 report, a committee of the National Research Council stated that

> there are many ways in which acid deposition might affect forests, but to date it is not clear from existing field evidence in North America that any of the mechanisms have caused changes in the forest or the growth of trees.[6]

As with other aspects of the acid rain issue, additional research will help clarify the nature and magnitude of the problem.

Accelerated chemical weathering of stone structures is another costly impact of acid precipitation. Buildings and statuary made of marble, limestone, and concrete are especially susceptible. The marble monument pictured in Figure 4–25, for example, has stood in Peking, China, for 500 years. Forty years ago, the inscriptions were legible. Now they are unreadable. It is believed that acid rain may have

---

[6] *Acid Deposition: Long-Term Trends* (Washington, D.C.: National Academy Press, 1986), pp. 17–18.

FIGURE 4–25
Acid rain accelerates the chemical weathering of stone structures. Forty years ago the inscriptions on this 500-year-old marble monument in Peking, China, were legible. Today they are no longer visible. (Photo by Roger Cheng)

contributed to the accelerated deterioration.[7] Metals also corrode at a more rapid rate when exposed to acid rain.

Although preliminary research indicated that acid rain may reduce crop yields, subsequent experiments have not yet shown this to be the case. One 2-year experiment in Illinois that involved soy beans and corn indicated that acid rain was not a major factor in reducing yields, nor did it have an adverse effect on grain quality. In fact, the study actually found that one particular variety of corn had a higher yield when treated with very acid (pH 3) rain.

Acid precipitation is a complex and multifaceted issue. Because current knowledge regarding many aspects of the problem is incomplete, additional research is certainly needed. Yet it appears to many who are familiar with the problem that our understanding is sufficient to begin taking corrective actions as well. These actions could take the form of stricter government regulations regarding emissions, greater conservation, and the development of technologies that effectively reduce emissions while avoiding additional environmental problems.

---

[7] For more on this problem see Roger J. Cheng, "Deterioration of Marble Structures: The Role of Acid Rain," *Analytical Chemistry*, 59, no. 2 (January 15, 1987).

REVIEW

1. Describe the movement of water through the hydrologic cycle.

2. The quantity of water lost to evaporation over the oceans is not equaled by precipitation. Why, then, does the sea level not drop?

3. Summarize the processes by which water changes from one state to another. Indicate whether heat energy is absorbed or liberated.

4. After examining Table 4–1, write a generalization relating temperature and the capacity of air to hold water vapor.

5. How do absolute humidity and mixing ratio differ? What do they have in common? How is relative humidity different from absolute humidity and mixing ratio?

6. Refer to Figure 4–6 and then answer the following questions.
   a. When is relative humidity highest during a typical day? When is it lowest?
   b. At what time of day would dew most likely form?
   Write a generalization relating air temperature and relative humidity.

7. If temperature remains unchanged and if the mixing ratio decreases, how will relative humidity change?

8. Explain the principle of the psychrometer; the hair hygrometer.

9. What are the disadvantages of the hair hygrometer? Does this instrument have any advantages over the psychrometer?

10. Using the standard tables (Tables 4–2 and 4–3), determine the relative humidity and dew-point temperature if the dry-bulb thermometer reads 22°C and the wet-bulb thermometer reads 16°C. How would the relative humidity and dew point change if the wet-bulb thermometer read 19°C?

11. As you drink an ice-cold beverage on a warm day, the outside of the glass or bottle becomes wet while the contents warm rapidly. Explain.

12. Why does air cool when it rises through the atmosphere?

13. If unsaturated air at 20°C were to rise, what would its temperature be at a height of 500 meters? If the dew-point temperature at the lifting condensation level were 11°C, at what elevation would clouds begin to form?

14. Why does the adiabatic rate of cooling change when condensation begins? Why is the wet adiabatic rate not a constant figure?

15. The contents of an aerosol can are under very high pressure. When you push the nozzle on such a can, the spray feels cold. Explain.

16. How does stable air differ from unstable air?

17. Explain the difference between the environmental lapse rate and adiabatic cooling.

18. How is the stability of air determined?

19. List some weather conditions that would lead you to believe that air is either stable or unstable.

20. How can the stability of air be altered?

21. Distinguish between subsidence and convergence. How might each influence the stability of the air?

22. How do orographic lifting and frontal wedging act to force air to rise?

23. Explain why the Great Basin area of the western United States is dry. What term is applied to such a situation?

24. What is the difference between primary and secondary pollutants? What is a photochemical reaction?

25. Why are air pollution problems more acute when winds are weak or calm?

26. How do temperature inversions influence air pollution?

27. Describe the formation of a surface inversion and compare it with an inversion that occurs aloft.

28. How much more acidic is a substance with a pH of 4 compared with a substance with a pH of 6?

29. How has the building of tall smokestacks contributed to interregional air pollution problems?

30. List some possible environmental effects of acid precipitation.

VOCABULARY REVIEW

hydrologic cycle

transpiration

calorie

latent heat

evaporation

condensation

melting

freezing

sublimation

deposition

humidity

saturation

saturation vapor pressure

absolute humidity

mixing ratio

relative humidity

dew point

hygrometer

psychrometer

adiabatic temperature change

dry adiabatic rate

lifting condensation level

wet adiabatic rate

stable

unstable

absolute stability

absolute instability

conditional instability

subsidence

convergence

orographic lifting

rain shadow desert

frontal wedging

primary pollutant

secondary pollutant

photochemical reaction

smog

mixing depth

temperature inversion

pH scale

acid precipitation

5

Forms
of Condensation
and Precipitation

C louds and fog, as well as rain, snow, sleet, and hail, are among the most conspicuous and observable aspects of the atmosphere and its weather. In this chapter we endeavor to gain a basic understanding of each of these phenomena. Although clouds of many types are familiar to all of us, their names usually are not. Chapter 5 begins with a look at the basic scheme for classifying and naming clouds. Next, we will discover that the formation of an average raindrop requires water from roughly one million cloud droplets. What mechanism fosters this development? As we shall see, more than one process is responsible for the formation of raindrops. Are scientists able to trigger these processes? Can modern weather-modification technology increase the amount of precipitation that falls from a cloud? The latter portions of this chapter will address these and other questions regarding our ability to control weather events.

CLOUDS

Cloud

Clouds are a form of condensation best described as visible aggregates of minute droplets of water or tiny crystals of ice. In addition to being prominent and sometimes spectacular features in the sky, clouds are of continual interest to meteorologists because they provide a visible indication of what is going on in the atmosphere. Anyone who observes clouds with the hope of recognizing different types often finds that there is a bewildering variety of these familiar white and gray masses streaming across the sky. Once the basic classification scheme for clouds is known, however, most of the confusion vanishes.

Cloud Classification

Prior to the beginning of the nineteenth century there were no generally accepted names for clouds. In 1803 Luke Howard, an English naturalist, published a cloud classification that met with great success and subsequently served as the basis of our present-day classification.

Clouds are classified on the basis of two criteria: appearance and height (Figure 5–1). Three basic cloud forms are recognized: cirrus, cumulus, and stratus. **Cirrus** clouds are high, white, and thin. They are separated or detached and form delicate veil-like patches or extended wispy fibers and often have a feathery appearance. The **cumulus** form consist of globular individual cloud masses. Normally they

Cirrus

Cumulus

Stratus

High Cloud

Middle Cloud

Low Cloud

Clouds of Vertical Development

exhibit a flat base and have the appearance of rising domes or towers. Such clouds are frequently described as having a cauliflower-like structure. **Stratus** clouds are best described as sheets or layers that cover much or all of the sky. Although there may be minor breaks, there are no distinct individual cloud units. All other clouds either reflect one of these three basic forms or are combinations or modifications of them.

Looking at the second aspect of cloud classification, height, we see that three levels are recognized: high, middle, and low. **High** clouds normally have bases above 6000 meters; **middle** clouds generally occupy heights from 2000 to 6000 meters; **low** clouds form below 2000 meters. The altitudes listed for each height category are not hard and fast. There is some seasonal as well as latitudinal variation. At high latitudes or during cold winter months in the midlatitudes, for example, high clouds are often found at lower altitudes.

Because of the low temperatures and small quantities of water vapor found at high altitudes, all high clouds are thin and white and are made up of ice crystals. Because much more water vapor is available at lower altitudes, middle and low clouds are denser and darker.

Layered clouds in any of these height ranges generally indicate that the air is stable. We might not normally expect clouds to grow or persist in stable air. Yet cloud growth of this type is common when air is forced to rise, as along a front or near the center of a cyclone where converging winds cause air to ascend. Such forced ascent of stable air leads to the formation of a stratified cloud layer that is large horizontally compared to its depth.

Some clouds do not fit into any one of these three height categories. Such clouds have their bases in the low height range and often extend upward into the middle or high altitudes. Consequently, these clouds are referred to as **clouds of vertical development.** They are all related to one another and are associated with unstable air. Although cumulus clouds are often connected with "fair weather," they may, under the proper circumstances, grow dramatically. Once upward movement is triggered, acceleration is powerful, and clouds with great vertical extent are formed. As the cumulus enlarges, its top leaves the low height range and it is called a *cumulus congestus*. Finally, when the cloud becomes even more towering and rain begins to fall, it becomes a *cumulonimbus*.

Definite weather patterns can usually be associated with certain clouds or certain combinations of cloud types, and so it is important to become familiar with cloud descriptions and characteristics. Table 5–1 lists the ten basic cloud types that are recognized internationally and gives some characteristics of each. The series of photos in Figure 5–2 depicts common forms of several cloud types.

In addition to the names given to the ten basic cloud types, various adjectives may also be used to describe certain clouds. For example, the term *uncinus*, meaning hook shaped, is applied to streaks of cirrus clouds that are shaped like a comma resting on its side. Cirrus uncinus are often precursors of bad weather. When stratus or cumulus clouds appear to be broken into smaller pieces, the adjective *fractus* may be used in their description. Lens-shaped clouds called *lenticular* clouds are sometimes common sights in areas that have rough or mountainous topography (Figure 5–3). Although lenticular clouds can form whenever the airflow

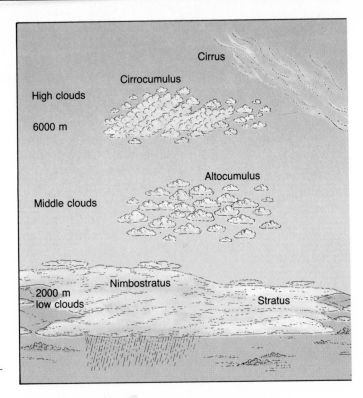

FIGURE 5–1
Classification of clouds according to height and form.
(After Ward's Natural Science Establishment, Inc., Rochester, N.Y.)

undulates sharply in the vertical, they most frequently form on the lee side of mountains. As air passes over mountainous terrain, a wave pattern develops as shown in Figure 5–3. Clouds form where the wavy flow causes air to ascend, whereas areas with descending air are cloud free.

Cloud Formation

As we learned earlier, condensation occurs when water vapor changes to a liquid. The result of this process may be dew, fog, or clouds. Although each type of condensation is very different, all have two properties in common. First, in order for any form of condensation to occur, the air must be saturated. Saturation occurs when the air is cooled below its dew point, which most commonly happens, or when water vapor is added to the air. Second, there generally must be a surface on which the water vapor may condense. When dew forms, objects at or near the ground serve this purpose. When condensation occurs in the air above the ground, tiny particles known as **condensation nuclei** serve as surfaces on which water vapor condenses. The importance of these nuclei should be noted because, if absent, a relative humidity well in excess of 100 percent is necessary to produce clouds. Condensation nuclei, such as microscopic dust, smoke, and salt particles, are profuse in the lower atmosphere. Consequently, the relative humidity rarely

Condensation Nuclei

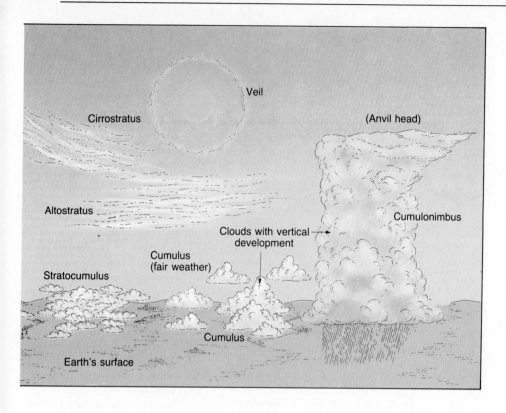

Veil

Cirrostratus

(Anvil head)

Altostratus

Cumulonimbus

Clouds with vertical development

Cumulus (fair weather)

Stratocumulus

Cumulus

Earth's surface

Hygroscopic Nuclei

exceeds 101 percent before condensation begins. Some particles, such as salt from the ocean, are particularly good nuclei because they absorb water. These particles are termed **hygroscopic** ("water-seeking") **nuclei.**

When condensation takes place, the initial growth rate of cloud droplets is rapid, but it diminishes quickly because the available water vapor is readily consumed by the large number of competing droplets. The result is the formation of a cloud consisting of billions of tiny water droplets, all so minute that they remain suspended in air. Even in very moist air the growth of these cloud droplets by additional condensation is slow. Furthermore, the immense size difference between cloud droplets and raindrops (it takes about one million cloud droplets to form a single raindrop) suggests that condensation alone is not responsible for the formation of drops large enough to fall as rain.

FORMATION OF PRECIPITATION

Although all clouds contain water, why do some produce precipitation and others drift placidly overhead? This seemingly simple question perplexed meteorologists for many years. First, cloud droplets are tiny, averaging less than 20 micrometers

TABLE 5–1 Basic Cloud Types

Cloud Family and Height	Cloud Type	Characteristics
High clouds— above 6000 m	Cirrus	Thin, delicate, fibrous ice-crystal clouds. Sometimes appear as hooked filaments called "mares' tails." (cirrus uncinus) (Figure 5–2a)
	Cirrocumulus	Thin, white ice-crystal clouds, in the form of ripples, waves, or globular masses all in a row. May produce a "mackerel sky." Least common of the high clouds. (Figure 5–2b)
	Cirrostratus	Thin sheet of white ice-crystal clouds that may give the sky a milky look. Sometimes produce halos around the sun or moon. (Figure 12–13)
Middle clouds— 2000–6000 m	Altocumulus	White to gray clouds often made up of separate globules; "sheepback" clouds.
	Altostratus	Stratified veil of clouds that are generally thin and may produce very light precipitation. When thin, the sun or moon may be visible as a "bright spot," but no halos are produced. (Figure 5–2c)
Low clouds— below 2000 m	Stratocumulus	Soft, gray clouds in globular patches or rolls. Rolls may join together to make a continuous cloud.
	Stratus	Low uniform layer resembling fog but not resting on the ground. May produce drizzle.
	Nimbostratus	Amorphous layer of dark gray clouds. One of the chief precipitation-producing clouds. (Figure 5–2d)
Clouds of vertical development	Cumulus	Dense, billowy clouds often characterized by flat bases. May occur as isolated clouds or closely packed. (Figure 5–2e)
	Cumulonimbus	Towering cloud sometimes spreading out on top to form an "anvil head." Associated with heavy rainfall, thunder, lightning, hail, and tornadoes. (Figure 5–2f)

FIGURE 5–2
(a) Cirrus. (b) Cirrocumulus. (c) Altostratus.

(a)

(b)

(c)

FIGURE 5–2 (*cont.*)
(d) Nimbostratus. (e) Fair weather cumulus.
(f) Cumulonimbus. (Photos by E. J. Tarbuck)

(d)

(e)

(f)

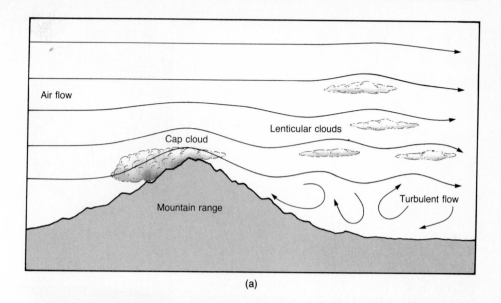

(a)

(b)

FIGURE 5–3

(a) This diagram depicts the formation of lenticular clouds in the turbulent flow which develops in the lee of a mountain range. (b) Lenticular clouds are relatively common in many mountainous areas. (Photo by E. J. Tarbuck)

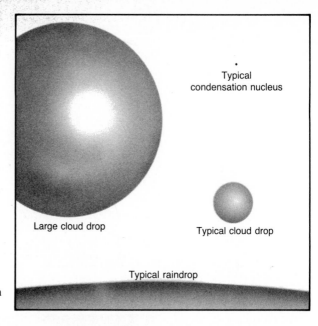

FIGURE 5–4
Comparative diameters of particles involved in condensation
and precipitation processes.

in diameter (Figure 5–4). For comparison, a human hair is about 75 micrometers
in diameter.[1] Because of their small size, the rate at which cloud droplets fall is
incredibly slow. An average cloud droplet falling from a cloud base at 1000 meters
would require about 48 hours to reach the ground. Of course, it would never
complete its journey. Even falling through humid air, a cloud droplet would evaporate before it fell a few meters below the cloud base. Secondly, clouds consist of
many billions of these droplets, all competing for the available water; thus their
continued growth via condensation is slow.

A raindrop large enough to reach the ground without evaporating contains
roughly a million times the water of a cloud droplet. Therefore in order for
precipitation to form, millions of cloud droplets must somehow coalesce (join
together) into drops large enough to sustain themselves during their descent.
Two mechanisms have been proposed to explain this phenomenon: the Bergeron
process and the collision-coalescence process.

Bergeron Process

Bergeron Process

The **Bergeron process** is named for its discoverer, the highly respected Swedish
meteorologist, Tor Bergeron.[2] It relies on two interesting properties of water.
First, cloud droplets do not freeze at 0°C as expected. In fact, pure water suspended
in air does not freeze until it reaches a temperature of nearly −40°C. Water in

---

[1] One micrometer equals 0.001 millimeter.
[2] After reading this section, read Duncan Blanchard's essay, "Science and Serendipity," p. 134, which
deals in a most interesting way with Bergeron's discovery of this important process.

TABLE 5–2 Relative Humidity with Respect to Ice when
Relative Humidity with Respect to Water is 100 Percent

| Temperature (°C) | Relative Humidity with Respect to: | |
	Water (%)	Ice (%)
0	100	100
−5	100	105
−10	100	110
−15	100	115
−20	100	121

Supercooled Water

the liquid state below 0°C is generally referred to as **supercooled.** Supercooled water will readily freeze if sufficiently agitated, which explains why airplanes collect ice when they pass through a liquid cloud made up of supercooled droplets. In addition, supercooled droplets will freeze on contact with solid particles that have a crystal form closely resembling that of ice. These materials have been

Freezing Nuclei

termed **freezing nuclei.** The need for freezing nuclei to initiate the freezing process is similar to the requirement for condensation nuclei in the process of condensation. In contrast to condensation nuclei, however, freezing nuclei are sparse in the atmosphere and do not generally become active until the temperature reaches −10°C or below. Thus at temperatures between 0 and −10°C clouds consist mainly of supercooled water droplets. Between −10 and −20°C, liquid droplets coexist with ice crystals, and below −20°C clouds are generally composed entirely of ice crystals.

This brings us to a second important property of water. The saturation vapor pressure over ice crystals is much lower than over supercooled liquid droplets. This situation occurs because ice crystals are solid, which means that the individual water molecules are held together more tightly than those forming a liquid droplet. As a result, it is easier for water molecules to escape from the supercooled liquid droplets. This fact accounts for the higher-saturation vapor pressure over supercooled liquid droplets compared with ice crystals. Consequently, when air is saturated (100 percent relative humidity) with respect to liquid droplets, it is supersaturated with respect to ice crystals. Table 5–2, for example, shows that at −10°C, when the relative humidity is 100 percent with respect to water, the relative humidity with respect to ice is about 110 percent.

With these facts in mind, we can now explain how the Bergeron process produces precipitation. Visualize a cloud at a temperature of −10°C where each ice crystal is surrounded by many thousands of liquid droplets. Because the air was initially saturated with respect to liquid water, it will be supersaturated with respect to the newly formed ice crystals. As a result of this supersaturated condition, the ice crystals collect more water molecules than they lose by sublimation. Thus continued evaporation from the liquid drops provides a source of water vapor for the growth of ice crystals (Figure 5–5).

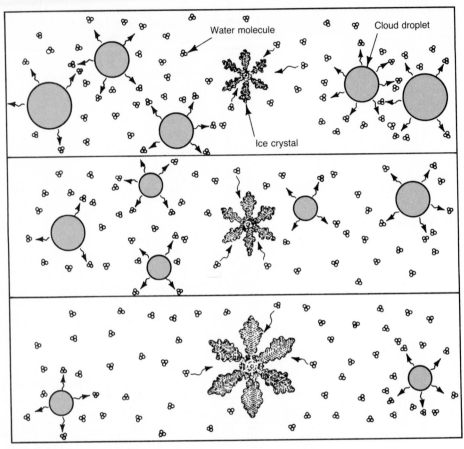

FIGURE 5–5
The Bergeron process. Ice crystals grow at the expense of cloud droplets until they are large enough to fall. The size of these particles has been greatly exaggerated.

Because the level of supersaturation with respect to ice can be great, the growth of ice crystals is generally rapid enough to generate crystals large enough to fall. During their descent these ice crystals enlarge as they intercept cloud drops that freeze on them. Air movement will sometimes break up these delicate crystals and the fragments will serve as freezing nuclei for other liquid droplets. A chain reaction develops and produces many ice crystals, which, by accretion, will form into larger crystals called snowflakes (Figure 5–6). When the surface temperature is above 4°C, snowflakes usually melt before they reach the ground and continue their descent as rain. Even a summer rain may have begun as a snowstorm in the clouds overhead.

Cloud seeding uses the Bergeron process. Adding freezing nuclei (commonly silver iodide) to supercooled clouds can markedly change the growth of these clouds. This process is discussed in greater detail later in the chapter.

(a)

(b)

FIGURE 5–6

(a) As this photograph illustrates, snow crystals are six-sided. This particular snow crystal was formed in the laboratory using a lead particle as the freezing nucleus. (Photo by Roger Cheng)

(b) Although snow crystals are always six-sided, they come in an infinite variety of forms. (Courtesy of NOAA)

SCIENCE AND SERENDIPITY

Duncan C. Blanchard[*]

Many important scientific discoveries are made as the result of serendipity, defined by Nobel Laureate Irving Langmuir as "the art of profiting from unexpected occurrences." In other words, if you are observing something or are doing an experiment and the entirely unexpected happens, and if you see in this happy accident a new and meaningful discovery, then you have experienced serendipity. Most nonscientists, some scientists, and, alas, many teachers, are not aware that many of the great discoveries in science are serendipitous.

One of the many excellent examples of serendipity in science occurred when Tor Bergeron, the great Swedish meteorologist, discovered the importance of ice crystals in the initiation of precipitation in supercooled clouds. This landmark discovery was presented in a classic paper at a meeting of the International Union of Geodesy and Geophysics in Lisbon in 1933, and published two years later in the proceedings of the meeting. In the paper Bergeron says he applied to supercooled clouds a principle of thermodynamics he had read about many years before in a book by the German scientist, Alfred Wegener. Wegener had pointed out that the vapor pressure of ice is less than the vapor pressure of water at the same temperature. Thus, said Bergeron, if ice crystals somehow appear in the midst of a cloud of supercooled droplets, they will grow rapidly as water molecules diffuse toward them from the evaporating droplets. This rapid growth causes the crystals to fall, to collide and stick to other crystals to form snowflakes which, depending on the air temperature beneath the cloud, fall to the ground as snow or rain.

I was curious about the story behind Bergeron's

discovery. Did he, after reading about the thermodynamic principle in Wegener's book, have a sudden flash of insight to connect it to the rapid growth of snow in clouds? Or did he make some serendipitous observation that made the application of the principle obvious? No hint can be found in his paper. In March 1971 I wrote to him and asked these questions. Three months later, Bergeron wrote back. In a fascinating letter he expressed his feelings about discovery, including the role played by subjectivity. In one of his German papers he said he wrote "Vor lauter Objektivität kein Fortschritt." (From pure objectivity, no progress.)

Included with his letter was a long account of the observations and ideas that led up to that day in Lisbon in 1933 when he presented the paper that outlined his "ice nucleus theory." He had indeed, while walking through the woods many years before, made a serendipitous observation that enabled him to make that brilliant conceptual leap to the role of ice crystals in the formation of precipitation.

After his death in 1977, I submitted his "Autobiographic Notes" to the *Bulletin of the American Meteorological Society*, where they were published. The following two paragraphs, taken from his Notes, describe the crucial observations that eventually led to his "ice nucleus theory."

> To return to my own problem. What I had read in Wegener's book evidently lay latent in my mind, when in February 1922, just before emigrating to Norway for a 13 years' stay there, I spent a couple of weeks for recreation at a health resort at an altitude of 430 m (1400 ft) on a hill near Oslo. On that hill we were often in a supercooled stratus layer, and when walking on a narrow road in the fir forest, parallel to the contours of the hillside, I noticed that the "fog" did *not* enter the "road tunnel" at, say, −5 to −10°C, but *did* enter it when the temperature was >0°C. The profile of the road, trees, and fog for the two temperature regimes is shown in the figure.
>
> My tentative explanation came immediately. At −10°C the rime-clad branches of the firs along

[*]Duncan C. Blanchard is a senior scientist at the Atmospheric Sciences Research Center at the State University of New York at Albany. His major research is in the area of air–water interaction, but he enjoys looking behind the research to the people who have made outstanding contributions to the science of meteorology. This essay is drawn from a longer piece by Dr. Blanchard that appeared in *Weatherwise*, 32, no. 6 (1979), 236–241, entitled "Science, Success and Serendipity."

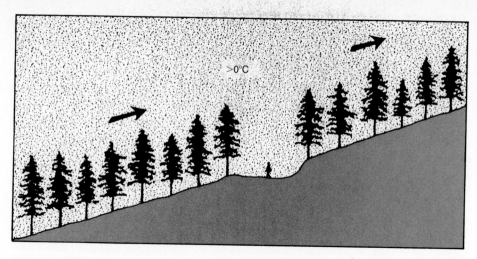

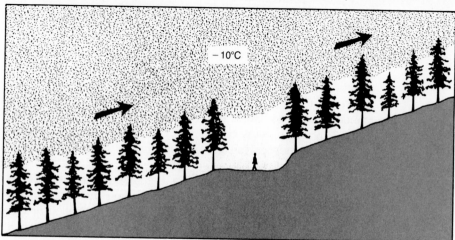

the windward side of the road by *diffusion transport* 'filtered away' so much of the water vapor in the air (because at −10°C the saturation vapor pressure over ice is less than that over water) that the fog droplets partly evaporated (some droplets were, of course, also captured directly by the 'netting' of fir needles). At temperatures >0°C the fog was seen everywhere between the trees and filled the whole 'road-tunnel.' So striking was the difference that it could not only be due to the mechanical capture of the droplets. Why should, indeed, the latter effect apparently fail utterly with the temperatures above freezing?

Serendipity not only played a role in Bergeron's discovery, but its influence can be seen in the entire realm of science. Can we, then, conclude that because a great many ideas and discoveries follow chance ob-servations, that anyone who makes these observations will necessarily make the discovery? Not at all. A perceptive and inquiring mind is required, a mind that has been groping for order in a labyrinth of facts. As Langmuir said, the unexpected occurrence is not enough. You must know how to profit from it. Louis Pasteur said it in another way . . . "In the field of observation, chance favors only the prepared mind." The discoverer of vitamin C, Nobel Laureate Albert Szent-Györgyi, was not defining serendipity, but might as well have been when he said that discoveries are made by those who ". . . see what everybody else has seen, and think what nobody else has thought." We need not be surprised that these noted scientists are saying the same thing, though with different words, for serendipity is at the heart of science itself.

Collision–Coalescence

Thirty years ago meteorologists believed that the Bergeron process was responsible for the formation of most precipitation except for light drizzle. Later it was discovered that copious rainfall is often associated with clouds located well below the freezing level (warm clouds), especially in the tropics. This finding led to the proposal of a second mechanism thought to produce precipitation, the **collision-coalescence process.**

Collision–Coalescence Process

Clouds made up entirely of liquid droplets must contain droplets larger than 20 micrometers if precipitation is to form. These larger droplets form when "giant" condensation nuclei are present and when hygroscopic particles, such as sea salt, exist. Recall that hygroscopic particles begin to remove water vapor from the air at relative humidities under 100 percent and can grow very large. Because the rate at which drops fall is size dependent, these "giant" droplets fall most rapidly. As such, they collide with the smaller, slower droplets and coalesce. Becoming larger in the process, they fall even more rapidly (or in an updraft they rise more slowly) and increase their chances of collision and rate of growth (Figure 5–7). After a million such collisions, they are large enough to fall to the surface without completely evaporating. Because of the number of collisions required for growth to raindrop size, droplets in clouds with great vertical thickness and abundant moisture have a better chance of reaching the required size. Updrafts also aid in this process because they allow the droplets to traverse the cloud repeatedly. Raindrops can grow to maximum size of 5 millimeters when they fall at the rate of 30 kilometers per hour. At this size and speed the water's surface tension, which holds the drop together, is surpassed by the drag imposed by the air, which, in turn, succeeds in pulling the drops apart. The resulting breakup of a large raindrop produces numerous small drops that begin anew the task of sweeping up cloud droplets. Drops less than 0.5 millimeter on reaching the ground are termed "drizzle" and require about 10 minutes to fall from a cloud 1000 meters overhead.

From the preceding discussion it should be apparent that large condensation nuclei are required if warm clouds are to generate any appreciable precipitation. This fact has led to attempts to "seed" these clouds by adding "giant" nuclei. One material used is common table salt. Water is also sprayed into clouds and acts as giant particles to initiate the collision–coalescence process.

The collision-coalescence process is not that simple, however. First, as the larger droplets descend, they produce an airstream around them similar to that produced by an automobile when driven rapidly down the highway. If an automobile is driven at night and we use the bugs that are often out as analogous to cloud droplets, it is easy to visualize how most cloud droplets are swept aside. The larger the cloud droplet (or bug), the better chance it will have of colliding with the giant droplet (or car).

Next, collision does not guarantee coalescence. Experimentation has indicated that the presence of atmospheric electricity may be the key to what holds these droplets together once they collide. If a droplet with a negative charge should collide with a positively charged droplet, their electrical attraction may bind them together.

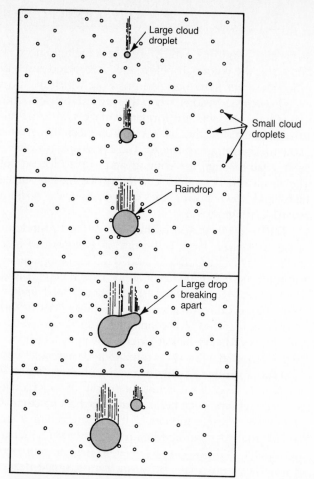

FIGURE 5–7
The collision-coalescence process. Because large cloud droplets fall more rapidly than smaller droplets, they are able to sweep up the smaller ones in their path and grow. Most cloud droplets are so small that the motion of the air keeps them suspended. Even if these small cloud droplets were to fall, they would evaporate before reaching the surface.

SLEET, GLAZE, AND HAIL

Rain and snow are the most common and familiar forms of precipitation, but other forms exist and are worthy of mention. Sleet, glaze, and hail fall into this category. Although limited in occurrence and sporadic in both time and space, these forms, especially glaze and hail, may on occasion cause considerable damage.

Sleet

Sleet is a wintertime phenomenon and refers to the fall of small, clear to translucent particles of ice. In order for sleet to be produced, a layer of air with temperatures above freezing must overlie a subfreezing layer near the ground.

Glaze

Hail

When the raindrops, which are often melted snow, leave the warmer air and encounter the colder air below, they freeze and reach the ground as small pellets of ice no larger than the raindrops from which they formed.

On some occasions, when the vertical distribution of temperatures is similar to that associated with the formation of sleet, freezing rain or **glaze** results instead. In such situations, the subfreezing air near the ground is not thick enough to allow the raindrops to freeze. The raindrops, however, do become supercooled as they fall through the cold air and turn to ice on colliding with solid objects. The result can be a thick coating of ice having sufficient weight to break tree limbs and down power lines as well as make walking or motoring extremely hazardous (Figure 5–8).

Hail is precipitation in the form of hard rounded pellets or irregular lumps of ice. Moreover, large hailstones often consist of a series of nearly concentric shells of differing densities and degrees of opaqueness (Figure 5–9). Usually, hailstones have a diameter of about 1 centimeter, but they may vary in size from 5 millimeters to more than 10 centimeters in diameter. The largest hailstone on record fell on Coffeyville, Kansas, September 3, 1970. With a 14-centimeter diameter and a circumference of 44 centimeters, this "giant" weighed 766 grams! The destructive effects of heavy hail are well known, especially to farmers whose crops can be devastated in a few short minutes and to people whose windows are shattered.

Hail is produced only in cumulonimbus clouds where updrafts are strong and where there is an abundant supply of supercooled water. Hailstones begin as small embryonic ice pellets that grow by collecting supercooled cloud droplets as they fall through the cloud. If they encounter a strong updraft, they may be carried upward again and begin the downward journey anew. Each trip though the supercooled portion of the cloud may be represented by an additional layer of ice. Hailstones may also form from a single descent through an updraft. In this case, a layered structure may result from variations in the rate at which supercooled droplets accumulate and freeze, which, in turn, is related to differences in the amount of supercooled water in different parts of the cumulonimbus tower.

In either case, hailstones grow by the addition of supercooled water on growing ice pellets. The ultimate size of the hailstone depends primarily on three factors: (1) the strength of the updrafts, (2) the concentration of supercooled water, and (3) the length of the path through the cloud.

PRECIPITATION MEASUREMENT

Standard Rain Gauge

The most common form of precipitation, rain, is probably the easiest to measure. Any open container that has a consistent cross section throughout can be a rain gauge. In general practice, however, more sophisticated devices are used in order to measure small amounts of rainfall more accurately as well as reduce losses resulting from evaporation. The **standard rain gauge** (Figure 5–10) has a diameter of about 20 centimeters at the top. Once the water is caught, a funnel conducts the rain through a narrow opening into a cylindrical measuring tube that has a

FIGURE 5–8

The results of an icing storm. (a) Close-up of glaze on a wire fence and branches; (b) a tree coated with shimmering ice. (Photographs by Jack Radgowski, courtesy of the *Peoria Journal Star*)

(a)

(b)

FIGURE 5–9
(a) A section through a hailstone photographed in polarized light through a microscope. Note the concentric layers of ice. (b) Like this large stone, many hailstones are irregular masses. (Photographs courtesy of the National Center for Atmospheric Research)

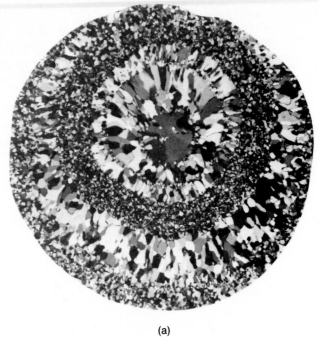

(a)

(b)

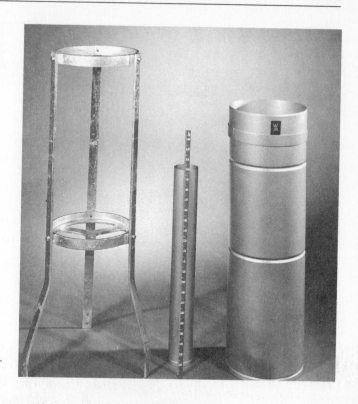

FIGURE 5–10
Standard rain gauge. (Courtesy of WeatherMeasure Corporation)

Trace of Precipitation

cross-sectional area only one-tenth as large as the receiver. Consequently, rainfall depth is magnified ten times, which allows for accurate measurements to the nearest 0.025 centimeter, while the narrow opening minimizes evaporation. When the amount of rain is less than 0.025 centimeter, it is generally reported as being a **trace** of precipitation.

In addition to the standard rain guage, several types of recording gauges are routinely used. These instruments not only record the amount of rain but also its time of occurrence and intensity (amount per unit of time). Two of the most common gauges are the tipping-bucket gauge and the weighing gauge.

Tipping-Bucket Gauge

As can be seen in Figure 5–11(a), the **tipping-bucket gauge** consists of two compartments, each one capable of holding 0.025 centimeter of rain, situated at the base of a 25-centimeter funnel. When one "bucket" fills, it tips and empties its water. Meanwhile, the other "bucket" takes its place at the mouth of the funnel. Each time a compartment tips, an electrical circuit is closed and the amount of precipitation is automatically recorded on a graph.

Weighing Gauge

The **weighing gauge,** as the name would indicate, works on a different principle. A shown in Figure 5–11(b), the precipitation is caught in a cylinder that rests on a spring balance. As the cylinder fills, the movement is transmitted to a pen that records the data.

No matter which rain gauge is used, proper exposure is critical. Errors arise when the gauge is shielded from obliquely falling rain by buildings, trees, or other high objects. Hence, the instrument should be at least as far away from

(a)

(b)

FIGURE 5–11
Two common types of recording precipitation gauges.
(a) An internal view of a tipping-bucket precipitation
gauge. (Courtesy of WeatherMeasure Corporation) (b)
A weighing gauge. (Courtesy of Belfort Instrument
Company)

such obstructions as the objects are high. Another cause for error is the wind. It has been shown that with increasing wind and turbulence, it becomes more difficult to collect a representative quantity of rain. To offset this effect, a windscreen is often placed around the instrument so that rain falls into the gauge and is not carried across it.

When snow records are kept, two measurements are normally taken: depth and water equivalent. Usually the depth of snow is measured with a calibrated stick. The actual measurement is not difficult, but choosing a representative spot often poses a dilemma. Even when winds are light or moderate, snow drifts freely. As a rule, it is best to take several measurements in an open place away from trees and obstructions and then average them. To obtain the water equivalent, samples may be melted and then weighed or measured as rain.

Sometimes large cylinders are used to collect snow. A major problem that hinders accurate measurement by snow gauges is the wind. Snow will blow around the top of the cylinder instead of falling into it. Therefore the amount caught by the gauge is generally less than the actual fall. As is often the practice with rain gauges, shields designed to break up wind eddies are placed around the snow gauge to ensure a more accurate catch.

The quantity of water in a given volume of snow is not constant. A general ratio of 10 units of snow to 1 unit of water is often used when exact information is not available, but the actual water content of snow may deviate widely from this figure. It may take as much as 30 centimeters of light and fluffy dry snow or as little as 4 centimeters of wet snow to produce 1 centimeter of water.

FOG

Fog

Fog is generally considered an atmospheric hazard. When it is light, visibility is reduced to 2 or 3 kilometers. When it is dense, visibility may be cut to a few tens of meters or less, making travel by any mode not only difficult but dangerous as well. Officially visibility must be reduced to 1 kilometer or less before it is reported. Although arbitrary, this figure does permit a more objective criterion for comparing fog frequencies at different locations.

Fog is defined as a cloud with its base at or very near the ground. Physically there is basically no difference between a fog and a cloud; the appearance and structure of both are the same. The essential difference is the method and place of formation. Whereas clouds result when air rises and cools adiabatically, fogs (with the exception of upslope fogs) are the consequence of radiation cooling or the movement of air over a cold surface. In other circumstances, fogs are formed when enough water vapor is added to the air to bring about saturation (evaporation fogs).

Fogs Formed by Cooling

Radiation Fog

Radiation fog, as the name implies, results from radiation cooling of the ground and adjacent air. It is a nighttime phenomenon that requires clear skies and a fairly high relative humidity. Under these circumstances the ground and the air immediately above will cool rapidly. Because the relative humidity is high, just a small amount of cooling will lower the temperature to the dew point. If the air is calm, the fog may be patchy and less than a meter deep. For radiation fog to be more extensive vertically, a light breeze of 3 to 4 kilometers per hour is necessary. Then the light wind creates enough turbulence to carry the fog upward 10 to 30 meters without dispersing it.

Because the air containing the fog is relatively cold and dense, it drains downslope. As a result, radiation fog is thickest in valleys, whereas the surrounding hills are clear (Figure 5–12). Normally these fogs dissipate within 1 to 3 hours after sunrise. Often the fog is said to "lift." However, it does not actually rise. Instead the sun warms the earth, which, in turn, heats the surface air first. Consequently, the fog evaporates from the bottom up, giving the impression of lifting, and the last vestiges may appear to be a low white cloud layer.

Advection Fog

When warm moist air is blown over a cold surface, it becomes chilled by contact and, to a certain extent, by mixing with the cold air associated with the cold surface below. If cooling is sufficient, the result will be a blanket of fog called **advection fog.** The term advection refers to air moving horizontally, and so such fogs are a consequence of air giving up heat to the surface below during horizontal movement.

A certain amount of turbulence is needed for proper development; thus winds between 10 and 30 kilometers per hour are usually associated with advection fog. Not only does the turbulence facilitate cooling through a thicker layer of air, but it also carries the fog to greater heights. Unlike radiation fogs, advection fogs are often thick (300 to 600 meters deep) and persistent.

Examples of such fogs are common. The foggiest location in the United States, and perhaps in the world, is Cape Disappointment, Washington. The name is indeed appropriate because the station averages 2552 hours of fog each year. The fog experienced at Cape Disappointment, as well as that at other West Coast locations during the summer and early autumn, is produced when warm, moist air from the Pacific Ocean moves over the cold California Current (Figure 5–13). It is then carried onshore by westerly winds or a local sea breeze. Advection

FIGURE 5–12
Early morning radiation fog in a valley. (Photograph by Jack L. Bradley, courtesy of the *Peoria Journal Star*)

FIGURE 5–13
Advection fog rolling in under the Golden Gate Bridge, San Francisco, California. (Photograph by Lee Blodget)

fogs also occur in the northern tier of states in the winter when warm air from the Gulf of Mexico moves over cold, often snow-covered surfaces.

Upslope Fog

As its name implies, **upslope fog** is created when relatively humid air moves up a gradually sloping plain or, in some cases, up the steep slopes of a mountain. Because of the upward movement, air expands and cools adiabatically. If the dew point is reached, an extensive layer of fog may form.

In the United States the Great Plains offers an excellent example. When humid easterly or southeasterly winds move westward from the Mississippi River toward the Rocky Mountains, the air gradually rises, resulting in an adiabatic decrease of about 13°C. When the difference between the air temperature and dew point of westward moving air is less than 13°C, an extensive fog often results in the western plains.

Evaporation Fogs

When cool air moves over warm water, enough moisture may evaporate from the water surface to produce saturation. As the rising water vapor meets the cold air, it immediately recondenses and rises with the air that is being warmed from below. Because the water has a steaming appearance, the phenomenon is

Steam Fog

called **steam fog** (Figure 5–14). It is a fairly common occurrence over lakes and rivers in the fall and early winter when the water may still be relatively warm

FIGURE 5–14
Steam fog rising from a small lake. (Courtesy of Ward's Natural Science Establishment, Inc., Rochester, N.Y.)

and when the air is rather crisp. Steam fog is often very shallow, for as the steam rises, it reevaporates in the unsaturated air above.

Steam fogs, however, can be dense. During the winter when cold air pours off the continents and ice shelves of the north into the open ocean, the air may be from 20 to 30°C colder than the water. Steaming is intense and it saturates a large volume of air. Because of its source and appearance, this type of steam fog is given the name *arctic sea smoke*.

When frontal wedging occurs, warm air is lifted over colder air. If the resulting clouds yield rain, and the cold air below is near the dew point, enough rain will evaporate to produce fog. A fog formed in this manner is called **frontal** or **precipitation fog.** The result is a more or less continuous zone of condensed water droplets reaching from the ground up through the clouds.

Frontal Fog

In summary, both steam fog and frontal fog result from the addition of moisture to a layer of air. As you saw, the air is usually cool or cold and already near saturation. Because air's capacity to hold water vapor at low temperatures is small, only a relatively modest amount of evaporation is necessary to produce saturated conditions and fog.

The frequency of occurrence of dense fog varies considerably from place to place. As might be expected, fog incidence is highest in coastal areas, especially where cold currents prevail, as along the Pacific and New England coasts. Relatively high frequencies are also found in the Great Lakes region and in the humid Appalachian Mountains of the East. In contrast, fogs are rare in the interior of the continent, especially in the arid and semiarid areas of the West.

INTENTIONAL WEATHER MODIFICATION

Weather Modification

Weather modification is deliberate human intervention to influence and improve the atmospheric processes and events that constitute the weather—that is, to aim the weather at human purposes. Certainly a desire to change the weather is far from modern. From earliest recorded times people have used prayer, wizardry, dancing, and even black magic in attempts to alter the weather. Until modern times, however, most attempts at weather modification remained largely in the realm of the mystic. By the nineteenth century such devices as smudge pots, sprinklers, and wind machines to fight frost were in use. During the U.S. Civil War observations that rainfall apparently increased following some battles led to experiments in which cannons were fired into clouds to bring more rain. Unfortunately, these experiments, as well as many others, proved unsuccessful.

Weather-modification strategies fall into three broad subdivisions. The first relies on the injection of energy by "brute force." The use of powerful heat sources or the intense mechanical mixing of the air (such as by helicopters) are both examples of techniques of fog dispersal attempted at some airports that fall into this category. The second subdivision involves the alteration of land and water surfaces in order to change their natural interactions with the lower atmosphere. One commonly discussed but still theoretical example of this technique is the blanketing of a land area with a dark substance. If it were done, the amount of heat absorbed by the surface would increase and lead to stronger upward air currents that, in turn, might aid cloud formation. Finally, the third subdivision involves triggering, intensifying, or redirecting the atmosphere's natural energies. The seeding of clouds with such agents as dry ice and silver iodide for many purposes, including precipitation enhancement, represents the primary example in this category. Because it offers a relatively inexpensive and easily used technique, **cloud seeding** has been and will probably continue to be the main focus of modern weather-modification technology.

Cloud Seeding

Cloud Seeding

The first scientific breakthrough in intentional weather modification came in 1946 when Vincent J. Schaefer discovered that dry ice dropped into a supercooled cloud spurred the growth of ice crystals. Recall from the discussion of precipitation formation that once ice crystals form, they grow larger at the expense of the remaining liquid cloud droplets and, on reaching a sufficient size, fall as precipitation. Just as Tor Bergeron's discovery of the role of ice crystals in the formation of precipitation was, at least in part, accidental, so too was Schaefer's discovery.[3] Although Schaefer was indeed trying to find something that would trigger the conversion of supercooled cloud droplets into ice crystals, his application of dry ice was not intended to produce the result. He spent weeks trying to find a substance that would convert the cloud of supercooled droplets in his cold box

---

[3] See Duncan Blanchard's essay in this chapter entitled "Science and Serendipity," p. 134.

(an ordinary home freezer lined with black velvet) into a cloud of ice crystals. He tested a large variety of materials by sprinkling them into the cloud, but nothing seemed to work. Then one hot July afternoon, after he noticed that the temperature of the air in his cold box was not as low as usual, Schaefer placed a block of dry ice into the chamber to lower its temperature. In a few moments ice crystals replaced the supercooled cloud. This method worked because when water droplets are cooled below −40°C, they will freeze with or without freezing nuclei.

Shortly after Schaefer's discovery it was learned that silver iodide could also be used for cloud seeding. The similarity in the crystalline structures of silver iodide and ice accounts for silver iodide's ability to initiate the growth of ice crystals. Thus unlike dry ice, silver iodide crystals act as freezing nuclei rather than as a cooling agent. This substance has an advantage over dry ice in that it can be supplied to clouds from burners on the ground as well as from aircraft (Figure 5–15). If either substance is to be successful, certain atmospheric conditions must exist. Clouds must be present, for seeding cannot generate them. In addition, at least the top portion of the cloud must be supercooled; that is, it must be made of liquid droplets having a temperature below 0°C.

Static Seeding

There are two basic cloud-seeding strategies. **Static seeding** is based on the assumption that cumulus clouds are deficient in freezing nuclei and that the addition of nuclei by cloud seeding will spur the formation of additional precipitation. The object, therefore, is to produce just the correct number of ice crystals. Overseeding will simply produce billions of minute ice crystals, none having

(b)

FIGURE 5–15
(a) Silver iodide can be supplied to clouds from burners on the ground.
(b) Cloud seeding using wing-mounted silver iodide flares. (Photos courtesy of Joe Montgomery, U.S. Department of the Interior, Bureau of Reclamation)

(a)

Dynamic Seeding

sufficient size to fall. **Dynamic seeding,** on the other hand, uses massive seeding in order to convert supercooled liquid droplets rapidly to ice during the active growth phase of the cloud. Scientists hypothesize that the resulting release of latent heat increases cloud buoyancy that causes the cloud to grow larger than it would have grown without seeding.

Static seeding has been the common strategy used in most cloud-seeding experiments thus far conducted. The seeding of winter clouds that form along mountain barriers has been one actively pursued example of this technique. Estimates are that only about 20 to 50 percent of the water that condenses in orographic clouds actually falls as precipitation; consequently, these clouds were believed to be good candidates for seeding to increase the percentage of water falling on mountain slopes. The idea is to increase winter snowpack that melts and runs off during warmer months and can be collected in reservoirs for use in irrigation and hydroelectric power generation. With such a goal in mind, experimental operations such as the Sierra Cooperative Pilot Project and the Colorado Orographic Seeding Program have been operating since the mid-1970s. Although these and similar projects have shown promise, their apparently positive results have not yet been confirmed to the satisfaction of most weather modification scientists. There are indications from statistical analyses that, under certain conditions, it is possible to increase precipitation from wintertime mountain storms in the western United States. However, statistical evidence alone is not sufficient to convince many atmospheric scientists of the effectiveness of cloud seeding. They require, in addition, firm knowledge of the physical mechanisms responsible for any apparent increase in precipitation. To date, such an understanding has remained elusive.[4]

Static seeding of nonorographic cumulus clouds is an even more difficult matter because of their great variability in time and space. Nevertheless, there have been many attempts at increasing rainfall by seeding such clouds, some of which appeared successful initially but many of which were failures. One of the first sophisticated operational cloud-seeding experiments was conducted over south-central Missouri in the 1950s. The net effect of this project, which spanned five summers, was a *decrease* in rainfall. Apparently these clouds had all the freezing nuclei they needed to produce precipitation, and seeding only created an excess of very small ice particles that tended to remain suspended in the clouds rather than fall as rain. Of the many experiments that appeared to have positive results, only a single project, one conducted in Israel, was able to show a statistically convincing confirmation of an increase in precipitation after cloud seeding. Results from two separate phases of testing indicated a 15 percent increase in rainfall. When the results became known, the Israeli government stopped experimenting and began a program of operational seeding.

Although first suggested in the 1940s, dynamic seeding was not attempted experimentally until the 1960s. Several early tests indicated that this method does lead to the increased growth of clouds that was predicted by models. Furthermore, some researchers believed that the dynamic seeding of certain summer cumulus

---

[4] For more about such projects see David W. Reynolds and Arnett S. Dennis, "A Review of the Sierra Cooperative Pilot Project," *Bulletin of the American Meteorological Society*, 67, no. 5 (May 1986), 513.

clouds could also increase rainfall. The Florida Area Cumulus Experiment (FACE) was designed to test this hypothesis. Experiments were conducted for five summers in the early 1970s. The results suggested that, under some conditions, dynamic seeding increases rainfall. However, a second phase of field experiments that were more statistically rigorous than the first failed to confirm the initial results.

In light of such uncertainty, even among the better-designed experiments, many scientists believe that a "return to basics" is necessary. Although there is little question that many weather events can be modified by human intervention, there is also no doubt that more basic knowledge of atmospheric processes is necessary before weather modification can be carried out with scientifically predictable results. Certainly a major result of the last four decades of work in cloud seeding has been the sobering realization that even the simplest of weather events are exceedingly complex and not yet fully understood. This fact is reflected in a policy statement on planned and inadvertent weather modification adopted by the American Meteorological Society.

> A critical need exists for a greater understanding of clouds and precipitation and their responses to various seeding agents. In all cases where indications of precipitation increases have been suggested, confirmatory experiments are required before any of the technologies can be considered scientifically proven. The establishment of the physical mechanisms active in any demonstrated modification effect is also needed to achieve general scientific acceptance.[5]

Fog and Cloud Dispersal

Perhaps the simplest of all scientific experiments in cloud seeding involve spreading dry ice pellets or silver iodide particles into layers of supercooled fog and low stratus clouds. Such applications trigger a transformation in cloud composition from water droplets into ice crystals. As Figure 5–16 illustrates, the results are easily observed and may be dramatic. Initially the ice crystals are too small to fall to the ground, but as they grow at the expense of the remaining water droplets and as turbulent air motions disperse them through nearby air, they can grow large enough to produce a snow shower. The amount of snowfall produced in this manner is minor, but visibility almost always improves and often a large hole is opened in the stratus clouds or supercooled fog. The U.S. Air Force has practiced this technology for many years at various airbases, and commercial airlines have used this method of weather modification at selected airports in the northwestern United States. Although the possibility of using this technology to open large holes in winter clouds so as to increase the amount of solar radiation reaching the ground has been discussed, it has received little research attention.

Unfortunately, most fog does not consist of supercooled water droplets. Such warm fog is more difficult and expensive to combat because seeding will not diminish it. Successful attempts at dispersing warm fogs have involved mechanical mixing of the fog with drier, warmer air from above or by heating the air. When

---

[5] *Bulletin of the American Meteorological Society*, 65, no. 12 (December 1984), 1322.

FIGURE 5–16
Effects produced by seeding a cloud deck with dry ice. Within 1 hour a hole developed over the seeded area. (Courtesy of General Electric)

the layer of fog is very shallow, helicopters have sometimes been used to disperse the fog. By flying just above the fog, the helicopter creates a strong downdraft that pulls drier air toward the surface and mixes it with the saturated air near the ground. If the air aloft is dry enough, sufficient evaporation will occur to improve visibility significantly. At some airports where warm fogs are a common hazard it has become more usual to heat and thus evaporate the fog. A sophisticated thermal fog dissipation system, called Turboclair, was installed in 1970 at Orly Airport in Paris. This system consists of eight jet engines located in underground chambers alongside the upwind edge of the runway. Although expensive to install, the system is capable of improving visibility for a distance of about 900 meters along the approach and touchdown zones.

Hail Suppression

Hail has been and continues to be a major cause of crop losses and property damage in many parts of the world (Figure 5–17). Consequently, hail-suppression efforts date from classical Greece to the present day. A major attempt in Europe near the turn of the twentieth century involved firing cannons at thunderstorms (Figure 5–18). This method was based on a hypothesis that hailstones could be prevented by injecting smoke particles into possible hail-producing clouds. Because early results appeared to be promising, the use of the method spread. After about a decade, however, the practice was judged ineffective and the use of cannons to suppress hail was abandoned.[6]

---

[6] An excellent examination of this effort and its relevance to modern weather modification is found in S. A. Changnon and J. Loreena Ivens, "History Repeated: The Forgotten Hail Cannons of Europe," *Bulletin of the American Meteorological Society*, 62, no. 3 (1981), 368–375.

FIGURE 5–17
A cornfield devastated by hail. (Courtesy of Illinois State Water Survey)

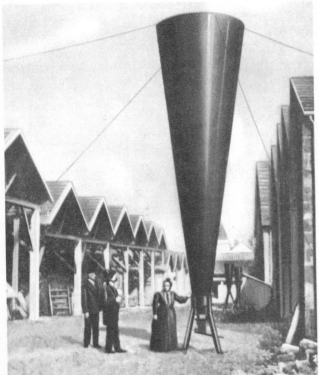

FIGURE 5–18
Near the turn of the 20th century hail shooting was a common activity in portions of Europe. It was thought that the formation of hailstones could be prevented by injecting smoke particles (to serve as condensation nuclei) by means of cannons fired at thunderstorms. The "cannons" were actually vertical-pointing muzzle-loading mortars that resembled very large upright megaphones. This large model weighed 9000 kilograms, was 9 meters long and pivoted in all directions. Hail shooting proved to be ineffective and the practice was abandoned after about 10 years. (Courtesy of J. Loreena Ivens, Illinois State Water Survey)

Hail is believed to form by the collecting and freezing of supercooled droplets around a nucleus. Updrafts of moist air in a cumulonimbus cloud are often so strong that small hailstones may be recirculated in the upper part of the cloud, collecting additional coats of water that freeze and thus cause the stone to enlarge. Normally only a small number of hail embryos exist, because freezing nuclei are relatively scarce; consequently, the embryos grow freely until they fall from the cloud. Modern attempts at hail suppression have been based largely on the idea that the introduction of silver iodide crystals into appropriate clouds will interrupt the formation and growth of hailstones. It is assumed that each of these crystals, acting as a freezing nucleus, attracts a portion of the cloud's water supply and thereby increases the competition for the available supercooled water in the upper portion of the cloud. Without ample supplies of supercooled water, hailstones cannot grow large enough to be destructive when they fall.

Because of dramatic crop and property losses, hail-suppression technology has been the focus of much interest and many field and laboratory studies. The results, however, have been mixed at best. In the Soviet Union during the 1960s scientists using rocket and artillery shells to carry freezing nuclei to the clouds claimed to have extraordinary success. Similar efforts attempted elsewhere, however, failed to show any reduction in hail. In the 1970s the federal government established the National Hail Research Experiment in northeastern Colorado. This experiment was a major effort that included a randomized seeding test stimulated by the USSR experience. An analysis of the data collected after three years of experimentation revealed that there was no statistically significant difference between the seeded and nonseeded incidence of hail. The planned five-year experiment was terminated after three years.

Commenting on current possibilities for successful hail suppression, one report states, "While certain positive results have been obtained, many important questions remain regarding the important chain of events leading to hail and to hypothesized hail suppression."[7] Certainly the current status of hail suppression must be described as one of uncertainty.

Frost Prevention

Frost

White Frost

Frost, the fruit growers' plight, occurs when the air temperature falls to 0°C or below. It may be accompanied by deposits of ice crystals commonly called **white frost.** This process, however, happens only if the air becomes saturated. White frost is not a requirement for crop damage.

Frost hazards exist when a cold air mass moves into a region or when ample radiation cooling occurs on a clear night. The conditions accompanying the invasion of cold air are characterized by low daytime temperatures, long periods of effective frost, strong winds, and widespread damage. Frost induced by radiation loss is a nighttime phenomenon associated with a surface temperature inversion and is much easier to combat.

---

[7] S. A. Changnon et al., "Review of the Tenth Conference on Planned and Inadvertent Weather Modification," *Bulletin of the American Meteorological Society*, 67, no. 12 (December 1986), 1502.

Several methods of frost prevention have been used with varying degrees of success. Generally these attempts are directed at reducing the amount of heat lost during the night or at adding heat to the lowermost layer of air. Heat conservation methods include covering plants with material having a low thermal conductivity, such as paper and cloth, and producing particles that, when suspended in air, reduce the rate of radiation cooling. Smudge fires have been used for particle production but have often proven unsatisfactory. In addition to the pollution problems created by the dense clouds of black smoke, the carbon particles impede daytime warming by reducing the amount of solar radiation that can reach the surface. This reduction in daytime warming may offset the benefits gained during the night.

Methods of warming include sprinkling, air mixing, and the use of orchard heaters. Sprinklers distribute water to the plants and add heat in two ways: first, from the warmth of the water and, second, but more important, from the latent heat of fusion that is released when the water freezes. As long as an ice-water mixture remains on the plant, the latent heat released will keep the temperature from dropping below 0°C. Air mixing is successful when the temperature at 15 meters above the ground is 5°C higher than the surface temperature. By using a wind machine, the warmer air aloft is mixed with the colder surface air (Figure 5–19a). Orchard heaters probably produce the most successful results (Figure 5–19b). Because as many as 30 or 40 heaters per acre are required, the fuel cost can be significant, but usually the effectiveness of this method seems to warrant the cost.

(a)

(b)

FIGURE 5–19
Two common frost-prevention methods: (a) wind machine and (b) orchard heater. (Photographs courtesy of Florida Department of Citrus)

REVIEW

1. What is the basis for the classification of clouds?

2. Why are high clouds always thin in comparison to low and middle clouds?

3. Which cloud types are associated with the following characteristics? Thunder, halos, precipitation, hail, mackerel sky, lightning, mares' tails.

4. What do layered clouds indicate about the stability of the air? What do clouds of vertical development indicate about the stability of air?

5. What is the importance of condensation nuclei?

6. Describe the steps in the formation of precipitation according to the Bergeron process. Be sure to include (a) the importance of supercooled cloud droplets, (b) the role of freezing nuclei, and (c) the difference in saturation vapor pressure between liquid water and ice.

7. How does the collision-coalescence process differ from the Bergeron process?

8. If snow is falling from a cloud, which process produced it? Explain.

9. Describe sleet and glaze and the circumstances under which they form. Why does glaze result on some occasions and sleet on others?

10. How does hail form? What factors govern the ultimate size of hailstones?

11. Although an open container can serve as a rain gauge, what advantages does a standard rain gauge provide?

12. How do recording rain gauges work? Do they have advantages over a standard rain gauge?

13. Describe some of the factors that could lead to an inaccurate measurement of rain or snow.

14. Distinguish between clouds and fog.

15. List five types of fog and discuss the details of their formation.

16. What actually happens when a radiation fog "lifts"?

17. Identify the fogs described in the following situations.
    a. You have stayed the night in a motel and decide to take an early morning swim. As you approach the heated swimming pool, you notice a fog over the water.
    b. You are located in the western Great Plains and the winds are from the east and fog is extensive.
    c. You are driving through hilly terrain during the early morning hours and experience fog in the valleys and clearing on the hills.

18. Why is there a relatively high frequency of dense fog along the Pacific Coast?

19. Why are silver iodide crystals often used to seed supercooled clouds?

20. If cloud seeding is to be successful (or have a chance of being successful), certain atmospheric conditions must exist. Name them.

21. Contrast static seeding and dynamic seeding.

22. How do frost and white frost differ?

23. Describe how smudge fires, sprinkling, and air mixing are used in frost prevention.

VOCABULARY REVIEW

clouds

cirrus

cumulus

stratus

high clouds

middle clouds

low clouds

clouds of vertical development

condensation nuclei

hygroscopic nuclei

Bergeron process

supercooled

freezing nuclei

collision–coalescence process

sleet

glaze

hail

standard rain gauge

trace (of precipitation)

tipping-bucket gauge

weighing gauge

fog

radiation fog

advection fog

upslope fog

steam fog

frontal or precipitation fog

weather modification

cloud seeding

static seeding

dynamic seeding

frost

white frost

6

Air Pressure and Winds

Of the various elements of weather and climate, changes in air pressure are the least noticeable. In listening to a weather report, generally we are interested in moisture conditions (humidity and precipitation), temperature, and perhaps wind. It is the rare person, however, who wonders about air pressure. Although the hour-to-hour and day-to-day variations in air pressure are not perceptible to human beings, they are very important in producing changes in our weather. Variations in air pressure from place to place are responsible for the movement of air (wind), besides being among the most significant factors in weather forecasting. As we shall see, air pressure is closely tied to the other elements of weather in a cause-and-effect relationship.

BEHAVIOR OF GASES

The concept of air pressure and the fact that it decreases with altitude can be more fully realized if we examine the behavior of gases and the principles that govern this behavior. Gas molecules, unlike those of the liquid and solid phases, are not "bound" to one another but are freely moving about, filling all space available to them. When two gas molecules collide, which happens frequently under the conditions normally experienced near the earth's surface, they bounce off each other as if they were very elastic balls. If a gas is confined to a container, this motion is restricted by the sides of the container, much the same way that the walls of a handball court redirect the motion of a handball. The continuous bombardment of gas molecules against the wall of the container exerts an outward push that we call air pressure. Although the atmosphere is without walls, it is confined from below by the earth's land-sea surface and from above by the force of gravity that prevents its outward escape. Here we define **air pressure** as the force exerted against a surface by the continuous collision of gas molecules.

Air Pressure

Two factors, temperature and density, largely determine the amount of pressure that a particular gas will exert. We first examine the effect of temperature on air pressure when density is kept constant by observing the behavior of a gas in a closed container (constant volume). As we shall see later, a change in temperature is nearly always accompanied by a change in density, which complicates the matter considerably. Nevertheless, when the density is kept constant and the temperature of the air is raised, the speed of the gas molecules, and hence their force, is increased. From this observation we conclude that an increase in temperature results in an increase in pressure and that, conversely, a decrease in temperature causes a decrease in pressure. The fact that air pressure is proportional to temperature is precisely why aerosol spray cans have a warning that cautions to keep them away from heat. Overheating of these containers can result in a dangerous explosion if the internal gas pressure exceeds the strength of the container.

It may appear from the preceding discussion that on warm days air pressure will be high and that on cold days it will be low. Such, however, is not necessarily the case. Over the continents in the midlatitudes, for example, the highest pressures are recorded in winter when temperatures are lowest, for air pressure is also proportional to density (the number of gas molecules involved) as well as temperature. On cold days the air molecules are more closely packed (greater density) than on warm days. Often the decrease in molecular motion associated with low temperatures is more than offset by the increased number of molecules exerting pressure. Thus low temperatures mean greater densities and often greater surface pressures. Conversely, when air is heated in the atmosphere, it expands (increases its volume) because of the increased speed of the gas molecules. Consequently, an increase in temperature is generally accompanied by a decrease in density and hence a decrease in pressure.

Ideal Gas Law

The relationship between pressure, temperature, and density described in this section can be expressed by the following equation, called the **ideal gas law:**[1]

$$\text{Pressure} = \text{density} \times \text{temperature} \times \text{constant}$$

In a verbal form the gas law states that the pressure exerted by a gas is proportional to its density and absolute temperature. Thus an increase in either temperature or density will cause an increase in pressure as long as the other variable (density or temperature) remains constant. Moreover, the gas law predicts the relationship between temperature and density observed earlier. When the pressure remains constant, a decrease in temperature results in increased density and vice versa.

Let us consider the decrease in pressure with altitude mentioned in Chapter 1. The relationship between air pressure and density largely explains the observed decrease. To illustrate, imagine a cylinder fitted with a movable piston as shown in Figure 6–1. If the temperature is kept constant and a weight is placed on the piston, the downward force exerted by gravity will begin to squeeze (compress) the gas molecules together. The result is to increase the density of the gas and hence the number of gas molecules (per unit area) bombarding the cylinder walls as well as the bottom of the piston. According to the gas law, the increase in density causes an increase in pressure. The piston will continue to squeeze the air molecules until the downward force is balanced by the ever-increasing gas pressure. When this balance is reached, the gas pressure within the cylinder will equal the weight of the piston and no further compression will occur. If more weight is added to the piston, the air will compress further until the pressure of the gas once again equals the new weight of the piston. Similarly, the pressure

---

[1] A mathematical treatment of the ideal gas law is provided in Appendix E.

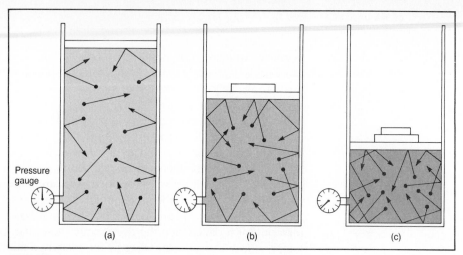

FIGURE 6–1
Schematic drawing showing the relationship between air pressure and density. As more weight is added (b and c) to the cylinder, the increased number of molecules per unit volume (density) causes an increase in the pressure exerted on the walls of the cylinder.

at any given altitude in the atmosphere is equal to the weight of the air directly above that point. At sea level a standard column of air weighs slightly more than 1 kilogram per square centimeter and as such exerts an equivalent amount of pressure. As we ascend through the atmosphere, we find that the air becomes less dense because of the lesser amount (weight) of air above, and as would be expected, the result is a corresponding decrease in pressure.

Recall from Chapter 1 that the rate at which pressure decreases with altitude is not a constant; that is, the rate of decrease is much greater near the earth's surface where pressure is high than aloft where air pressure is low. The "normal" decrease in pressure experienced with increased altitude is provided by the U.S. Standard Atmosphere in Table 6–1. We see from Table 6–1 that pressure is reduced by approximately one-half for each 5-kilometer increase in altitude. So at 5 kilometers the pressure is one-half its sea-level value, at 10 kilometers it is one-fourth, at 15 kilometers it is one-eighth, and so forth. Thus at the altitude at which commercial jets fly (10 kilometers), the air exerts a pressure equal to only one-fourth that at sea level.

In summary, the gas law explains how air pressure is affected by density and temperature. Still, it must be applied to the atmosphere with caution, for pressure depends on both variables. Recall that an increase in temperature is often accompanied by an increase in volume and hence a decrease in density and vice versa; therefore it is not always possible to conclude how a change in temperature will influence pressure.

MEASURING AIR PRESSURE

In measuring atmospheric pressure, meteorologists use the unit of force used in physics called the **newton.** One newton is the force that would accelerate 1

Newton

TABLE 6–1 U.S. Standard Atmosphere

Height (km)	Temperature (°C)	Pressure (mb)
0	15	1013.2
.5	12	954.6
1.0	9	898.8
1.5	5	845.6
2.0	2	795.0
2.5	−1	746.9
3.0	−4	701.2
3.5	−8	657.8
4.0	−11	616.6
5.0	−17	540.4
6.0	−24	472.2
7.0	−30	411.0
8.0	−37	356.5
9.0	−43	308.0
10.0	−50	265.0
12.0	−56	194.0
14.0	−56	141.7
16.0	−56	103.5
18.0	−56	75.65
20.0	−56	55.29
25.0	−51	25.49
30.0	−46	11.97
35.0	−36	5.75
40.0	−22	2.87
50.0	−2	0.798

Millibar

Mercurial Barometer

kilogram of mass 1 meter per second squared. At sea level the standard atmosphere exerts a force of 101,325 newtons per square meter. To simplify this large number, the U.S. National Weather Service has adopted the **millibar** (mb), which equals 100 newtons per square meter. Thus standard sea-level pressure is given as 1013.25 millibars.[2] The millibar has been the unit of measure on all U.S. weather maps since January 1940.

Although millibars are used almost exclusively in meteorology, you might be better acquainted with the expression "inches of mercury" that is used by the media to describe atmospheric pressure. In the United States the National Weather Service converts millibar values to inches of mercury for public and aviation use. This expression dates from 1643 when Torricelli, a student of the famous Italian scientist Galileo, invented the **mercurial barometer.** Torricelli correctly described the atmosphere as a vast ocean of air that exerts pressure on

---

[2] The standard unit of pressure in the SI system is the pascal, which is the name given to a newton per square meter (N/m^2). In this notation a standard atmosphere has a value of 101,325 pascals or 101.325 kilopascals. When the National Weather Service officially converts to the metric system, it will probably adopt this unit.

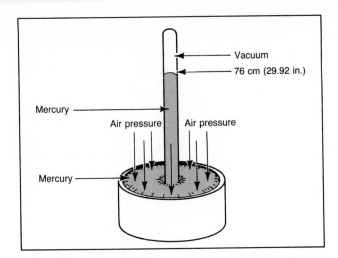

FIGURE 6–2
Simple mercurial barometer. The weight of the column of mercury is balanced by the pressure exerted on the dish of mercury by the air above. If the pressure decreases, the column of mercury falls; if the pressure increases, the column rises.

us and all things about us. To measure this force, he filled a glass tube that was closed at one end with mercury. The tube was then inverted into a dish of mercury (Figure 6–2). Torricelli found that the mercury flowed out of the tube until the weight of the column was balanced by the pressure exerted on the surface of the mercury by the air above. In other words, the weight of the mercury in the column equaled the weight of a similar diameter column of air that extended from the ground to the top of the atmosphere. Torricelli noted that when air pressure increased, the mercury in the tube rose; conversely, when air pressure decreased, so did the height of the column of mercury. The length of the column of mercury, therefore, became the measure of the air pressure. With some refinements, the mercurial barometer invented by Torricelli is still the standard pressure-measuring instrument used today. Standard atmospheric pressure at sea level equals 29.92 inches of mercury.

Aneroid Barometer

    The need for a smaller and more portable instrument for measuring air pressure led to the development of the **aneroid** (without liquid) **barometer.** Using a principle different from that used in the mercurial barometer, this instrument consists of partially evacuated metal chambers that have a spring inside, keeping them from collapsing. The metal chambers, being very sensitive to variations in air pressure, change shape, compressing as the pressure increases and expanding as the pressure decreases. Aneroids are often used in making **barographs,** instruments that continuously record pressure changes (Figure 6–3). Another important adaptation of the aneroid has been its use as an *altimeter* in aircraft. Recall that air pressure decreases with altitude and that the pressure distribution with height is well established. By marking an aneroid in meters instead of millibars, we have an altimeter. For example, we see in Table 6–1 that a pressure of 265 millibars "normally" occurs at a height of 10,000 meters and as such would indicate that altitude. Because of temperature variations, the standard pressures usually differ slightly from actual conditions; therefore altimeter corrections are made when accurate altitudes are required. Above 10 kilometers, where commercial jets fly and pressure changes are more gradual, these corrections cannot be made accurately enough. Consequently, such aircraft have their altimeters set by the standard

Barograph

FIGURE 6–3
Aneroid barograph. (Courtesy of WeatherMeasure Corporation)

atmosphere and fly paths of constant pressure instead of constant altitude. Stated another way, when an aircraft flies a constant altimeter setting, a pressure variation will result in a change in height. When pressure increases along a flight path, the plane will climb and it will descend with decreasing pressure. There is little risk of midair collisions because all aircraft will be adjusting their altitude a comparable amount. Large commercial aircraft also use radio altimeters to measure heights above the terrain. The time required for a radio signal to reach the surface and return accurately determines height. Nevertheless, this system has its drawbacks. A knowledge of the elevation of the land surface is required, which can pose a major problem in very rugged terrain.

As we shall see later, meteorologists are most concerned with pressure differences that occur horizontally across a given region. In order to compare readings obtained at various weather stations, compensation must be made for the elevation of the recording station. It is done by reducing all pressure measurements to sea-level equivalents. The process requires that meteorologists determine the pressure that would be exerted by an imaginary column of air equal in height to the elevation of the recording station and adding it to the pressure reading. Because temperature greatly affects the density, and hence the weight of this imaginary column, it must be considered in the calculations. Thus the corrected reading would give the pressure, at that time, as if it were taken at sea level under the same conditions.[3] A comparison of pressure records taken over the globe reveals that horizontal variations in pressure are rather small. Extreme pressure readings

---

[3] Appendix C explains how barometer corrections may be computed.

are rarely greater than 30 millibars (1 inch of mercury) over the standard pressure or less than 60 millibars (2 inches) below the standard pressure. Occasionally severe storms, such as hurricanes, are associated with even lower pressures.

FACTORS AFFECTING WIND

Wind

We discussed the upward movement of air and its importance in cloud formation. As important as vertical motion is, far more air is involved in horizontal movement, the phenomenon we call **wind.** Although we know that air will move vertically if it is warmer and thus more buoyant than surrounding air, what causes air to move horizontally? Simply stated, wind is the result of horizontal differences in air pressure. Air flows from areas of higher pressure to areas of lower pressure. You may have experienced this condition when opening a vacuum-packed can of coffee. The noise you hear is caused by air rushing from the higher pressure outside the can to the lower pressure inside. Wind is nature's attempt to balance similar inequalities in air pressure. Because unequal heating of the earth's surface generates these pressure differences, solar radiation is the ultimate driving force of wind.

If the earth did not rotate and if there were no friction, air would flow directly from areas of higher pressure to areas of lower pressure. But because both factors exist, wind is controlled by a combination of forces:

1. The pressure gradient force
2. The Coriolis effect
3. Centrifugal force
4. Friction
5. Gravity

Discussions of these factors follow.

PRESSURE GRADIENT FORCE

To get anything to accelerate (change its velocity) requires a net unbalanced force in one direction.[4] The force that drives the winds results from horizontal pressure differences. If air is subjected to greater pressure on one side than on another, this imbalance will produce a net force from the region of higher pressure toward the area of lower pressure. Thus pressure differences cause the wind to blow, and the greater these differences, the greater the wind speed.

Variations in air pressure over the earth's surface are determined from barometric readings taken at hundreds of weather stations. These pressure data are

---

[4] A definition of force as stated by Newton's second law is given in Appendix D.

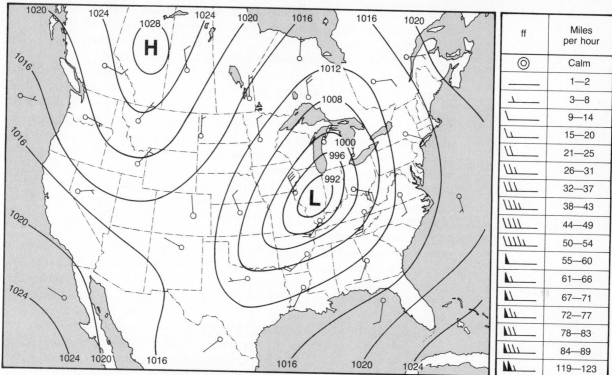

ff	Miles per hour
◎	Calm
—	1—2
⌐	3—8
⌐	9—14
⌐	15—20
⌐	21—25
⌐	26—31
⌐	32—37
⌐	38—43
⌐	44—49
⌐	50—54
◣	55—60
◣	61—66
◣	67—71
◣	72—77
◣	78—83
◣	84—89
◣	119—123

FIGURE 6–4

Isobars, which are lines connecting places of equal barometric pressure, are used to show the distribution of pressure on daily weather maps. The lines usually curve and often join where cells of high and low pressure exist. The "arrows" indicate the expected airflow surrounding cells of high and low pressure and are plotted as "flying" with the wind. Wind speed is indicated by flags and feathers as shown along the right-hand side of this diagram.

Isobar

Pressure Gradient

shown on surface weather maps by means of **isobars,** which are lines connecting places of equal air pressure (Figure 6–4). The spacing of the isobars indicates the amount of pressure change occurring over a given distance and is expressed as the **pressure gradient.** You might find it easier to visualize the concept of a pressure gradient if you think of it as being analogous to the slope of a hill. A steep pressure gradient, like a steep hill, causes greater acceleration of a parcel than does a weak pressure gradient. Thus the relationship between wind speed and the pressure gradient is rather simple. Closely spaced isobars indicate a steep pressure gradient and strong winds; widely spaced isobars indicate a weak pressure gradient and light winds. Figure 6–4 illustrates the relationship between the spacing of isobars and wind speed.

The factors that contribute to the pressure differences observed on the daily weather map are generally complex, but the underlying cause of these differences is simply unequal heating of the earth's land-sea surface. To illustrate how temperature differences can generate a pressure gradient and thereby create winds, we examine a common example, the *sea breeze.* Figure 6–5(a) shows a vertical cross section of a coastal location just before sunrise. At this time we are assuming that temperatures and pressures do not vary horizontally at any level. This assumption is shown in Figure 6–5(a) by the horizontal orientation of the lines (isobars) that indicate the pressure at various heights. Because there is no horizontal variation in pressure (zero-pressure gradient), there would be no wind. After sunrise, however, the unequal rates at which land and water heat will initiate airflow. Recall from Chapter 3 that surface temperatures over the ocean change only slightly on a daily basis. On the other hand, land surfaces and the air above can be substantially warmed during a single daylight period. As air over the land warms, it expands, causing the isobars to bend upward as shown in Figure 6–5(b). Although this warming does not by itself produce a surface pressure change, the pressure aloft does become higher over the land than at comparable altitudes over the ocean. The resultant pressure gradient aloft causes the air there to move from over the land toward the ocean. The mass transfer of air seaward creates a surface high over the ocean where the air is collecting and a surface low over the land. The surface circulation that develops from this redistribution of mass aloft is from the sea toward the land (sea breeze) as shown in Figure 6–5(c). Thus a simple thermal circulation has been developed with flow seaward aloft and landward at the surface. Note that vertical movement is required to make the circulation complete.

An important relationship exists between pressure and temperature, as we saw in the preceding discussion. Temperature variations create pressure differences—and hence wind—and the greater these temperature differences, the stronger the pressure gradient and resultant wind. Daily temperature differences and the pressure gradients so generated are confined to a rather shallow layer of the atmosphere. On a global scale, however, latitudinal variations in solar incidence similarly generate the much larger general atmospheric circulation. This is the topic of the next chapter.

In summary, the horizontal pressure gradient is the driving force of wind. It has both magnitude and direction. Its magnitude is determined from the spacing of isobars, and the direction of force is always from areas of higher pressure to areas of lower pressure and at right angles to the isobars.[5] Once the air starts to move, the Coriolis force and friction come into play but only to modify the movement, not to produce it.

CORIOLIS EFFECT

Figure 6–4 shows the typical air movements associated with surface high- and low-pressure systems. As expected, the air moves out of the regions of higher pressure and into the regions of lower pressure. The wind does not cross the isobars at right angles as the pressure gradient force directs, however. This deviation is the result of the earth's rotation and has been named the **Coriolis effect**

Coriolis Effect

---

[5] The mathematical expression of the pressure gradient force is provided in Appendix D.

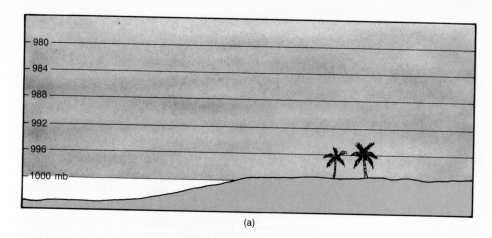

(a)

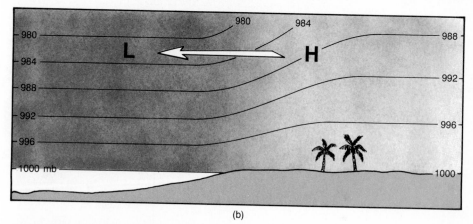

(b)

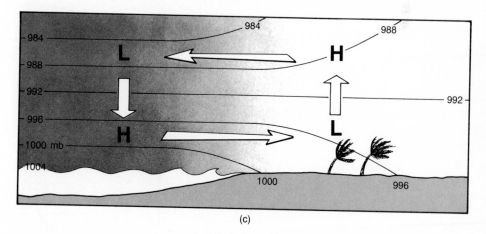

(c)

FIGURE 6–5
Cross-sectional view illustrating the formation of a sea breeze.

after the French scientist who first expressed its magnitude quantitatively. All free-moving objects, including wind, are deflected to the right of their path of motion in the northern hemisphere and to the left in the southern hemisphere. The reason for this deflection can be illustrated by imagining the path of a rocket launched from the North Pole toward a target located on the equator (Figure 6–6). If the rocket took an hour to reach its target, the earth would have rotated 15 degrees to the east during its flight. To someone standing on the earth, it would look as if the rocket veered off its path and hit the earth 15 degrees west of its target. The true path of the rocket was straight and would appear so to someone out in space looking down at the earth. It was the earth turning under the rocket that gave it its apparent deflection. Note that the rocket was deflected to the right of its path of motion because of the counterclockwise rotation of the northern hemisphere. Clockwise rotation produces a similar deflection in the southern hemisphere, but to the left of the path of motion.

FIGURE 6–6
The Coriolis effect. During the rocket's flight from the North Pole to point x, the earth's rotation carried point x to a position denoted by point x_1. The earth's rotation gives the rocket's trajectory a curved path when plotted on the earth's surface.

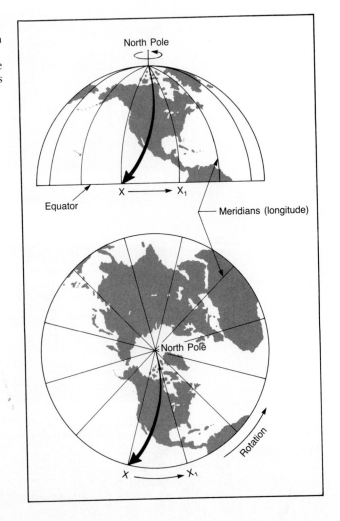

Although it is usually easy for people to visualize the Coriolis effect when the motion is from north to south, as in our rocket example, it is not so easy to see how a west-to-east flow would be deflected. Figure 6–7 illustrates this situation by examining a wind that is blowing eastward along the fortieth parallel. Several hours later, what started as a west wind will be coming from a more northwesterly direction. Note that it was really the observer's orientation that changed, not the air movement.

It is of interest to point out that any "free-moving" object will experience a similar deflection. This fact was dramatically discovered by our navy in World War II. During target practice long-range guns on battleships continually missed their targets by as much as several hundred yards until ballistic corrections were made for the changing position of a seemingly stationary target. Over a short distance the Coriolis effect can be ignored. The Roman legions, for example, did not have to account for it when they used their catapults.

For convenience, we attribute this apparent shift in wind direction to the Coriolis force. It is hardly a "real" force but rather the effect of the earth's rotation on a moving body. This deflecting force (1) is always directed at right angles to the direction of airflow; (2) affects only wind direction, not wind speed; (3) is

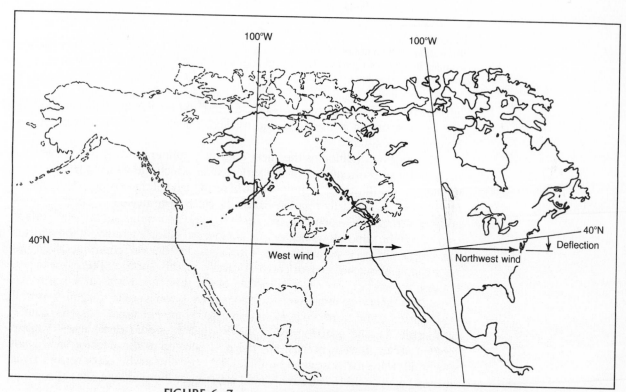

FIGURE 6–7
Coriolis deflection of a west wind. After a few hours the earth's rotation changes the position of the surface over which the wind blows, thus causing the apparent deflection.

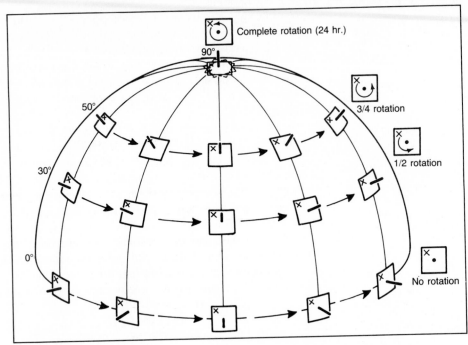

FIGURE 6–8
Illustration of the amount of rotation of a horizontal surface about a vertical axis at various latitudes in a 24-hour period. The magnitude of the Coriolis force is proportional to the amount of rotation experienced.

affected by wind speed (the stronger the wind, the greater the deflecting force); and (4) is strongest at the poles and weakens equatorward, where it eventually becomes nonexistent.[6]

The fourth condition is the most difficult to understand. It results because the amount of reorientation of the surface over which the wind blows depends on latitude. To illustrate, at the poles, where the land surface is perpendicular to the earth's axis, the daily rotation causes the horizon to make one complete turn about a vertical axis each day (Figure 6–8). Restated more simply, the surface over which the wind blows makes one complete rotation each day. At the equator the earth's surface is parallel to the earth's axis of rotation; consequently, it has no rotation about an axis oriented vertically to this surface. Therefore at the equator the surface over which the air blows does not rotate in a horizontal sense. This difference might be easier to visualize if you were to imagine a vertical post situated at the North Pole and one situated at the equator. During the course of a single day the post located at the North Pole would make one complete rotation about its vertical axis, but the post situated at the equator would not spin at all. Here the post would tumble end over end as the earth rotates. Posts situated between these extremes would experience intermediate rates of rotation

---

[6] A mathematical treatment of the Coriolis effect is provided in Appendix D.

about their vertical axis each day. Consequently, because the horizontal orientation (rotation about a vertical axis) of the earth's surface would change more rapidly at high latitudes than at low latitudes, the Coriolis force would be greater there as well. To summarize, the magnitude of the Coriolis force is dependent on latitude; it is strongest at the poles, and it weakens equatorward, where it eventually becomes nonexistent.

Although the magnitude of the Coriolis force varies considerably with both wind speed and latitude, calculations of its strength indicate that at reasonable wind speeds it can approximate the strength of the pressure gradient force. This fact is the basis of the following section.

THE GEOSTROPHIC WIND

Earlier we stated that the pressure gradient force is the primary driving force of the wind. As an unbalanced force, it causes air to accelerate from regions of higher pressure to regions of lower pressure. Thus wind speeds should continually increase (accelerate) for as long as this imbalance exists. But we know from our own personal experiences that winds do not get faster indefinitely. So some other force, or forces, must oppose the pressure gradient force in order to moderate airflow. From our everyday experiences we know that friction acts to slow a moving object. Although friction significantly influences airflow near the earth's surface, its effect is negligible above a height of a few kilometers. For this reason, we will divide our discussion of surface winds where friction is significant from that of flow aloft where its effect is small. This section will deal only with airflow above a few kilometers, where the effects of friction are significantly small enough to be neglected.

Aloft the Coriolis force is responsible for balancing the pressure gradient force and thereby directing airflow.[7] Figure 6–9 helps to show how a balance is reached between these opposing forces. For illustration only, we assume a nonmoving parcel of air at the starting point in Figure 6–9. It should be remembered that air is rarely stationary in the "real" atmosphere. Because our parcel of air has no motion, the Coriolis force exerts no influence; only the pressure gradient force can act. Under the influence of the latter force, which is always directed perpendicularly to the isobars, the parcel begins to accelerate directly toward the area of low pressure. As soon as the flow begins, the Coriolis force causes a deflection to the right for winds in the northern hemisphere. As the parcel continues to accelerate, the Coriolis force intensifies (recall that the magnitude of the Coriolis force is proportional to wind speed). Thus the increased speed results in further deflection. Eventually the wind will be turned so that it is flowing parallel to the isobars, with the pressure gradient force directed toward the area of low pressure and opposed by the Coriolis force, which is directed toward the region of high pressure (Figure 6–9). When these two opposing forces are equal, the wind will

---

[7] The magnitude of the pressure gradient force and the magnitude of the Coriolis effect are compared in Appendix D.

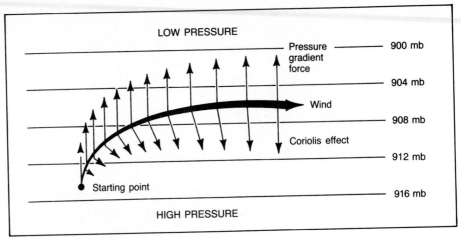

FIGURE 6–9
The geostrophic wind. Upper-level winds are deflected by the Coriolis force until the
Coriolis force just balances the pressure gradient force. Above 600 meters, where friction
is negligible, these winds will flow nearly parallel to the isobars and are called geostrophic
winds.

continue to flow parallel to the isobars at a constant speed. This situation occurs
because when the airflow is parallel to the isobars, the pressure gradient is oriented
perpendicularly to the flow, where it no longer causes further acceleration; and
because the Coriolis force is proportional to wind speed, it too will remain constant.
Consequently, the balance that exists between these forces will be maintained
and airflow will continue along the isobars. Under these idealized conditions
the wind can be considered to be coasting (not accelerating or decelerating)
along a pathway defined by the isobars.

So far we have assumed that these two opposing forces just balance each
other. In order for this assumption to be true, the pressure gradient force must
produce a wind with the exact velocity needed to generate a Coriolis force equal
to itself. But there is no reason why it should, at least not on the first attempt.
Theory predicts that the speed of the wind generated in the manner described
would result in a Coriolis force that exceeds the pressure gradient force, thereby
causing the wind to be deflected beyond the middle ground and to cross the
isobars toward the region of high pressure (see Figure 6–9). Or stated another
way, the winds would move against the pressure gradient force. Yet doing so
would result in an immediate weakening of the wind and a reduced Coriolis
force. So the wind would be deflected back toward the region of low pressure
by the pressure gradient force and again accelerated. Instead of flowing directly
along the isobars, the wind is constantly adjusting its speed and direction in an
attempt to reach a balance between these opposing forces.

The preceding notwithstanding, the pressure gradient and Coriolis forces
can be considered to reach a state of equilibrium, called the geostrophic balance.
The winds generated by this balance are called **geostrophic winds** and flow
nearly parallel to the isobars with velocities proportional to the pressure gradient

Geostrophic Wind

force.[8] A steep pressure gradient creates strong winds that generate an equally strong Coriolis force.

Referring back to Figure 6–9, we see that for geostrophic flow the pressure gradient force is directed at right angles to the wind toward the area of low pressure and is opposed by the Coriolis force that is directed toward the area of high pressure. This rather simple relationship between wind direction and pressure distribution was first formulated by a Dutch meteorologist, Buys Ballott, in 1857.

Buys Ballott's Law

Essentially **Buys Ballott's law** states: If you stand with your back to the wind in the northern hemisphere, low pressure will be to your left and high pressure will be to your right. In the southern hemisphere the situation is reversed, for the Coriolis deflection is to the left. Although Buys Ballott's law holds for airflow aloft, it must be used with some caution when considering surface winds. Numerous geographic effects can generate local disturbances that interfere with the larger circulation.

In the real atmosphere the winds are never purely geostrophic. Nonetheless, the importance of the idealized geostrophic flow lies in the fact that it gives a useful approximation of the actual winds aloft. Thus by measuring the pressure field (orientation and spacing of isobars) aloft, meteorologists can determine both wind direction and speed. The idealized geostrophic flow predicts winds parallel to the isobars with speeds that depend on isobaric spacing (Figure 6–10). Of even greater use to meteorologists is that the same method can be used in reverse to determine the distribution of pressure from airflow measurements. This interrelationship between pressure and winds greatly enhances the reliability of upper-air weather charts by providing checks and balances. In addition, it minimizes the number of direct observations required to describe adequately the conditions aloft, where accurate data are most expensive and difficult to obtain.

In summary, winds above a few kilometers can be considered geostrophic; that is, they flow parallel to the isobars at speeds that can be calculated from the pressure gradient. The major discrepancy from true geostrophic winds involves the flow along highly curved paths, a topic considered next.

CURVED FLOW AND THE GRADIENT WIND

Even a casual glance at a weather map reveals that the isobars (height contours) are not generally straight; instead they make broad sweeping curves. Occasionally the isobars connect to form roughly circular cells of either high or low pressure. The curved nature of the pressure field is one factor that tends to modify true geostrophic flow primarily by affecting wind speed but not direction, which remains roughly parallel to the isobars. Because the geostrophic wind is a reasonable approximation of the airflow aloft, we will first examine the flow around pressure centers as if it were geostrophic; that is, we will ignore the effect of curvature.

---

[8] The speed of geostrophic flow is also inversely proportional to the magnitude of the Coriolis force. When influenced by equal pressure gradients, geostrophic flow is faster near the equator, where the Coriolis force is weak, than at higher latitudes where it is stronger.

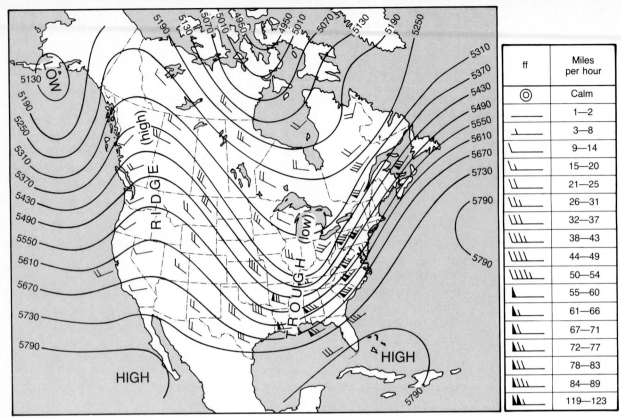

FIGURE 6–10
This simplified weather chart shows the direction and speed of the upper-air winds. Note that the airflow is almost parallel to the contours. These isolines are height contours measured in meters for the 500-millibar level.

The wind speed legend table:

ff	Miles per hour
◎	Calm
	1—2
	3—8
	9—14
	15—20
	21—25
	26—31
	32—37
	38—43
	44—49
	50—54
	55—60
	61—66
	67—71
	72—77
	78—83
	84—89
	119—123

Cyclone

Cyclonic Flow

Anticyclone

Anticyclonic Flow

Trough

Ridge

Figure 6–11 shows that, for geostrophic flow around a center of low pressure, the pressure gradient force is directed inward and is balanced by the outward-directed Coriolis force. In the northern hemisphere, where the Coriolis force deflects the flow to the right, the resultant wind blows counterclockwise about a low. Conversely, around a high-pressure cell the outward-directed pressure gradient force that is balanced by the inward-directed Coriolis force results in clockwise flow. Because the Coriolis force deflects the winds to the left in the southern hemisphere, the flow is reversed there—clockwise around low-pressure centers and counterclockwise around high-pressure centers.

It is common practice to call all centers of low pressure **cyclones** and the flow around them cyclonic. **Cyclonic flow** has the same direction of rotation as the earth: counterclockwise in the northern hemisphere and clockwise in the southern hemisphere. Centers of high pressure are called **anticyclones** and have **anticyclonic flow** (opposite that of the earth's rotation). Whenever the isobars curve to form elongated regions of low and high pressure, these areas are called **troughs** and **ridges,** respectively. The flow about a trough is cyclonic; the flow around a ridge is anticyclonic.

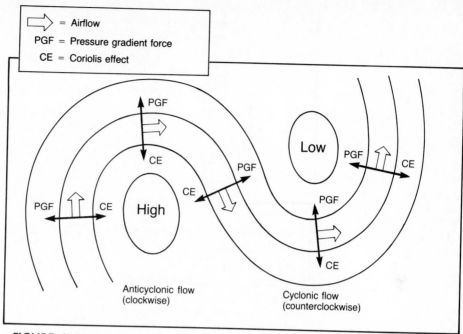

FIGURE 6–11

Idealized illustration showing expected airflow aloft around high- and low-pressure centers.

Next let us consider the effects of curvature on cyclonic and anticyclonic flow. According to Newton's first law of motion, a moving object will continue to travel in a straight line unless acted upon by an opposing force. You have experienced the effect of this law of motion when the auto you were riding in made a sharp turn and your body continued to move straight ahead. The tendency of a particle to travel in a straight line creates an outward force we call **centrifugal force.** Unlike geostrophic flow, which is straight and represents a balance between the pressure gradient force and the Coriolis force, curved flow is also influenced by centrifugal force. An examination of upper-level flow charts (see Figure 6–10) reveals that curved airflow patterns generally parallel the isobars (height contours). Since this is the case, we can conclude that a balance between these forces is usually achieved. Referring to Figure 6–12, it can be seen that airflow around a high-pressure center represents a balance between the inward-directed Coriolis force on the one hand and the outward-directed pressure gradient force plus centrifugal force on the other. Conversely, it is apparent that the flow around a low-pressure center represents a balance between the inward-directed pressure gradient force and the outward-directed Coriolis force plus centrifugal force. The curved airflow pattern that results from this balance is called the **gradient wind.** As Figure 6–12 illustrates, the gradient wind is a curved flow that travels in a cyclonic manner around a low-pressure center (counterclockwise in the northern hemisphere) and in an anticyclonic manner around a high-pressure center (clockwise in the northern hemisphere).

Although the magnitude of centrifugal force is small compared with the other

Centrifugal Force

Gradient Wind

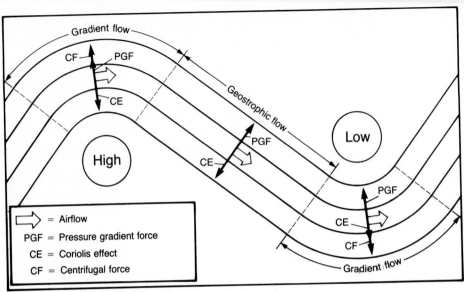

FIGURE 6–12
Schematic showing the gradient wind which represents a balance between the pressure gradient force, Coriolis force, and centrifugal force.

forces involved, its influence does affect upper-level airflow in a very important way. To visualize this effect, examine Figure 6–10 and notice that over the Plains states the winds aloft are geostrophic, following a path parallel to the isobars. However, in other regions the flow curves sharply around an upper-level ridge (high pressure) and an upper-level trough (low pressure). As soon as a straight airstream starts to exhibit a curved flow, centrifugal force comes into play and disrupts the balance that existed between the pressure gradient force and the Coriolis force. It is important to note that this disruption in the pressure field is temporary, and that these three forces will quickly establish a new balance which we call the gradient wind. Nevertheless, during the period of realignment, the effect of centrifugal force is to direct the flow away from the center of curvature. In the case of a trough the effect of the outward-directed centrifugal force is to cause a net outflow of air from the center of low pressure. This loss of air serves to intensify the area of low pressure. In a similar manner centrifugal force causes a net outflow around a ridge of high pressure. In this situation the loss of mass serves to weaken the central high pressure. Figure 6–10 illustrates the resulting upper-level airflow caused by this net outflow of air. Notice that the isobars are more closely spaced on the downwind side of the trough than they are on the downwind side of the ridge. By examining the wind arrows we see that the wind speeds are much greater around the trough than around the ridge. As we shall see later, the strong airflow normally found downwind from an upper-level trough plays a major role in the development of surface weather systems.

Despite the importance of centrifugal force in establishing curved flow aloft, near the surface, friction comes into play and greatly overshadows the much weaker centrifugal force. Consequently, except for rapidly rotating storms such as tornadoes and hurricanes, the effect of centrifugal force is negligible and thus will not be considered in the discussion of surface circulation.

FRICTION LAYER WINDS

Friction as a factor affecting wind is important only within the first few kilometers of the earth's surface. We know that it acts to slow the movement of air; consequently, it also alters wind direction. Recall that the Coriolis force is proportional to wind speed. By lowering the wind speed, friction reduces the Coriolis force. Because the pressure gradient force is not affected by wind speed, it wins the tug of war against the Coriolis force (Figure 6–13). The result is the movement of air at an angle across the isobars toward the area of lower pressure. The roughness of the terrain determines the angle at which the air will flow across the isobars as well as the speed at which it will move. Over the relatively smooth ocean surface, where friction is low, air moves at an angle of 10 to 20 degrees to the isobars and at speeds roughly two-thirds that of geostrophic flow. Over rugged terrain, where friction is high, the angle could be as great as 45 degrees from the isobars, with wind speeds reduced by as much as 50 percent.

By changing the direction of airflow, friction plays a major role in redistributing air within the atmosphere. This fact is especially noticeable when considering the motion around surface cyclones and anticyclones, two of the most common features on surface weather maps. In the preceding section we learned that above the friction layer in the northern hemisphere winds blow counterclockwise around a cyclone and clockwise around an anticyclone with winds nearly parallel to the isobars. When we add the effects of friction, we notice that the airflow will cross the isobars at varying angles, depending on the roughness of the terrain, but

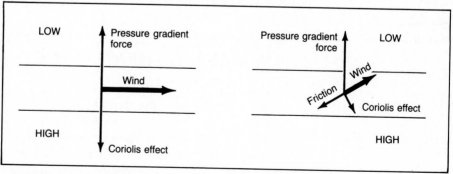

(a) Upper-level wind (no friction)   (b) Surface wind (effect of friction)

FIGURE 6–13

Comparison between upper-level winds and surface winds showing the effects of friction on airflow. Friction slows surface wind speed, which weakens the Coriolis force, causing the winds to cross the isobars.

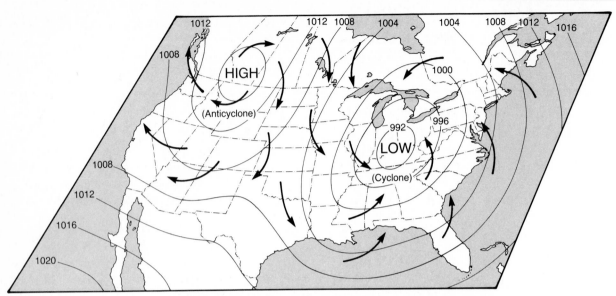

FIGURE 6–14
Cyclonic and anticyclonic winds in the northern hemisphere. Arrows show the winds blowing in and counterclockwise about a low and out and clockwise about a high.

always from higher to lower pressure. In a cyclone, in which pressure decreases inward, friction causes a net flow toward its center (Figure 6–14), but in an anticyclone just the opposite is the case; the pressure decreases outward and the resultant winds blow into and counterclockwise about a surface cyclone (Figure 6–15) and outward and clockwise about a surface anticyclone. Of course, in the southern hemisphere the Coriolis force deflects the winds to the left and reverses the direction of flow. In whatever hemisphere, however, friction causes a net inflow **(convergence)** around a cyclone and a net outflow **(divergence)** around an anticyclone. This very important relationship between cyclonic flow and convergence and anticyclonic flow and divergence will be considered again.

Convergence

Divergence

HOW WINDS GENERATE VERTICAL MOTION

So far we have discussed the factors that affect wind without regard to how the movement of air might alter the state of the atmosphere. In other words, how might wind in one zone of the atmosphere affect airflow elsewhere? Of particular importance is the question, How do winds relate to vertical flow? It should be remembered that although vertical transport is small compared to horizontal motion, it is very important as a weather maker. Rising air is associated with cloudy conditions and precipitation, whereas subsidence produces adiabatic heating and clearing conditions. In this section we will discern how the movement of air (dynamic effect) can itself create pressure change and hence generate winds. On doing so, we will examine the interrelationship between horizontal and vertical flow and its effects on the weather.

FIGURE 6–15
This view of a hurricane from space shows the storm's cyclonic flow. By examining the cloud pattern, the inward and counterclockwise circulation is revealed. (Courtesy of NASA)

Let us first consider the situation around a surface low-pressure system in which the air is spiraling inward. Here the net inward transport of air causes a shrinking of the area occupied by the air mass, a process called horizontal convergence. Whenever air converges horizontally, it must pile up; that is, it must increase in height to allow for the decreased area it now occupies. This process generates a "taller" and therefore heavier air column; yet a surface low can exist only as long as the column of air above remains light. We seem to have encountered a paradox—low-pressure centers cause a net accumulation of air, which increases their pressure. Consequently, a surface cyclone should quickly eradicate itself in a manner not unlike what happens to the vacuum in a coffee can on being opened.

Based on the preceding discussion, it should be apparent that in order for a surface low to exist for any reasonable time, compensation must occur at some layer aloft. Surface convergence could be maintained, for example, if divergence (spreading out) aloft occurred at a rate equal to the inflow below. Figure 6–16(a) diagrammatically shows the relationship between surface convergence (inflow) and divergence (outflow) aloft that is needed to maintain a low-pressure center. Note that surface convergence about a cyclone causes a net upward movement. The rate of this vertical movement is slow, generally less than 1 kilometer per day. Nevertheless, because rising air often results in cloud formation and precipita-

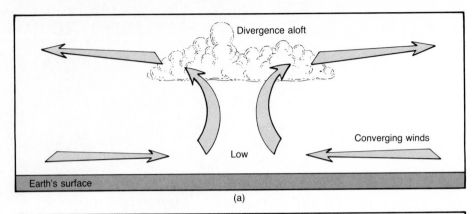

(a)

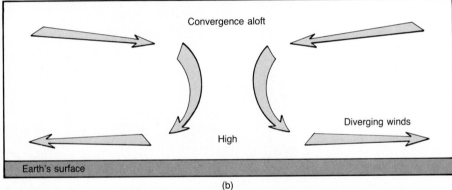

(b)

FIGURE 6–16
Schematic of the airflow associated with cyclones and anticyclones. (a) Converging winds and rising air are associated with a low, or cyclone. (b) Highs, or anticyclones, are associated with descending air and diverging winds.

tion, the passage of a low is generally related to unstable conditions and "stormy" weather. Divergence aloft may occasionally even exceed surface convergence, thereby resulting in intensified surface inflow and increased vertical motion. Thus divergence aloft can intensify these storm centers as well as maintain them. On the other hand, inadequate divergence aloft permits surface flow to "fill" and weaken the accompanying cyclone. As often as not, it is divergence aloft that first creates the surface low. Spreading out (divergence) aloft initiates upflow in the layer directly below, and it eventually works its way to the surface where inflow is encouraged.

Like their cyclonic counterparts, anticyclones must also be maintained from above. The mass outflow near the surface is accompanied by convergence aloft and general subsidence of the air column (Figure 6–16b). Because descending air is compressed and warmed, cloud formation and precipitation are unlikely in an anticyclone and "fair" weather can usually be expected with the approach of a high.

For reasons that should now be obvious, it has been common practice to write the words "stormy" at the low end and "fair" at the high end on barometers

Pressure Tendency

intended for household use. By noting whether the pressure is rising, falling, or steady, we have a good indication of what the forthcoming weather will be. Such a determination, called the **pressure** or **barometric tendency,** is a useful aid in short-range weather prediction. The generalizations relating cyclones and anticyclones to the weather conditions just considered are stated rather poetically in the rhyme that follows. Note that glass refers to the barometer.

> When the glass falls low,
> Prepare for a blow;
> When it rises high,
> Let all your kites fly.

Because of the close tie between winds and weather systems, we will consider some factors that contribute to horizontal convergence and divergence. Earlier we mentioned the effect of friction on curved flow, causing the winds to cross the isobars toward the area of low pressure. To recapitulate, cyclonic flow encourages convergence and uplifting and anticyclonic flow encourages divergence and subsidence. Friction can also cause mass convergence when the flow is straight. When air moves from the relatively smooth ocean surface onto land, for instance, the increased friction causes an abrupt drop in wind speed. This reduction of wind speed downstream results in a pile up of air upstream. Thus converging winds and ascending air accompany flow off the ocean. This effect contributes to the cloudy conditions over the land often associated with a sea breeze in a humid region like Florida. As expected, general divergence and subsidence accompany the flow of air seaward because of increasing wind speeds.

Mountains, which also hinder the flow of air, cause divergence and convergence in yet another way. As air passes over a mountain range, it must shrink vertically, which produces horizontal spreading (divergence) aloft. On reaching the lee side of the mountain, the air experiences vertical stretching, which causes horizontal convergence. This effect greatly influences the weather in the United States east of the Rocky Mountains, as we shall examine later. When air flows equatorward, where the Coriolis force is weakened, divergence and subsidence prevail; during poleward migration convergence and slow uplift are favored.

In conclusion, you should now be better able to understand why local television weather broadcasters emphasize the positions and projected paths of cyclones and anticyclones. The "villain" on these weather programs is always the cyclone that produces "bad" weather in any season. Lows move in roughly a west-to-east direction across the United States and require as little as a few days to more than a week for the journey. Because their paths can be somewhat erratic, accurate prediction of their migration is difficult and yet essential for short-range forecasting. Meteorologists must also determine if the flow aloft will intensify an embryo storm or act to suppress its development. As a result of the close tie between surface conditions and those aloft, a great deal of emphasis has been placed on the importance of understanding the total atmospheric circulation, especially in the midlatitudes. Once we have examined the workings of the general circulation, we will again consider the development and structure of the cyclone in light of these findings.

WIND MEASUREMENT

Wind Vane

Two basic wind measurements—direction and speed—are particularly significant to the weather observer. Winds are always labeled by the direction *from* which they blow. A north wind blows from the north toward the south; an east wind blows from the east toward the west. One instrument that is commonly used to determine wind direction is the **wind vane** (Figure 6–17). This instrument, which is a common sight on many buildings, always points into the wind. Often the wind direction is shown on a dial that is connected to the wind vane. The dial indicates the direction of the wind either by points of the compass—that is, N, NE, E, SE, and so on—or by a 0- to 360-degree scale. On the latter scale 0 degrees or 360 degrees is north, 90 degrees is east, 180 degrees is south, and 270 degrees is west. When the wind consistently blows more often from one direction than from any other, it is termed a **prevailing wind.**

Prevailing Wind

Cup Anemometer

Wind speed is often measured with a **cup anemometer** (Figure 6–17). The wind speed is read from a dial much like the speedometer of an automobile. Sometimes an **aerovane** is used instead of a wind vane and cup anemometer. As can be seen in Figure 6–18, this instrument resembles a wind vane with a

Aerovane

FIGURE 6–17
Wind vane and cup anemometer. (Courtesy of Weather-Measure Corporation)

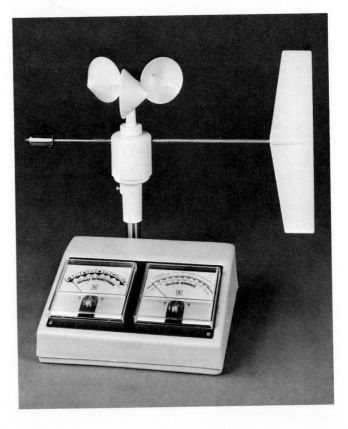

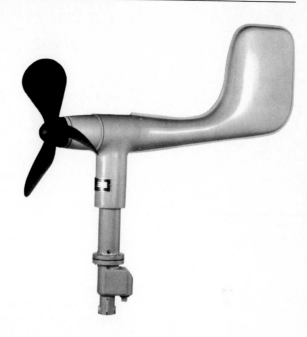

FIGURE 6–18
An aerovane. (Courtesy of the Belfort Instrument Company)

TABLE 6–2 Beaufort Scale

Beaufort Number	m/s	mph	International Description	Specifications
0	< 1	< 1	Calm	Calm; smoke rises vertically
1	1	1–3	Light air	Direction of wind shown by smoke drift but not by wind vanes
2	2	4–7	Light breeze	Wind felt on face; leaves rustle, vanes moved by wind
3	4	8–12	Gentle breeze	Leaves and small twigs in constant motion; wind extends light flag
4	7	13–18	Moderate	Raises dust, loose paper; small branches moved
5	10	19–24	Fresh	Small trees in leaf begin to sway; crested wavelets form on inland waters
6	12	25–31	Strong	Large branches in motion; whistling heard in telegraph wires; umbrellas used with difficulty
7	15	32–38	Near gale	Whole trees in motion; inconvenience felt walking against wind
8	18	39–46	Gale	Breaks twigs off trees; impedes progress
9	20	47–54	Strong gale	Slight structural damage occurs
10	26	55–63	Storm	Trees uprooted; considerable damage occurs
11	30	64–72	Violent storm	Widespread damage
12	≥ 33	> 73	Hurricane	

FIGURE 6–19
Giant wind turbine with windmill in background located near Medicine Bow, Wyoming. (Photo by Brooks Martner, University of Wyoming)

Beaufort Scale

propeller at one end. The fin keeps the propeller facing into the wind, allowing the blades to rotate at a rate that is proportional to the wind speed. This instrument is commonly attached to a recorder in order to keep a continuous record of wind speed and direction. When instruments are not available, wind speed can be estimated by using the **Beaufort scale** (Table 6–2).

If you are aware of the locations of low-pressure centers and high-pressure centers, you can predict the changes in wind direction that will be experienced as the pressure centers move past. Changes in wind direction often bring changes in temperature and moisture conditions; therefore the ability to predict the winds can be useful. In the Midwest, for example, a north wind may bring cool, dry air from Canada, whereas a south wind may bring warm, humid air from the Gulf of Mexico. Sir Francis Bacon summed it up nicely when he wrote, "Every wind has its weather."

In addition to understanding how important winds are in weather forecasting, researchers are also interested in winds as a source of energy. Windmills have been used for many years to pump water. More recent attempts to use wind power have concentrated on the generation of electrical energy (Figure 6–19). Presently, however, wind generators are usually not able to produce electricity at costs that are competitive with other sources. The primary disadvantage of wind generators is the fact that winds at most locations (except mountainous areas) are not strong enough or steady enough to produce sufficient quantities of electricity to make the venture profitable. Moreover, even when winds are strong, the problem of efficient storage has not been solved.

REVIEW

1. When density remains constant and the temperature is raised, how will the pressure of a gas change?

2. When gases in the atmosphere are heated, air pressure normally decreases. Based on your answer to the first question, explain this apparent paradox.

3. What is standard sea-level pressure in millibars? in inches of mercury?

4. Describe the principles of the mercurial barometer and the aneroid barometer.

5. What force is responsible for generating wind?

6. Write a generalization relating the spacing of isobars to the speed of wind.

7. Although vertical pressure differences may be great, such variations do not generate strong vertical currents. Explain.

8. Temperature variations create pressure differences, which, in turn, produce winds. On a small scale, the sea breeze illustrates this principle nicely. Describe how a sea breeze forms.

9. Briefly describe how the Coriolis effect modifies the movement of air.

10. Which two factors influence the magnitude of the Coriolis effect?

11. Explain the formation of a geostrophic wind.

12. Answer the following question by applying Buys Ballott's law: If you are facing north and the wind is at your back, in what direction is the low pressure?

13. Unlike winds aloft, which blow nearly parallel to the isobars, surface winds generally cross the isobars. Explain why this difference exists.

14. Sketch a diagram (isobars and wind arrows) showing the winds associated with surface cyclones and anticyclones in both the northern and southern hemispheres.

15. In order for surface low pressure to exist for an extended period of time, which condition must exist aloft?

16. Describe the general weather conditions to be expected when the pressure tendency is rising and when the pressure tendency is falling.

17. Converging winds and ascending air are often associated with the flow of air from the oceans onto land. Conversely, divergence and subsidence often accompany the flow of air from land to sea. What causes this convergence over land and divergence over the ocean?

18. A southwest wind is blowing from the _____ (direction) toward the _____ (direction).

19. The wind direction is 315 degrees. From what compass direction is the wind blowing?

VOCABULARY REVIEW

air pressure	wind
ideal gas law	isobar
newton	pressure gradient
millibar	Coriolis effect
mercurial barometer	geostrophic wind
aneroid barometer	Buys Ballott's law
barograph	cyclone

cyclonic flow

anticyclone

anticyclonic flow

trough

ridge

centrifugal force

gradient wind

convergence

divergence

pressure or barometric tendency

wind vane

prevailing wind

cup anemometer

aerovane

Beaufort scale

7

Global Circulation

$\mathbf{T}$hose of us who live in the United States are familiar with the term "westerlies" to describe our winds. But all of us have experienced winds from the south and north and even directly from the east. You may even recall being in a storm when shifts in wind direction and speed came in such rapid succession that it was impossible to determine the wind's direction. With such variations, how can we describe our winds as westerly? The answer lies in our attempt to simplify descriptions of the atmospheric circulation by sorting out events according to size. On the scale of a weather map, for instance, where observing stations are spaced about 150 kilometers apart, small whirlwinds that carry dust skyward are far too small to show up. Instead these maps reveal larger-scale wind patterns, such as those associated with traveling cyclones and anticyclones. Not only do we separate winds according to the size of the system, but equal consideration is given to the time frame in which they occur. A small eddy may last only a few moments, whereas a larger vortex like a hurricane may last for days.

In Chapter 7 an important goal is to gain an understanding of the earth's complex circulation system. Although the main focus will be on global pressure and wind patterns, we will consider local wind systems as well. The chapter concludes with a discussion of global precipitation patterns. As we shall see, the distribution of rainfall is closely linked to the pattern of atmospheric pressure.

SCALES OF ATMOSPHERIC MOTIONS

Macroscale Wind Systems

The time and space scales we will use for atmospheric motions are provided in Table 7–1. The largest-scale wind pattern is called **macroscale wind systems** and is exemplified by long waves in the westerlies. These *planetary-scale* flow patterns extend around the entire globe and often remain essentially unchanged for weeks at a time. A somewhat smaller type of macroscale circulation is commonly called the *synoptic scale* and consists mainly of individual traveling cyclones and anticyclones. These vortices are an important part of the circulation in the middle latitudes. The airflow in these less than global-sized macroscale circulations consists primarily of horizontal flow, with only modest amounts of vertical motion. In

Mesoscale Winds

contrast, **mesoscale** and **microscale** winds influence smaller areas and exhibit extensive vertical flow, which can be rapid, as in a developing thunderstorm.

Microscale Winds

Although the primary emphasis of this chapter is on developing an understanding

TABLE 7–1 Time and Space Scales for Atmospheric Motions

Name of Scale	Time Scale	Length Scale	Examples
Macroscale			
Planetary Scale	Weeks to years	1000 to 40,000 km	Waves in the Westerlies
Synoptic Scale	Days to weeks	100 to 5000 km	Cyclones, anticyclones, and hurricanes
Mesoscale	Minutes to days	1 to 100 km	Land-sea breeze, thunderstorms, and tornadoes
Microscale	Seconds to minutes	<1 km	Turbulence

of the largest type of macroscale circulation, global wind patterns, we also deal with smaller local wind systems, which are classed as mesoscale phenomena.

Although it is common practice to divide atmospheric motions according to scale, it is important to remember that this flow is complex—much like a turbulent river with smaller eddies within larger eddies within still larger eddies. Also, like the flow of a turbulent river, each scale of motion is related to the other. Let us examine the flow associated with the hurricanes that form over the North Atlantic, for example. When we view one of these tropical cyclones on satellite photographs, the storm appears as a large whirling cloud migrating slowly across the ocean (Figure 7–1). From this perspective (synoptic scale) the general counterclockwise rotation can be easily seen and the storm's path, which persists for a week or so, can be easily followed. However, when we average the winds of a hurricane, we find that they have a net motion from east to west, thereby indicating that these rather large eddies are embedded in a still larger flow (the general circulation) that is moving westward across the tropical portion of the North Atlantic. When we examine a hurricane more closely by flying an airplane through it, some of the small-scale aspects of the storm become noticeable. As the plane approaches the outer edge of the system, it becomes evident that the large rotating cloud that we saw in the satellite images is made of numerous towering cumulonimbus clouds. Each of these mesoscale phenomena (thunderstorms) lasts for only a few hours and must be continually replaced by new ones if the hurricane is to persist. When we fly into these storms, we quickly realize that the individual clouds are also made up of even smaller-scale turbulences. The small thermals of rising air that compose these clouds make for a rather rough trip. Thus a typical hurricane exhibits several different scales of motion, including many mesoscale thunderstorms, which, in turn, consist of numerous microscale turbulences, whereas the counterclockwise circulation of the hurricane is itself part of the larger general circulation in the tropics.

FIGURE 7–1
Satellite view of Hurricane Ione (*left*) and Hurricane Kirsten (*right*).

IDEALIZED GLOBAL CIRCULATION

Our knowledge of global winds comes from both the observed distribution of the pressure-wind regimes and theoretical studies of fluid motion. We begin our discussion by first considering the classical model of global circulation that was developed largely from the mean worldwide pressure distribution. We then add to the idealized circulation some of the more recently discovered aspects of the atmosphere's complex motions.

One of the first contributions to the classical model of general circulation came from George Hadley in 1735. Hadley was well aware of the fact that solar energy drives the winds. He proposed that the large temperature contrast between the poles and the equator would create a thermal circulation similar to that of the sea breeze discussed in Chapter 6. As long as the earth's surface is heated unequally, air will move in an attempt to balance the inequalities. Hadley suggested that on a nonrotating earth the air movement would take the form of one large **convection cell** in each hemisphere as shown in Figure 7–2. The more intensely heated equatorial air would rise and move poleward. Eventually this upper-level flow would reach the poles, where it would sink and spread out at the surface and return to the equator. As the cold polar air approached the equator, it would be reheated and rise again. Thus the proposed Hadley circulation for a nonrotating earth has upper-level air flowing poleward and surface air flowing equatorward. When we add the effect of the earth's rotation, the Coriolis effect causes the surface flow to become more or less easterly (toward the west) and the flow aloft to be westerly.

There are several reasons why the simple convectional circulation developed by Hadley cannot exist. Perhaps the easiest to understand hinges on the fact that the atmosphere is connected to the earth by the force of gravity and therefore must rotate with the earth. In the Hadley model the surface winds blow toward the west and would, because of friction, oppose the earth's rotation, which is toward the east. Because the atmosphere is attached to the earth, however, it cannot, over the long haul, slow the earth's rotation. Thus easterly flow at one latitude must be balanced by westerly flow at another. Furthermore, Hadley's simple convective system did not fit the observed pressure distribution that was subsequently established for the earth. Consequently, the popular Hadley model was replaced by a version that better fit the observed situation.

In the 1920s a three-cell circulation in each hemisphere was proposed to carry on the task of maintaining the earth's heat balance. Although this model has been modified to fit recent upper-air observations, it is nonetheless still useful.

Convection Cell

FIGURE 7–2
Global circulation on a nonrotating earth. A simple convection system is produced by unequal heating of the atmosphere on a nonrotating earth.

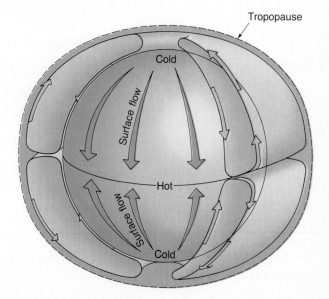

Figure 7–3 illustrates the three-cell model and the surface winds that result. Note that the surface flow resulting from the three-cell model has a much greater east-west component than a north-south component.

In the zone between the equator and roughly 30 degrees latitude the circulation closely resembles the convective model used by Hadley for the whole earth; so the name **Hadley cell** is generally applied to it. Here the surface flow is equator-ward, whereas the flow aloft is poleward. Near the equator the warm rising air that releases latent heat during the formation of cumulus towers is believed to provide the energy to drive this cell. These clouds also provide the rainfall that maintains the lush vegetation of the equatorial rain forests. As the upper flow in this cell moves poleward, it begins to subside in a zone between 20 and 35 degrees latitude. Two factors are thought to contribute to the general subsidence found here. First, as this flow moves away from the stormy equatorial region where the release of latent heat of condensation keeps the air warm and buoyant, radiation cooling would result in increased density of the air aloft. This factor could account for the subsidence. Satellites that monitor the radiation emitted in the upper troposphere have supplied data to support this mechanism for subsidence. Second, because the Coriolis effect becomes stronger with increasing dis-

Hadley Cell

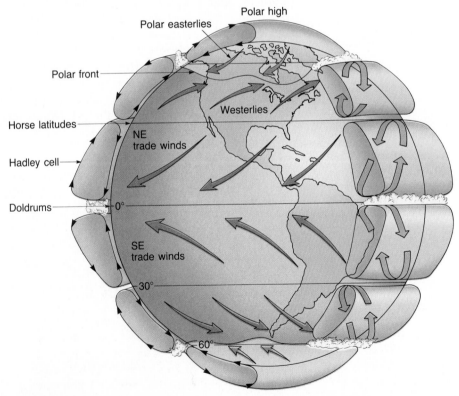

FIGURE 7–3
Idealized global circulation proposed for the three-cell circulation model.

tance from the equator, the once poleward-directed winds are deflected into a nearly west-to-east flow by the time they reach 25 degrees latitude. A restricted poleward flow of air ensues. Stated another way, the Coriolis force produces a blocking effect that causes a general pileup of air (convergence) aloft. In either case, general subsidence occurs in the zone between 20 and 35 degrees latitude. This subsiding air is relatively dry, for it has released its moisture near the equator. In addition, the effect of adiabatic heating during descent further reduces the relative humidity of the air. Consequently, this subsidence zone is the site of the world's tropical deserts. Winds are generally weak and variable near the center of this zone of descending air. Because of this, the region was popularly called

Horse Latitudes

the **horse latitudes.** The name is believed to have been coined by Spanish sailors, who, while crossing the Atlantic, were sometimes becalmed in these waters and reportedly were forced to throw horses overboard when they could no longer water or feed them. From the center of the horse latitudes the surface flow splits into a poleward branch and an equatorward branch. The equatorward flow is

Trade Winds

deflected by the Coriolis force and forms the reliable **trade winds.** In the northern hemisphere the trades are from the northeast, where they provided the sail power to explore the New World; in the southern hemisphere the trades are from the southeast. The trade winds from both hemispheres meet near the equator in a region that has a weak pressure gradient. This region is called the **doldrums.**

Doldrums

Here light winds and humid conditions provide the monotonous weather that is the basis for the expression "down in the doldrums."

In the three-cell model the circulation between 30 and 60 degrees latitude is just opposite that of the Hadley cell. The net surface flow is poleward and, because of the Coriolis effect, the winds have a strong westerly component. These

Prevailing Westerlies

prevailing westerlies were known to Benjamin Franklin, perhaps the first American weather forecaster, who noted that storms migrated eastward across the colonies. Franklin also observed that the westerlies were much more sporadic and unreliable than the trades for sail power. We now know that it is the migration of cyclones and anticyclones across the midlatitudes that disrupts the general westerly flow at the surface. Again, referring to the three-cell model in Figure 7–3, we can see that the flow aloft in the midlatitudes is equatorward, from which the Coriolis force would produce an east wind. Since World War II, however, numerous observations have indicated that a general westerly flow exists aloft in the midlatitudes as well as at the surface. Consequently, the center cell in this model does not completely fit observations. Because of this complication and the importance of the midlatitude circulation in maintaining the earth's heat balance, we will consider the westerlies in more detail in a later section.

Relatively little is known about the circulation in high latitudes. It is generally believed that subsidence near the poles produces a surface flow that moves equator-

Polar Easterlies

ward and is deflected into the **polar easterlies** of both hemispheres. As these cold polar winds move equatorward, they eventually encounter the warmer westerly flow of the midlatitudes. The region where these contrasting flows clash has been

Polar Front

named the **polar front.** This important region will be considered in future discussions.

OBSERVED DISTRIBUTION OF SURFACE PRESSURE AND WINDS

As we would expect, the planetary circulation considered earlier is accompanied by a compatible distribution of surface pressure. Let us now consider the relationship between the mean surface winds and this pressure distribution. To simplify the discussion, we will first examine the idealized pressure distribution that would be expected if the earth's surface were uniform. Under this restriction, four latitudinally oriented belts of either high or low pressure emerge (Figure 7–4a). Near the equator, the converging air from both hemispheres is associated with the pressure zone known as the **equatorial low,** a region marked by abundant precipitation. Because it is the region where the trade winds meet, it is also referred to as the **intertropical convergence zone (ITC).** In the belts about 20 to 35 degrees on either side of the equator, where the westerlies and trade winds originate and go their separate ways, are located the **subtropical high**-pressure zones. These are regions of subsidence and divergent flow. Yet another surface pressure region is situated at about 50 to 60 degrees latitude in a position corresponding to the polar front. Here the polar easterlies and westerlies meet to form a convergent zone known as the **subpolar low.** Finally, near the earth's poleward extremes are the **polar highs** from which the polar easterlies originate.

Up to this point we have considered the surface pressure systems to be continuous belts around the earth. However, the only true zonal distribution of pressure

Equatorial Low

Intertropical Convergence Zone (ITC)

Subtropical High

Subpolar Low

Polar High

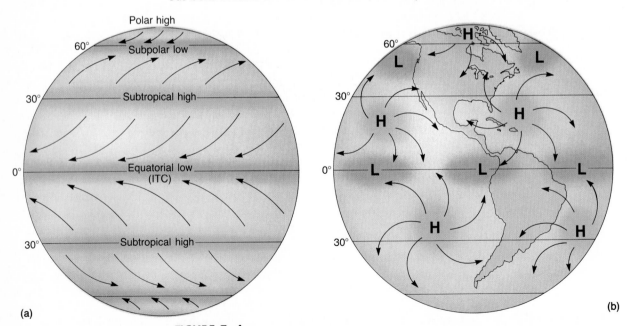

FIGURE 7–4
Idealized zonal pressure patterns for (a) a uniform earth and (b) disruptions to this zonal pattern caused by large landmasses.

exists in the region of the subpolar low in the southern hemisphere, where the ocean is continuous. To a lesser extent, the equatorial low is also zonal. At other latitudes, particularly in the northern hemisphere where there is a higher proportion of land, this zonal pattern is replaced by semipermanent cells of high and low pressure (Figure 7–4b). This pattern is further complicated by large seasonal temperature changes, which serve to either strengthen or weaken these pressure cells. As a consequence, the earth's pressure patterns vary considerably during the course of a year. A better approximation of global pressure patterns and resulting winds is found in Figure 7–5.

Notice on these pressure maps that, for the most part, the observed pressure regimes are cellular (or elongated) instead of zonal. The most prominent features on both maps are the subtropical highs. These systems are centered between 20 and 35 degrees latitude over all the larger oceans. It should also be noted that the subtropical highs are situated toward the east side of these oceans, particularly in the North and South Pacific and the North Atlantic. This fact greatly affects the west coast climates of the adjacent continents. When we compare Figure 7–5(a) and (b), we see that some pressure cells are more or less permanent features, like the subtropical highs, and can be seen on both the January and July charts. Others, however, are seasonal, such as the low situated over the southwestern United States in July, which can be seen on only one map. The seasonal change in pressure is much more noticeable in the northern than in the southern hemisphere.

Relatively little pressure variation occurs from midsummer to midwinter in the southern hemisphere, a fact we attribute to the dominance of water in that hemisphere. The most noticeable changes here are the seasonal 5- to 10-degree shifts of the subtropical highs that move with the position of the sun's vertical rays.

Numerous departures from the idealized zonal pattern shown in Figure 7–4a are evident in the northern hemisphere. The main cause of these variations is the seasonal temperature fluctuations experienced over the landmasses, especially those in the middle and higher latitudes. In January we find a very strong high-pressure center situated over frigid Siberia and a somewhat weaker high-pressure system existing over the chilled North American continent. These cold anticyclones are consistent with the greater density of this frigid air during this season. As these highs develop over the land, a weakening is observed in the subtropical anticyclones over the northern oceans. Although these subtropical highs do retain their identity during the winter, the circulation over the oceans is dominated by two intense cell-like cyclones, the **Aleutian** and **Icelandic lows.** These two semipermanent lows are statistical averages that represent the great number of cyclonic storms that migrate eastward across the globe and converge in these areas.

Aleutian Low

Icelandic Low

High surface temperatures over the continents in the summer generate lows that replace the winter highs. These are thermal lows in which warm outflow aloft induces inflow at the surface. The strongest of these low-pressure centers develops in northern India. It is this system, along with the northward migration of the ITC, which creates the well-known summer monsoon that is experienced in India. A weak thermal low is also generated in the southwestern United States.

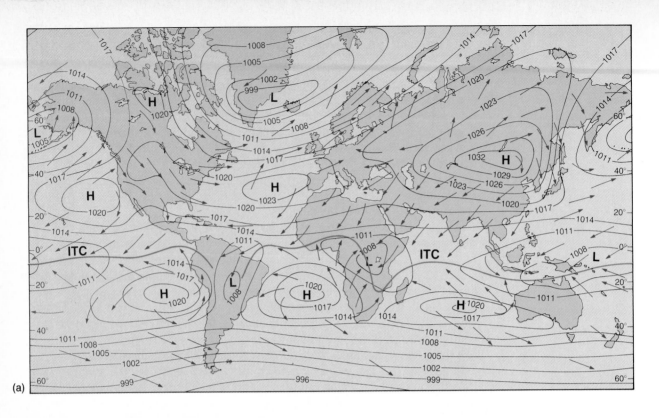

(a)

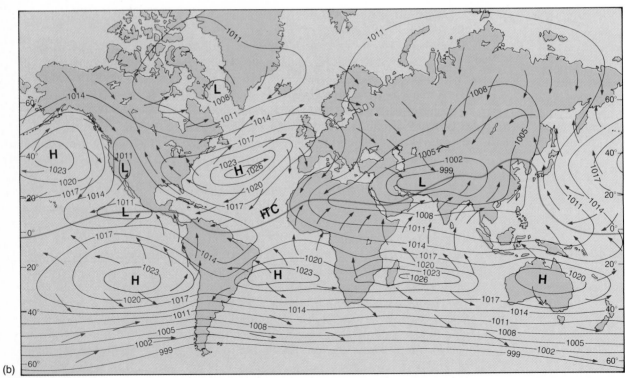

(b)

FIGURE 7–5
Average surface pressure and associated global circulation for (a) January and (b) July.

196

In addition, notice in Figure 7–5 that during summer months the subtropical highs in the northern hemisphere are more intense than during winter months. These large pressure systems dominate the summer circulation over the oceans and aid the general influx of air onto the continents during that season.

Having examined the distribution of pressure in summer and winter, let us now turn to the airflow associated with these pressure patterns. Although the wind patterns shown in Figure 7–5 are highly idealized, they do depict the mean flow and are useful in studying the effects of seasonal changes in pressure on the flow pattern. These charts clearly show the dominance of the subtropical anticyclones in generating the surface flow. Also notice the seasonal change in the circulation surrounding these subtropical highs.

Still another important seasonal change evident in Figure 7–5 is the migration of the intertropical convergence zone, which is indicated on these diagrams by a blue line. Along with the migration of the ITC, we can detect the well-known **Siberian High** monsoon circulation of Asia. In January the strong **Siberian high** produces flow off the Asian continent and across the equator toward the ITC. This outflow of cool, dry continental air produces the dry winter monsoon for much of southern and southeastern Asia. With the onset of summer, the low that develops over northern India draws warm moist air from the adjacent oceans into Asia, producing the wet summer monsoon. Many other regions around the globe experience a somewhat similar seasonal wind shift caused by the seasonal fluctuation in the global pressure regimes. In fact, if you examine the flow over the United States, you will notice a much greater surge of warm moist air from the Gulf of Mexico during the summer when the high-pressure centers over the North Atlantic are most intense. Although the seasonal wind shift over the United States is sometimes referred to as a monsoon, it is not nearly as abrupt or dramatic a change as that experienced in Asia, where even the most unobservant person cannot help but notice the change.

THE WESTERLIES

Prior to World War II upper-air observations were scarce. Since then data collected from aircraft and a network of radiosonde stations have aided in filling this gap. One of the most important pieces of information discovered was that at most latitudes, except near the equator where the Coriolis force is weak, the airflow in the middle and upper troposphere is westerly.

Why Westerlies?

Let us consider the reason for the predominance of westerly flow aloft. Recall from the gas law that cold air is more dense than warm air. Therefore air pressure decreases more rapidly in a column of cool air than in a column of warm air. Figure 7–6 shows the resulting pressure distribution with height. This figure is a vertical cross section through the northern hemisphere. The warm equatorial region is on the right side of the drawing, and the cold polar region is on the

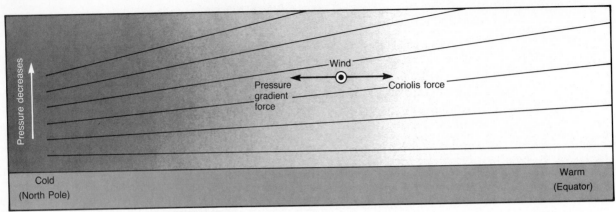

FIGURE 7–6
Cross section showing the poleward-directed pressure gradient that is responsible for generating the westerlies of the middle latitudes.

left. The lines represent the normal pressure distribution with height, and to simplify matters we have assumed the same surface pressure at all latitudes. Note that over the equator, where temperatures are highest, the pressure decreases more slowly than over the colder polar regions. Consequently, above the surface higher pressure exists over the tropics and lower pressure exists above the poles. Thus the resulting pressure gradient is directed from the equator toward the poles. Adding the effect of the Coriolis deflection, which directly opposes the pressure gradient force, we generate a wind with a strong westerly component—that is, a wind that is 90 degrees to the right of the northward-directed pressure gradient force. In our diagram the flow is into the page. Because the equator-to-pole temperature gradient shown in Figure 7–6 is typical over the globe, a westerly flow aloft should be expected; on most occasions it is observed. It can also be seen that the pressure gradient increases with altitude; as a result, so should wind speeds. This increase in wind speed continues only to the tropopause. Here the temperature gradients are reversed (warmer over the poles); so, the westerlies reach a maximum at the top of the troposphere and decrease upward into the stratosphere.

Jet Streams

It should be evident from our previous discussion that greater temperature contrasts at the surface will produce greater pressure gradients aloft and hence faster upper-air winds as well. In the winter it would not be unusual to have a warm balmy day in southern Florida and near-freezing temperatures in Georgia, only a few hundred kilometers to the north. Because of large wintertime temperature contrasts in the midlatitudes, like the one just mentioned, we would expect faster westerly flow at that time of year. Observations substantiate such expectations. In the flow directly above these regions of large temperature contrasts, very fast currents of air do indeed exist. These fast streams of air have been considered analogous to **Jet Stream**   jets of water and thus are called **jet streams.**

Although predicted earlier, the existence of jet streams was first dramatically illustrated during World War II. U.S. bombers heading toward Japanese-occupied islands occasionally made little headway. On abandoning their missions, the planes experienced tail winds that sometimes exceeded 300 kilometers per hour during the return flight. Even today commercial aircraft use these strong tail winds to increase their ground speed when making eastward flights across the United States. Naturally, attempts are made on westward flights to fly away from these fast currents of air.

The cause of jet streams was alluded to earlier. Recall that large temperature contrasts over a short distance produce strong pressure gradients and strong winds. These large temperature contrasts occur along regions called fronts. In the midlatitudes a jet is found in association with the polar front. Remember that the surface position of the polar front is situated between the cool polar easterlies and warm westerlies. Because other jet streams have been discovered, this midlatitude jet is called the **polar jet stream.** Instead of flowing in a nearly straight west-to-east manner, the polar jet stream often meanders wildly, on occasion flowing almost due north and south. Sometimes it splits into two jets that may or may not rejoin. Like the polar front, this jet is not continuous around the globe.

Polar Jet Stream

The polar jet stream plays a very important role in the weather of the midlatitudes. In addition to supplying energy to the circulation of surface storms, it also directs their paths of movement. Consequently, determining changes in the location and flow pattern of the polar jet is an important part of modern weather forecasting.

Occasionally jet stream speeds of over 500 kilometers per hour have been recorded. On the average, however, the polar jet travels at rates of 125 kilometers per hour in the winter and roughly half that speed in the summer (Figure 7–7). This seasonal difference is due to the much stronger temperature gradient that exists in the middle latitudes during the wintertime. During the cold winter months the polar jet stream may extend as far south as 30°N latitude (Figure 7–7). With the coming of spring the jet begins a gradual northward migration. By midsummer its average position is usually about 50°N latitude. As the polar jet migrates northward, there is a corresponding change in the regions where outbreaks of severe thunderstorms and tornadoes occur. For example, in February most tornadoes occur in the states bordering the Gulf of Mexico. By midsummer, the center of this activity shifts to the northern plains and Great Lakes states. We shall return to this important relationship between the polar jet and midlatitude weather in later chapters.

Several other jet streams are known to exist, but none has been studied in as much detail as the polar jet stream. An almost permanent jet exists over the subtropics centered at 25 degrees latitude. This westerly jet is located about 13 kilometers above the surface in the region where the Hadley cell and the middle latitude cell meet in Figure 7–3. The convergence of cool and warm air at this location generates a front aloft; consequently, the appropriate conditions for the development of a jet are met. There also exists a jet that flows from east to west at about 15°N latitude over the region occupied by the Himalayas. This jet exists only in the summer when the ITC has shifted northward.

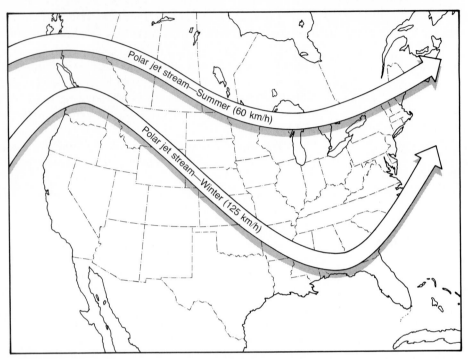

FIGURE 7–7
The position and speed of the polar jet stream changes with the seasons. Shown are flow patterns that are common for summer and winter.

Waves in the Westerlies

It is important to remember that the polar jet stream is an integral part of the westerlies. In fact, the jet stream can more accurately be described as the fast core of the overall westerly flow. Studies of upper-level wind charts reveal that the westerlies follow wavy paths that have rather long wavelengths. Much of our knowledge of these large-scale motions is attributed to C. G. Rossby, who first explained the nature of these waves. The longest wave patterns (called *Rossby waves*) have wavelengths of 4000 to 6000 kilometers, so that three to six waves will fit around the earth. Although the air flows eastward along this wavy path, these long waves tend to remain in the same position or to move slowly. In addition to Rossby waves, shorter waves also occur in the middle and upper troposphere. These shorter waves are often associated with cyclones at the surface and, like these storms, the waves travel eastward around the globe at rates of up to 15 degrees per day.

To get a better understanding of this wavy flow, let us examine the pattern on an upper-level weather chart. Figure 7–8 is a simplified 500-millibar height contour chart provided daily by the National Weather Service. Like most upper-air charts, this one shows the height at which a given pressure (500 millibars) is found instead of giving the pressure at a given height, as do surface maps. But

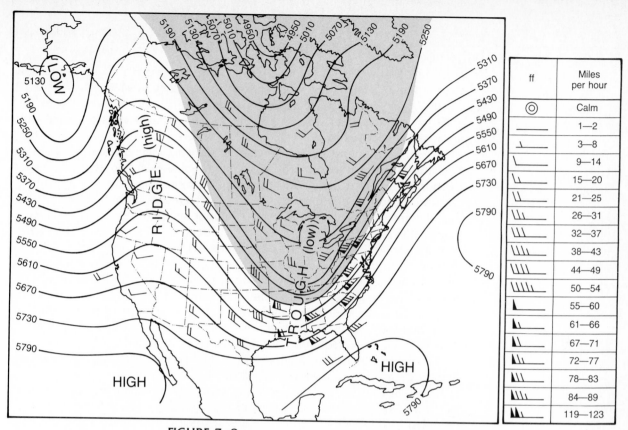

FIGURE 7–8

Simplified 500-millibar height contour chart. Note the position of the trough (low) and ridge (high). The position of the jet stream core is found by reading the wind speeds.

do not let that bother you, for there is a simple relationship between height contours and pressure. Places where the same pressure occurs at higher altitudes are experiencing higher pressures than places where the height contours indicate lower altitudes. Thus higher contour readings indicate higher pressures and vice versa. Notice in Figure 7–8 that the height of the 500-millibar level decreases poleward as well. This fact is in agreement with the pressure gradient we concluded must exist in the upper troposphere in order to produce westerly winds there. Recall that the winds aloft are nearly geostrophic. So the flow is nearly parallel to the contours, having wind speeds proportional to the contour spacing. Notice the large, sweeping wavy pattern of the contour lines, which generally outline the flow pattern as well. Although this chart is well below the altitude of the jet stream core, the position of the jet at 500 millibars can be estimated from the indicated wind speeds.

Now let us return to the wind's function of transporting heat from the equator toward the poles. In Chapter 2 we showed that the equator receives more solar radiation than it radiates to space, whereas the poles experience the reverse situation. Thus the equator has excess heat while the poles experience a deficit. Although the flow near the equator is somewhat meridional (north to south), at most other latitudes the flow is zonal (west to east). The reason for the zonal flow, as we have seen, is the Coriolis force. The question we now consider is, How can wind with a west-to-east flow transfer heat from south to north?

In Figure 7–8 the shaded area represents cold air that is bounded by the polar front to the south. We can also see that the polar front is displaced with the wavy flow of the jet. It should be remembered, however, that the surface winds and temperature gradients will be somewhat different from those aloft. Notice in Figure 7–8 that where the jet bends equatorward an elongate low or **trough** is produced that allows colder air to move southward. Conversely, a poleward bend in the jet produces a **ridge** of high pressure that draws warmer air poleward. Also notice that along one limb of a meander warm air is directed poleward and along the other limb of the meander cold air is directed equatorward. Thus the wavy flow of the westerlies provides an important mechanism for heat transfer across the midlatitudes. In addition, vortices (cyclones and anticyclones) help in this redistribution of energy. Imagine the counterclockwise circulation around a cyclone in the northern hemisphere; the east side draws warm air northward, whereas the west side pulls cool air southward. Although the amount of air carried southward is roughly equal to the amount of air carried northward, more heat is transported northward than is transported southward.

Laboratory experiments using rotating fluids to duplicate the earth's circulation support the existence of waves and eddies to carry on the task of heat transfer in the middle latitudes. In these studies, called *dishpan experiments*, a large circular pan is heated around the outer edge to represent the equator while the center is cooled to duplicate the poles. Colored particles are added so that the flow can be easily observed and photographed. When the pan is heated, but not rotated, a simple convection cell forms to redistribute the heat. This cell is similar to the Hadley cell considered earlier. When the pan is rotated, however, the simple circulation breaks down and the flow develops a wavy pattern with eddies embedded between meanders, as seen in Figure 7–9(a). These experiments indicate that changing the rate of rotation and varying the temperature gradient largely determine the flow pattern produced. Such studies have added greatly to our understanding of global circulation. It is now known that the wavy flow aloft largely determines surface pressure patterns. During periods when the wavy flow aloft is relatively flat (small-amplitude waves), little cyclonic activity is generated at the surface. In contrast, when the flow acquires large-amplitude waves having short wavelengths, vigorous cyclonic storms are experienced. This important relationship between the flow aloft and cyclonic storms is considered in more detail in Chapter 9.

Trough

Ridge

LOCAL WINDS

Having examined the large macroscale circulation for the earth, let us now turn to some mesoscale winds. Remember that all winds are produced for the same reason: temperature differences that arise because of unequal heating of the earth's surface. In turn, these differences generate pressure differences. Local winds are

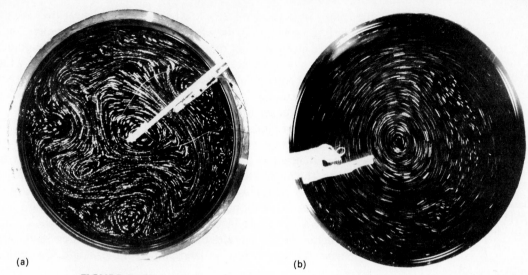

(a)   (b)

FIGURE 7–9
Photographs obtained from a "dishpan" experiment simulating (a) wavy and (b) flat flow
of the general circulation. (Courtesy of D. H. Fultz, University of Chicago Hydrodynamics
Laboratory)

simply small-scale winds produced by a locally generated pressure gradient. Although many winds are given local names, some are actually part of the general circulation just described. The "norther" of Texas, for instance, is a cold southward flow produced by the circulation around anticyclones that invade the United States from Canada in the winter. Because these winds are not locally generated, they cannot be considered true local winds. Others, like those described in the following discussion, are truly mesoscale and are caused either by topographic effects or variations in surface composition in the immediate area.

Land and Sea Breezes

Sea Breeze

Land Breeze

The daily temperature contrast between the land and the sea, plus the pressure pattern that generates a **sea breeze,** was discussed in the preceding chapter (see Figure 6–5). Recall that land is heated more intensely during daylight hours than is an adjacent body of water. As a result, the air above the land surface heats and expands, creating an area of low pressure. A sea breeze then develops, blowing cooler air off the water and onto the land. At night the reverse may take place; the land cools more rapidly than the sea and a **land breeze** develops.

The sea breeze has a significant moderating influence on the temperatures in coastal areas. Shortly after the breeze begins, land temperatures may drop by as much as 5 to 10°C. However, the cooling effect of these breezes generally reaches a maximum of only 100 kilometers inland in the tropics and often only half that distance in the middle latitudes. These cool sea breezes generally begin shortly before noon and reach their greatest intensity, about 10 to 20 kilometers per hour, in the midafternoon. Smaller-scale sea breezes can also develop along the shores of large lakes. People who live in a city near the Great Lakes, such as Chicago, recognize the lake effect, especially in the summer. These people are reminded daily by weather reports of the cool temperatures near the lake compared with warmer outlying areas. In many places sea breezes also affect the amount of cloud cover and rainfall (see Figure 4–13). The peninsula of Florida, for example, experiences a summer precipitation maximum that is caused partly by the convergence of sea breezes from both the Atlantic and Gulf coasts.

The scale of land and sea breezes depends on the location and the time of year. Tropical areas where intense solar heating is continuous throughout the year experience more frequent and stronger sea breezes than do midlatitude locations. The most intense sea breezes develop along tropical coastlines adjacent to cool ocean currents. In the middle latitudes sea breezes are most common during the warmest months of the year, but their counterpart, the land breeze, is often missing, for the land does not always cool below the ocean temperature. In the higher middle latitudes the frequent migration of pressure systems generally dominates the circulation; so land and sea breezes are less noticeable in these areas.

Mountain and Valley Breezes

Valley Breeze

A daily wind similar to land and sea breezes occurs in many mountainous regions. Here during the daylight hours the air along the slopes of the mountains is heated more intensely than the air at the same elevation over the valley floor. This warm air glides up along the slope and generates a **valley breeze.** The occurrence of these daytime upslope breezes can often be identified by the cumulus clouds that develop over adjacent mountain peaks. The late afternoon thundershowers so common on warm summer days in the mountains can also be attributed to this phenomenon. After sunset the pattern is reversed. Rapid radiation heat loss along the mountain slopes results in cool air drainage into the valley below and causes the so-called **mountain breeze.** The same type of cool air drainage can occur in regions that have little slope. The result is that the coldest pockets of air are usually found in the lowest spots. Consequently, low areas are the first to experience radiation fog and are also the most likely spots for frost damage to crops.

Mountain Breeze

Like many other winds, mountain and valley breezes have seasonal preferences. Although valley breezes are most common during the warm season when solar heating is most intense, mountain breezes tend to be more dominant in the cold season.

Chinook (Foehn) Winds

Chinook

Foehn

Warm, dry winds sometimes move down the east slopes of the Rockies, where they are called **chinooks,** and the Alps, where they are called **foehns.** Such winds are often created when a pressure system, such as a cyclone, is situated on the leeward side of the mountains where it pulls air over these imposing barriers. As the air descends the leeward slopes of the mountain, it is heated substantially by compression. Because condensation may have occurred as the air ascended the windward side, releasing latent heat, the air descending the leeward side will be warmer and drier than at a similar elevation on the windward side. Although the temperature of these winds is generally less than 10°C, which is not particularly high, they usually occur in the winter and spring when the affected area may be experiencing below-freezing temperatures. Thus, by comparison, these dry, warm winds often bring a drastic change. Within a few minutes after the arrival of a chinook, the temperature normally climbs 20°C. When the ground has a snow cover, these winds are known to melt it in short order. The word chinook is an Indian word that literally means "snoweater."

The chinook is viewed by some as beneficial to ranchers east of the Rockies, for their grasslands are kept clear of snow during much of the winter, but this benefit is offset by the loss of moisture that the snow would bequeath to the land had it remained until the spring melt.

Santa Ana

Another chinooklike wind that occurs in the United States is the **Santa Ana.** Found in southern California, these hot, desiccating winds greatly increase the threat of fire in this already dry area.

Katabatic Winds

Katabatic Wind

In the winter season areas adjacent to highlands may experience a local wind called a **katabatic** or **fall wind.** These winds originate when cold air, situated over a highland area, such as the ice sheets of Greenland or Antarctica, is set in motion. Under the influence of gravity the cold air cascades over the rim of a highland like a waterfall. Although the air is heated adiabatically, as are chinooks, the initial temperatures are so low that the wind arrives in the lowlands still colder and more dense than the air it displaces. In fact, it is a requirement that this air be colder than the air it invades, for it is the air's greater density that causes it to descend. Occasionally as this frigid air descends, it is channeled into narrow valleys where it acquires high velocities capable of great destruction.

Mistral

A few of the better-known katabatic winds have been given local names. The most famous is the **mistral** that blows from the Alps over France toward the Mediterranean Sea. Another is the **bora** that originates in the mountains of Yugoslavia and blows to the Adriatic Sea.

Bora

GLOBAL DISTRIBUTION OF PRECIPITATION

Even a casual glance at Figure 7–10(a) reveals the complex nature of the global distribution of precipitation. Nevertheless, the gross features of the earth's precipitation regimes shown in this figure can be explained by using our knowledge of the global wind and pressure systems. In general, regions influenced by high pressure, with its associated subsidence and divergent winds, experience rather dry conditions. On the other hand, regions under the influence of low pressure and its converging winds and ascending air receive ample precipitation. If the

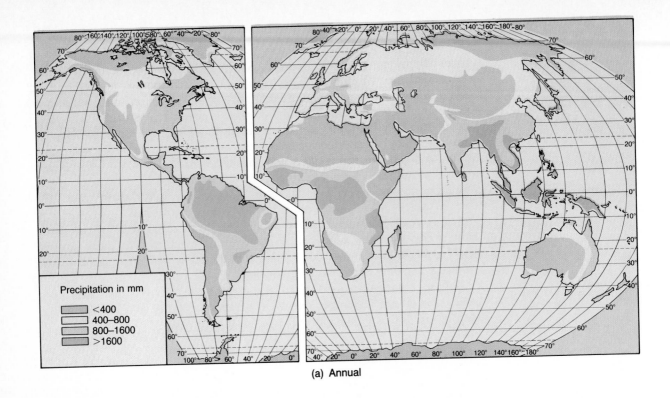

(a) Annual

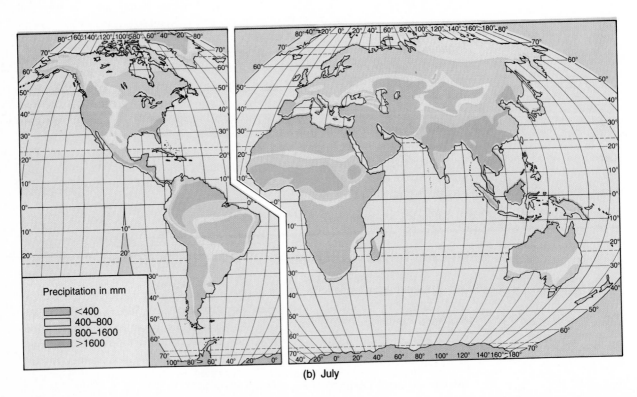

(b) July

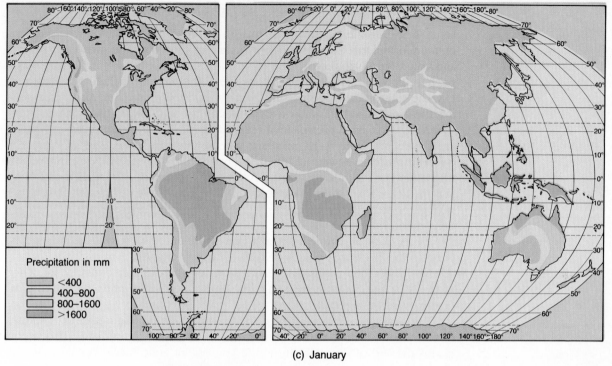

(c) January

FIGURE 7–10
Global distribution of precipitation. (a) Annual, (b) July, and (c) January.

wind-pressure regimes were the only control of precipitation, the pattern shown in Figure 7–10(a) would be much simpler. The inherent nature of the air involved, however, is also an important factor in determining the potential for precipitation. Because cold air has a very low capacity for moisture compared with warm air, we would expect a latitudinal variation in precipitation, with low latitudes receiving the greatest amounts of precipitation and high latitudes receiving the least amounts. An examination of Figure 7–10(a) does indeed reveal heavy precipitation in equatorial regions and meager precipitation near the landmasses poleward of 60 degrees latitude. A noticeably arid region, however, is also found in the subtropics. This situation can be explained by examining the global wind-pressure regimes.

In addition to the latitudinal variations in precipitation, the distribution of land and water complicates the precipitation pattern found over the earth. Large landmasses in the middle latitudes commonly experience decreased precipitation toward their interiors. For example, central North America and central Eurasia receive considerably less precipitation than coastal regions at the same latitude. Furthermore, the effects of mountain barriers alter the idealized precipitation regimes we would expect from the global wind systems. Windward mountain slopes receive abundant precipitation, whereas leeward slopes and adjacent lowlands are usually deficient in moisture.

Zonal Distribution of Precipitation

Let us first examine the zonal distribution of precipitation that we would expect on a uniform earth and then add the variations caused by the influences of land and water. Recall from our earlier discussion that on a uniform earth four major pressure zones emerge in each hemisphere. These zones include the equatorial low, the subtropical high, the subpolar low, and the polar high. Also, remember that these pressure belts show a marked seasonal shift toward the summer hemisphere. The idealized precipitation regimes expected from these pressure systems are shown in Figure 7–11, in which we can see that the equatorial regime is centered over the equatorial low throughout most of the year. So in this region where the trade winds converge (ITC), heavy precipitation is experienced in all seasons. Poleward of the equatorial low in each hemisphere lies the belt of subtropical high pressure. In these regions subsidence contributes to the rather dry conditions found here throughout the year. Between the wet equatorial regime and the dry subtropical regime lies a zone that is influenced by both pressure systems. Because the pressure systems migrate with the sun, these transitional regions receive most of their precipitation in the summer when they are under the influence of the ITC. They experience a dry season in the winter when the subtropical high moves equatorward.

The midlatitudes receive most of their precipitation from traveling cyclones that frequent this region and generate the subpolar low (Figure 7–12). This region is also the site of the polar front, the convergent zone between cold polar air

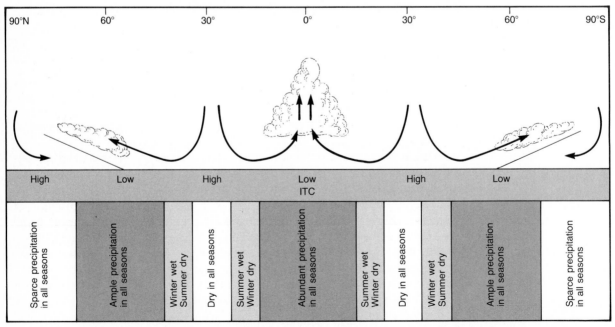

FIGURE 7–11
Schematic illustration of zonal precipitation patterns.

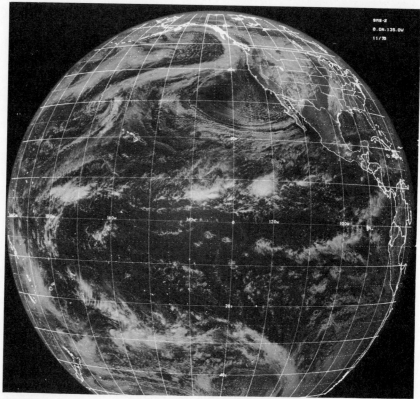

FIGURE 7–12
Satellite images of the earth's cloud cover reveal the global wind systems. The curved cloud patterns found poleward of 30 degrees latitude in both hemispheres are associated with traveling cyclones. The ITC over the Pacific is visible as a band of clouds located at 5 degrees north latitude. (Courtesy of NOAA)

and the warmer westerlies. It is along the polar front that cyclones often form. Because the position of the polar front migrates freely between approximately 30 and 70 degrees latitude, most midlatitude areas receive ample precipitation. But the mean position of this zone also moves with the sun on a seasonal basis, so that a narrow belt between 30 and 40 degrees latitude experiences a marked seasonal fluctuation in precipitation. In winter this zone experiences numerous cyclones with associated precipitation as the position of the polar front moves equatorward. During summer, however, the circulation of this region is dominated by subsidence associated with the subtropical high; thus rather dry conditions prevail. Compare the July and January precipitation data for the west coast of North America by referring to Figure 7–10(b) and (c).

The polar regions are dominated by cold air that holds little moisture. Throughout the year this region experiences only meager amounts of precipitation. Even in the summer, when temperatures rise, these areas of ice and snow are dominated by high pressure that blocks the movement of the few cyclones that do travel poleward.

Distribution of Precipitation over the Continents

The zonal pattern outlined in the previous section roughly approximates the major aspects of global precipitation. Abundant precipitation is found in the equatorial and midlatitude regions, whereas substantial portions of the subtropics and polar areas are relatively dry. Yet numerous exceptions to this idealized zonal pattern can also be found in Figure 7–10(a). For example, several arid areas are found in the midlatitudes. The desert of Patagonia, located along the southeast sector of South America, is one example. For the most part, midlatitude deserts like that of Patagonia are found on the leeward (rain shadow) side of a mountain barrier or in the interior of a continent cut off from a source of moisture. Most other departures from the true zonal scheme result from the effects of the distribution of continents and oceans on global circulation.

The most notable breakdown in the zonal distribution of precipitation occurs in the subtropics. Here we find not only many of the world's largest deserts but also regions of abundant rainfall. This pattern results because the high pressure centers that dominate the circulation in these latitudes have different characteristics on their eastern and western flanks. Subsidence is most pronounced on the eastern side of these oceanic highs, and a strong temperature inversion is encountered very near the surface and results in stable atmospheric conditions. The upwelling of cold water along the west coasts of the adjacent continents cools the air from below and adds to the stability on the eastern sides of these highs. Because these anticyclones tend to crowd the eastern side of an ocean, we find that the western sides of the continents adjacent to these subtropical highs are arid. Centered at approximately 25°N or S latitude on the western side of their respective continents, we find the Sahara Desert of North Africa, the Namib of southwest Africa, the Atacama of South America, the deserts of the Baja peninsula of North America, and the Great Desert of Australia. On the western side of these highs, however, subsidence is less pronounced, and convergence with associated rising air appears to be more prevalent. In addition, as this air travels over a large expanse of warm water, it acquires moisture through evaporation that acts to enhance its instability. Consequently, the eastern regions of subtropical continents generally receive abundant precipitation all year round. Southern Florida is a good example.

When we consider the influence of land and water on the distribution of precipitation, the general pattern illustrated in Figure 7–13 emerges. This figure is a highly idealized precipitation scheme for a hypothetical landmass in the northern hemisphere. By rotating this diagram 180 degrees around the line representing the equator, the pattern for the southern hemisphere is established. Although this diagram contains the most salient features of the global precipitation pattern, frequent reference should also be made to Figure 7–10 as you read the following discussion. To make this diagram more useful, Figure 7–14 provides precipitation data for locations in selected precipitation regimes.

First, notice in Figure 7–13 that the precipitation regime for the western section of the hypothetical continent is the same as the zonal pattern depicted earlier for a uniform earth. Precipitation data for San Francisco and Mazatlán illustrate the marked seasonal variations found in regimes 3 and 5, respectively.

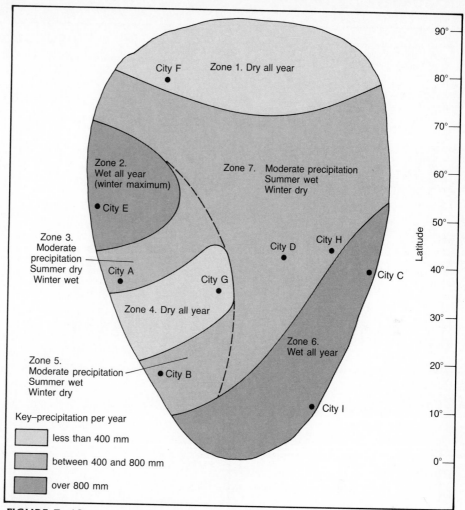

FIGURE 7–13
Idealized precipitation regimes for a hypothetical landmass in the northern hemisphere.
Refer also to Figure 7–14.

Recall that it is the seasonal migration of the pressure systems that causes these fluctuations. Second, notice that precipitation regime 2 shows a strong maximum in the winter, as illustrated by data for Juneau. This seasonal variation results because cyclonic storms are more prevalent in the winter.

The eastern segment of the continent shows a marked contrast to the zonal pattern described earlier. Only the dry polar regime is similar in size and position to the western seaboard. The most noticeable departure from the zonal pattern is the absence of the arid subtropical region on the east coast. As noted earlier this difference is caused by the predominance of convergence and lifting on the western side of the oceanic anticyclones and subsidence on the eastern side. Also notice that as one moves inland from the east coast, a decrease in precipitation is experienced. This decrease, however, does not hold true for mountainous re-

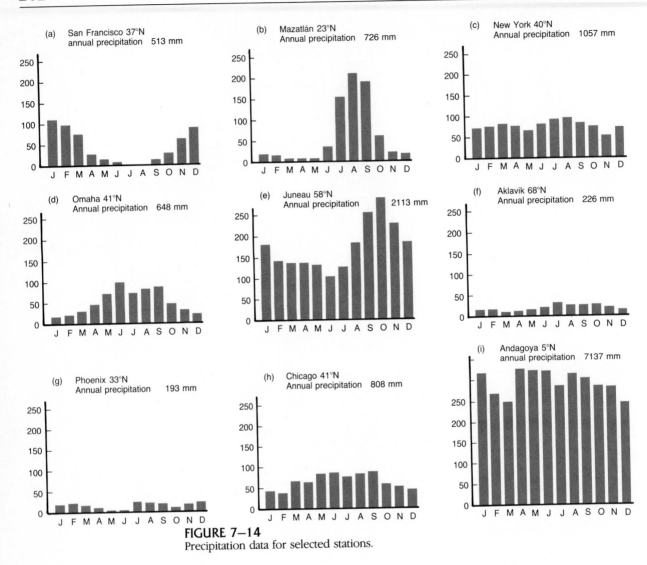

FIGURE 7–14
Precipitation data for selected stations.

gions; even relatively small ranges like the Appalachians are able to extract more than their share of precipitation.

Another variation is seen with latitude. As we move poleward along the eastern section, we note a decrease in precipitation. This is the decrease we would expect with cooler air temperatures and corresponding lower moisture capacities. But we do not find a similar decrease in precipitation in the higher middle latitudes along the west coast. Because this region lies in the westerly wind belt, the west coasts of midlatitude regions, such as Canada and Norway, receive abundant moisture. In fact, some of the rainiest regions on earth are found in such settings. In contrast, the east coasts of continents in the midlatitudes experience temperature and moisture regimes more typical of continental locations, particularly in the winter when the flow is off the land. Consider next the interior of continents, which shows a somewhat different precipitation pattern, especially in the middle latitudes. Because cyclones are more common in the winter, we would expect midlatitude locations also to experience maximum precipitation then. The precipitation regimes of the interior, however, are affected to a degree by the so-called monsoon circulation that offsets this picture. Cold winter temperatures and dominance of flow off the land in the winter make it the dry season. This pattern holds true, for example, for our own Great Plains. Although cyclonic storms frequent these regions in the winter, they do not become precipitation producers until they have drawn in abundant moist air from the Gulf of Mexico. Thus most of the winter precipitation falls in the eastern one-third of the United States. The lack of winter precipitation in the Plains states compared with the East can be seen in the data for Omaha and New York City. In the summer the inflow of warm moist air from the ocean is aided by the thermal low that develops over the land and causes a general increase in precipitation over the midcontinent. The dominance of summer monsoon-type precipitation is evident when we compare the patterns for India in summer and winter by using Figure 7–10(b) and (c). In the United States areas in the southeastern portion of the midlatitudes begin their rainy season in the early spring, but more northerly and westerly locations do not reach their peak precipitation periods until summer or early fall.

EL NIÑO AND GLOBAL WEATHER

El Niño

In 1982 and 1983 an event called **El Niño** made news headlines and was blamed for weather extremes of many types in many parts of the world. Heavy rains and flooding in these years plagued normally dry portions of Ecuador and Peru. Some locations that normally receive only 10 to 13 centimeters of rain had as much as 3.5 meters of precipitation. At the same time severe drought beset Australia, Indonesia, and the Philippines. Huge crop losses, property damage, and much human suffering were recorded. Farther north, one of the warmest winters on record was followed by one of the wettest springs for much of the United States. The ferocious storms that struck the California coast brought unprecedented beach erosion, landslides, and floods. Heavy snows in the Sierra Nevada and the mountains of Utah and Colorado led to mudflows and flooding in Utah, Nevada, and along the Colorado River in the spring of 1983. Neither was the Gulf of Mexico spared. Excessive rains brought floods to the Gulf states and Cuba. Exactly what is El Niño and how is it related to these diverse weather events?

Near the end of each year a weak, warm ocean current begins to flow southward along the coast of Ecuador and Peru. Many years ago local residents named this annual current El Niño, which means "the Child," because it usually appeared during the Christmas season. In some years a much larger and more extensive warming of the ocean occurs, affecting waters far out into the central Pacific. Today the name *El Niño* is applied primarily to these major episodes of ocean warming, even though the area covered and the physical processes that produce them differ greatly from the relatively weak phenomenon that originally bore the name. The El Niño of 1982–1983 was such a major episode. In fact, it is believed to be the strongest warming of the equatorial Pacific in this century. Unlike its predecessors, it began to develop in the spring of 1982, long before the normal Christmastime onset of earlier events.

Major El Niño events are intimately related to large-scale changes of atmospheric circulation in the Pacific. Each time an El Niño occurs, the barometric pressure over large portions of the southeastern Pacific drops, whereas across the Pacific, near Indonesia and northern Australia, the pressure rises. Moreover, about the time a major El Niño event ends, the pressure difference between these two regions swings back in the opposite direction. This seesaw pattern of atmospheric **Southern Oscillation** pressure between the eastern and western Pacific is termed the **Southern Oscillation.** It is an inseparable part of the warmings that occur in the central and eastern Pacific every 3 to 10 years.

The winds in the lower portion of the atmosphere are the link between the Southern Oscillation and the extensive ocean warming associated with El Niño. Typically the Pacific trade winds converge near the equator and move westward into a semipermanent region of low pressure near Indonesia (Figure 7–15a). This steady westward flow of air literally drags the warm surface waters to the west. The result is a "piling up" of a thick layer of warm surface water and a higher sea level in the western Pacific. Meanwhile, strong coastal upwelling of cold water and a lower sea level characterize the eastern Pacific.

When, for unknown reasons, the Southern Oscillation occurs, the normal situation just described changes dramatically. Barometric pressure rises in the Indonesian region, causing the pressure gradient along the equator to weaken or, as occurred in 1982–1983, actually to reverse. As a consequence, the once-steady easterly winds in the western Pacific diminish and then change direction. This reversal creates a major change in the equatorial current system, with warm water moving toward the east. With time, water temperatures in the eastern Pacific increase and sea level in the region rises (Figure 7–15b). A major El Niño event is now fully developed.

The reader should now be better able to understand the meteorological consequences of the events outlined at the beginning of this section. The droughts in the western Pacific must be related to the pressure seesaw of the Southern Oscillation; that is, the development of higher pressure in the region is largely responsible for the dry conditions. Moreover, as long as the elevated ocean temperatures in the eastern Pacific persist, enormous quantities of additional heat and moisture are added to the overlying air. This additional energy acts to intensify the winds of the tropical jet stream and provides the necessary ingredients for additional storminess and rainfall over the western hemisphere. Referring to this fact, J. Murray Mitchell, a noted researcher at the National Oceanic and Atmospheric Administration, stated that the

> miserably wet weather that beset the sunbelt of the United States this past winter and spring along with a host of other weather anomalies, among them the huge

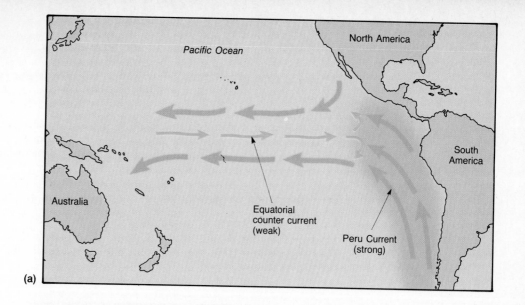

(a)

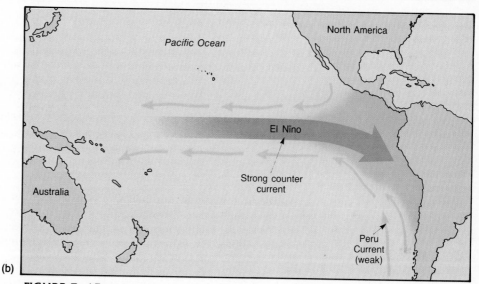

(b)

FIGURE 7–15

The relationship between the Southern Oscillation and El Niño is illustrated on these simplified maps. (a) Normally the flow of air and surface water is toward the west, and upwelling of cold water occurs along the west coast of South America. (b) When the Southern Oscillation develops, the trade winds diminish and then reverse, leading to an eastward movement of warm water along the equator. As a result, the surface waters of the central and eastern Pacific warm.

mountain snowpack that recently brought the Colorado River to flood stage, can be traced back to the strong El Niño event of recent months.[1]

Thus the shifting pressure pattern associated with the Southern Oscillation, together with the temperature increase in the waters of the eastern Pacific called

[1] "El Niño: The Global Weather Connection," *Weatherwise*, 36, no. 4 (August 1983), 168.

El Niño, is believed to have played an important role in generating many extreme weather events in 1982–1983. Mitchell puts it this way:

> . . . there is no place on Earth where the weather is indifferent to air and ocean conditions in the tropical Pacific: the Southern Oscillation and El Niño events associated with it are now understood to have a significant influence on the state of weather and climate almost everywhere.[2]

Although these powerful events were relatively unpredictable in the past, it is hoped that the extensive monitoring of the 1982–1983 episode, along with future monitoring efforts, will provide enough data to allow scientists eventually to predict these events a year or more in advance. If so, more accurate long-range forecasts will result.

REVIEW

1. Distinguish between macroscale, mesoscale, and microscale winds. Give an example of each.

2. If you were to view a weather map of the entire world for any single day of the year, would the global pattern of winds likely be visible? Explain your answer.

3. Briefly describe the idealized global circulation proposed by George Hadley. Did subsequent observations confirm Hadley's proposal?

4. Which factors are believed to cause air to subside in the latitude zone 20 to 35 degrees?

5. Referring to the idealized three-cell model of atmospheric circulation, most of the United States is situated in which belt of prevailing winds?

6. Briefly explain each of the following statements that relate to the global distribution of surface pressure:
   a. The only true zonal distribution of pressure exists in the region of the subpolar low in the southern hemisphere.
   b. The subtropical highs are more continuous in the winter hemisphere.
   c. The subpolar low in the northern hemisphere is represented by individual oceanic cells that are strongly seasonal in character.
   d. There is a strong high-pressure cell over Eurasia in the winter.

7. Why is the flow aloft predominantly westerly?

8. At what time of year should we expect the fastest westerly flow? Explain.

9. Describe the situation in which jet streams were first encountered.

10. Describe the manner in which pressure distribution is shown on upper-air charts. How are high- and low-pressure areas depicted on these charts?

11. What were the "dishpan experiments" and what have we learned from them?

12. Why is the well-known Texas "norther" not a true local (mesoscale) wind?

13. The most intense sea breezes develop along tropical coasts adjacent to cool ocean currents. Explain.

14. What are katabatic (fall) winds? Name two examples.

---

[2] Ibid.

15. Relying only on your knowledge of the zonal distribution of precipitation, describe the seasonal pattern of precipitation at the following representative locations: equator, 15°N latitude, Tropic of Cancer, 40°N latitude, 80°N latitude.

16. What factors, other than global wind and pressure systems, exert an influence on the world distribution of precipitation?

17. Describe the relationship between the Southern Oscillation and a major El Niño event (see Figure 7–15).

VOCABULARY REVIEW

macroscale circulation	Siberian high
mesoscale winds	jet stream
microscale winds	polar jet stream
convection cell	trough
Hadley cell	ridge
horse latitudes	sea breeze
trade winds	land breeze
doldrums	valley breeze
prevailing westerlies	mountain breeze
polar easterlies	chinook
polar front	foehn
equatorial low	Santa Ana
intertropical convergence zone (ITC)	katabatic or fall wind
subtropical high	mistral
subpolar low	bora
polar high	El Niño
Aleutian low	Southern Oscillation
Icelandic low	

8

Air Masses

Any person living in the middle latitudes has experienced hot, "sticky" summer heat waves and frigid winter cold waves. In the first case, after several days of sultry weather the spell may come to a dramatic end that is marked by thundershowers and followed by several days of relatively cool relief. In the second case, thick stratus clouds and snow may replace the clear skies that had prevailed and temperatures may climb to values that seem mild compared with what preceded them. In both examples, what was experienced was a period of generally uniform weather conditions followed by a relatively short period of change and the subsequent reestablishment of a new set of weather conditions that remained for perhaps several days before changing again.

Air Mass

The weather patterns just described are the result of the movements of large bodies of air, called air masses. An **air mass,** as the term implies, is an immense body of air, usually 1600 kilometers or more across and perhaps several kilometers thick, which is characterized by homogeneous physical properties (in particular, temperature and moisture content) at any given altitude. When this air moves out of its region of origin, it will carry these temperatures and moisture conditions elsewhere, eventually affecting a large portion of a continent.

The horizontal uniformity of an air mass is not complete because it may extend through 20 degrees or more of latitude and cover hundreds of thousands to millions of square kilometers. Consequently, small differences in temperature and humidity from one point to another at the same level are to be expected. Still, the differences observed within an air mass are small in comparison to the rapid rates of change experienced across the boundaries between air masses. Because it may take several days for an air mass to traverse an area, the region under its influence will probably experience generally constant weather conditions, a situation called **air-mass weather.** Certainly some day-to-day variations may exist, but the events will be very unlike those in an adjacent air mass.

Air-Mass Weather

The air-mass concept is an important one because it is closely related to the study of atmospheric disturbances. Most disturbances in the middle latitudes originate along the boundary zones that separate different air masses.

SOURCE REGIONS

Where do air masses form? What factors determine the nature and degree of uniformity of an air mass? These two basic questions are closely related, for the site where an air mass forms vitally affects the properties that characterize it.

Source Region

Areas in which air masses originate are termed **source regions.** Because the atmosphere is heated chiefly from below and gains its moisture by evaporation from the earth's surface, the nature of the source region largely determines the initial characteristics of an air mass. An ideal source region must meet two essential criteria. First, it must be an extensive and physically uniform area. A region having highly irregular topography or one that has a surface consisting of both water and land is not satisfactory. The second criterion is that the area be characterized by a general stagnation of atmospheric circulation so that air will stay over the region long enough to come to some measure of equilibrium with the surface; in general, it means regions dominated by stationary or slow-moving anticyclones with their extensive areas of calms or light winds. Regions under the influence of cyclones are not likely to produce air masses because such systems are characterized by converging surface winds. The winds in lows are constantly bringing air with unlike temperature and humidity properties into the area. Because the time involved is not long enough to eliminate these differences, steep temperature gradients result, and air-mass formation is precluded.

Figure 8–1 shows the source regions that produce the air masses that most

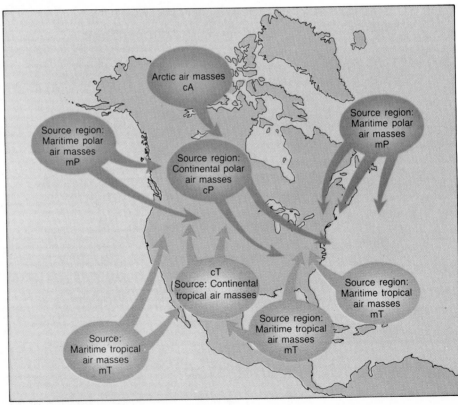

FIGURE 8–1
Air-mass source regions for North America. (Courtesy of Ward's Natural Science Establishment, Inc., Rochester, N.Y.)

often influence North America. The waters of the Gulf of Mexico and Caribbean Sea and similar regions in the Pacific west of Mexico yield warm air masses, as does the land area that encompasses the southwestern United States and northern Mexico. In contrast, the North Pacific, the North Atlantic, and the snow- and ice-covered areas comprising northern North America and the adjacent Arctic Ocean are major source regions for cold air masses.

Notice that the major source regions are not found in the middle latitudes but instead are confined to tropical and polar locations. The fact that the middle latitudes are the site where cold and warm air masses clash, often because the converging winds of a traveling cyclone draw them together, means that this zone lacks the conditions that are necessary for a source region. Almost every air mass found in the middle latitudes had its source elsewhere.

CLASSIFYING AIR MASSES

The classification of an air mass depends on the latitude of the source region and the nature of the surface in the area of origin—ocean or continent. The latitude of the source region indicates the temperature conditions within the air mass and the nature of the surface below strongly influences the moisture content of the air.

Polar (P) Air Mass

Arctic (A) Air Mass

Tropical (T) Air Mass

Equatorial (E) Air Mass

Maritime (m) Air Mass

Continental (c) Air Mass

Air masses are identified by two-letter codes. With reference to latitude (temperature), air masses are placed into one of four categories: **polar** (P), **arctic** (A), **tropical** (T), and **equatorial** (E). The differences between polar and arctic and between tropical and equatorial are usually small and simply serve to indicate the degree of coldness or warmness of the respective air masses. The lowercase letter m (for **maritime**) or the lowercase letter c (for **continental**) is used to designate the nature of the surface in the source region and hence the humidity characteristics of the air mass. Because maritime air masses form over oceans, they have a relatively high water vapor content compared to continental air masses that originate over landmasses.

When this classification scheme is applied, the following air masses may be identified:

cA	continental	arctic
cP	continental	polar
cT	continental	tropical
mT	maritime	tropical
mP	maritime	polar
mE	maritime	equatorial

Notice that the list includes neither mA (maritime arctic) nor cE (continental equatorial). These air masses are not listed because they seldom, if ever, form. Although arctic air masses form over the Arctic Ocean, this water body is ice covered throughout the year. Consequently, the air masses that originate here consistently have the moisture characteristics associated with a continental source region. By contrast the region of the trade winds produces warm and humid air masses almost exclusively. Continental equatorial air masses do not generally form because this area is dominated by the oceans. In the latitude belt that includes the region 10 degrees on either side of the equator, more than 75 percent of the surface is ocean. Furthermore, the land areas adjacent to the equator are warm and wet tropical rain forests. So if an air mass were to originate over such locales, its moisture content would be relatively high.

AIR-MASS MODIFICATION

After an air mass forms, it normally migrates from the area where it acquired its distinctive properties to a region with different surface characteristics. Once the air mass moves from its source region, it not only modifies the weather of the area it is traversing, but it is also gradually modified by the surface over which it is moving. Warming or cooling from below, the addition or subtraction of moisture, and vertical movements all act to bring about changes in an air mass. The amount of modification may be relatively small or, as the following example illustrates, the changes may be profound enough to alter completely the original identity of the air mass.

When cA or cP air moves over the ocean in winter, it undergoes considerable change. Evaporation from the water surface rapidly transfers large quantities of moisture to the once dry continental air. Furthermore, because the underlying water is warmer than the air above, the air is also heated from below. This factor leads to instability and vertically ascending currents that rapidly transport heat and moisture to higher levels. In a matter of days, cold, dry, and stable continental air is transformed into an unstable mP air mass.

When an air mass is colder than the surface over which it is passing, as in the preceding example, the lowercase letter k is added after the air-mass symbol. If, however, an air mass is warmer than the underlying surface, the lowercase letter w is added. It should be remembered that the k or w suffix does not mean that the air mass itself is cold or warm. It means only that the air is relatively cold or warm in comparison with the underlying surface over which it is traveling. For example, an mT air mass from the Gulf of Mexico is usually classified as mTk as it moves over the southeastern states in summer. Although the air mass is warm, it is still cooler than the highly heated continent over which it is passing.

The k or w designation gives an indication of the stability of an air mass and hence the weather that might be expected. An air mass that is colder than the surface is obviously going to be warmed in its lower layers. This fact causes an increased lapse rate and greater instability that favor the ascent of the heated lower air and create the possibility of cloud formation and precipitation. Indeed, a k air mass is often characterized by cumuliform clouds, and should precipitation occur, it will be of the shower or thunderstorm variety. Also, visibility is generally good (except in rain) because of the stirring and overturning of the air. Conversely, when an air mass is warmer than the surface over which it is moving, its lower layers are chilled. A surface inversion that increases the stability of the air mass often develops. This condition does not favor the ascent of air, and so it opposes cloud formation and precipitation. Any clouds that do form will be stratiform, and precipitation, if any, will be light to moderate. Moreover, because of the lack of vertical movements, smoke and dust often become concentrated in the lower layers of the air mass and cause poor visibility. During certain times of the year fogs, especially the advection type, may also be common in some regions.

In addition to modifications resulting from temperature differences between an air mass and the surface below, vertical movements induced by cyclones and anticyclones or topography may also affect the stability of an air mass. Such modifications are often called mechanical or dynamic and are usually independent of the changes caused by surface cooling or heating. For example, significant modification can result when an air mass is drawn into a low. Here convergence and lifting dominate and the air mass is rendered more unstable. Conversely, the subsidence associated with anticyclones acts to stabilize an air mass. Similar alterations in stability occur when an air mass is lifted over highlands or descends the leeward side of a mountain barrier. In the first case, the air's stability is reduced; in the second case, the air becomes more stable.

PROPERTIES OF NORTH AMERICAN AIR MASSES

Air masses are continually passing over us, which means that the day-to-day weather that we experience depends primarily on the temperature, stability, and moisture content of these large bodies of air. In the following section we briefly examine the properties of the principal North American air masses. In addition, Table 8–1 serves as a summary.

Continental Polar (cP) and Continental Arctic (cA) Air Masses

Continental polar and continental arctic air masses are, as their classification implies, cold and dry. Continental polar air originates over the snow-covered interior regions of Canada and Alaska, poleward of the fiftieth parallel; continental arctic air forms farther north over the Arctic Basin and the Greenland ice cap (Figure 8–1). Continental arctic air is distinguished from cP air by its generally lower temperatures, although at times the differences may be slight. In fact, some meteorologists do not differentiate between cP and cA.

During the winter both air masses are bitterly cold and very dry. Since the winter nights are long, the earth radiates heat that is, for the most part, not replenished by incoming solar energy. Hence, because of the prolonged earth radiation, the surface reaches very low temperatures and the air near the ground is gradually chilled to heights of perhaps 1 kilometer or more. The result is a strong and persistent temperature inversion with the lowest temperatures near the ground. Marked stability is, therefore, the rule. Since the air is very cold and the surface below is frozen, the mixing ratio of these air masses is necessarily low, ranging

TABLE 8–1 Weather Characteristics of North American Air Masses

Air Mass	Source Region	Temperature and Moisture Characteristics in Source Region	Stability in Source Region	Associated Weather
cA	Arctic basin and Greenland ice cap	Bitterly cold and very dry in winter	Stable	Cold waves in winter
cP	Interior Canada and Alaska	Very cold and dry in winter Cool and dry in summer	Stable entire year	a. Cold waves in winter b. Modified to cPk in winter over Great Lakes bringing "lake-effect" snow to leeward shores
mP	North Pacific	Mild (cool) and humid entire year	Unstable in winter Stable in summer	a. Low clouds and showers in winter b. Heavy orographic precipitation on windward side of western mountains in winter c. Low stratus and fog along coast in summer; modified to cP inland
mP	Northwestern Atlantic	Cold and humid in winter Cool and humid in summer	Unstable in winter Stable in summer	a. Occasional "northeaster" in winter b. Occasional periods of clear, cool weather in summer
cT	Northern interior Mexico and southwestern U.S. (summer only)	Hot and dry	Unstable	a. Hot, dry, and clear, rarely influencing areas outside source region b. Occasional drought to southern Great Plains
mT	Gulf of Mexico, Caribbean Sea, western Atlantic	Warm and humid entire year	Unstable entire year	a. In winter it usually becomes mTw moving northward and brings occasional widespread precipitation or advection fog b. In summer, hot and humid conditions, frequent cumulus development and showers or thunderstorms
mT	Subtropical Pacific	Warm and humid entire year	Stable entire year	a. In winter it brings fog, drizzle, and occasional moderate precipitation to N.W. Mexico and S.W. United States b. In summer this air mass occasionally reaches the western United States and is a source of moisture for infrequent convectional thunderstorms

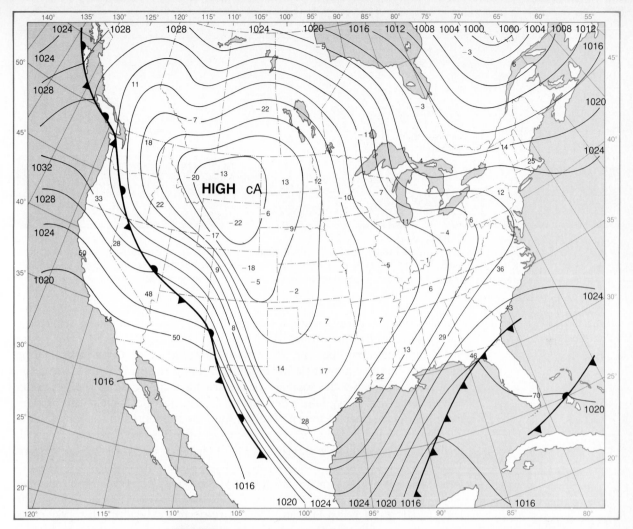

FIGURE 8–2

An intense winter cold wave depicted on the January 10, 1962, surface weather map. This outbreak of continental arctic air brought subfreezing temperatures as far south as the Gulf of Mexico. As shown here, these frigid air masses are generally associated with high pressure in the winter. (Courtesy of NOAA)

from perhaps 0.1 gram per kilogram in cA up to 1.5 gram per kilogram in some cP air.

As wintertime cP or cA air moves outward from its source region, it carries its cold, dry weather to the United States, normally entering between the Great Lakes and the Rockies. There are no major barriers to their movement between their source regions and the Gulf of Mexico; consequently, cP and cA air masses can sweep rapidly and with relative ease far southward into the United States. The winter cold waves experienced in much of the central and eastern United States are closely associated with such polar outbreaks. One such cold wave, which was among the most intense and widespread in this century, is depicted on the weather map in Figure 8–2. Usually the last freeze in spring and the first in autumn can be correlated with outbreaks of polar or arctic air.

225

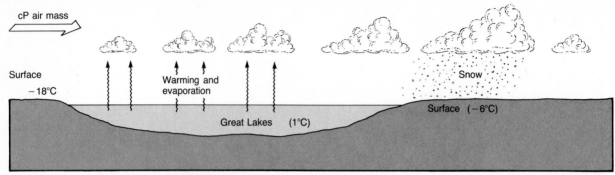

FIGURE 8–3
As continental polar air crosses the Great Lakes in winter, it acquires moisture and is made less stable because of warming from below. "Lake-effect" snow showers on the lee side of the lakes is often the consequence of this air-mass modification.

Lake-Effect Snow

Although cP air masses are not, as a rule, associated with heavy precipitation, those that cross the Great Lakes often bring snow to the leeward shores. Figure 8–3 depicts the movement of a cP air mass across one of the Great Lakes. During its journey the air acquires heat and moisture from the relatively warm lake surface. By the time it reaches the southern or eastern shore, this cPk air is quite humid and unstable, and heavy **lake-effect snow** is characteristic. Such storms account for a high percentage of the snowfall on the leeward shores of the Great Lakes. This fact is demonstrated nicely by examining Figure 8–4 and comparing snowfall totals on the eastern (leeward) shore of Lake Michigan with totals on the western shore. Table 8–2 serves as another example. It lists the ten snowiest major metropolitan areas in the United States for the 10-year period 1975–1985. With snowfall totals of more than 2500 centimeters, Buffalo and Rochester, New York, on the leeward shores of Lakes Erie and Ontario respectively, head the list. Lake-effect snows also contribute to the totals for the other Great Lakes cities on the list.

Because cA air is present principally in the winter, only cP air has any influence on our summer weather, and this effect is considerably reduced when compared with winter. During summer months the properties of the source region for cP air are very different from those during winter. Instead of being chilled by the ground, the air is warmed from below as the long days and higher sun angle warm the snow-free land surface. Although summer cP air is warmer and has a higher moisture content than its wintertime counterpart, the air is still cool and relatively dry compared with areas farther south. Summer heat waves in the northern portions of the eastern and central United States are often ended by the southward advance of cP air, which for a day or two brings cooling relief and bright, pleasant weather.

Maritime Polar (mP) Air Masses

Maritime polar air masses form over oceans at high latitudes. As the classification would indicate, mP air is cool to cold and humid, but compared with cP and cA

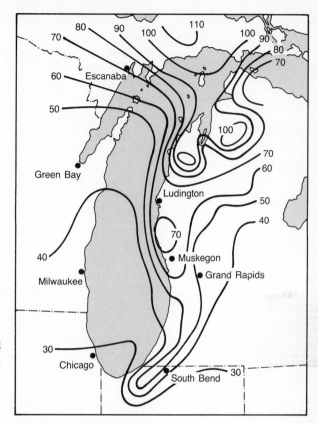

FIGURE 8–4

Average annual snowfall (in inches) over Lake Michigan and environs. Snowfall totals are much higher along the eastern shore of Lake Michigan because of the contribution of lake-effect storms. (After Stanley A. Changnon, *Precipitation Scavenging of Lake Michigan Basin*, Bulletin 52, Urbana, Ill.: Illinois State Water Survey, 1968)

TABLE 8–2 The Ten Snowiest Metropolitan Areas, 1975–1985*

City	Ten-Year Snowfall Totals (centimeters)
1. Buffalo, New York	2710.9
2. Rochester, New York	2635.8
3. Salt Lake City, Utah	1622.6
4. Minneapolis–St. Paul, Minnesota	1605.5
5. Albany, New York	1604.3
6. Cleveland, Ohio	1592.6
7. Denver, Colorado	1483.4
8. Milwaukee, Wisconsin	1400.3
9. Detroit, Michigan	1176.8
10. Chicago, Illinois	1114.8

* Official snowfall statistics for the 50 largest metropolitan areas as identified by the Census Bureau were used to compile the list.

air masses in winter, mP air is relatively mild because of the higher temperatures of the ocean surface as contrasted to the colder continents. Two regions are important sources for mP air that influences the United States: the North Pacific and the northwestern Atlantic from Newfoundland to Cape Cod (Figure 8–1). Because of the general west-to-east circulation in the middle latitudes, mP air masses from the North Pacific source region have a more profound influence on North American weather than mP air masses generated in the northwestern Atlantic. Whereas air masses that form in the Atlantic generally move eastward toward Europe, mP air from the North Pacific has a strong influence on the weather along the West Coast of North America, especially in the winter.

During the winter mP air masses from the Pacific usually begin as cP air in Siberia. Although air rarely stagnates over this area, the source region is extensive enough to allow the air moving across it to acquire its characteristic properties. As the air progresses eastward over the relatively warm water, active evaporation and heating in the lower levels occur. Consequently, what was once a very cold, dry, and stable air mass is changed into one that is mild and humid near the surface and relatively unstable. As this mP air arrives at the West Coast of North America, it is often accompanied by low clouds and shower activity. When the mP air advances inland against the western mountains, orographic uplift produces heavy rain or snow on the windward slopes of the mountains (see Figure 4–15).

As is the case for mP air from the Pacific, air masses forming in the northwestern Atlantic source region were originally cP air masses that moved from the continent and were transformed over the ocean. However, unlike air masses from the North Pacific, mP air from the Atlantic seldom affects the weather of North America. Nevertheless, this air mass does have an effect when the northeastern United States is on the northern or northwestern edge of a passing low. On these occasions, cyclonic winds draw mP air into the region. Its influence is generally confined to the area east of the Appalachians and north of Cape Hatteras, North Carolina. The weather associated with an invasion of mP air from the Atlantic is known

Northeaster

locally as a **northeaster.** Strong northeast winds, freezing or near freezing temperatures, high relative humidity, and the possibility of precipitation make this weather phenomenon a most unwelcome event.

During summer months the influence of mP air masses from the North Pacific on North American weather changes. Then the water is cooler than the surrounding continents. In addition, the Pacific anticyclone lies off the West Coast of the United States (see Figure 7–5). Consequently, there is an almost continuous southward flow of air having moderate temperatures. Although the air near the surface may often be conditionally unstable, the presence of the Pacific high means that there is subsidence and stability aloft. Thus low stratus clouds and summer fogs characterize much of the West Coast. Once summer mP air from the Pacific moves inland, it is heated at the surface over the hot and dry interior. The heating and resulting turbulence act to reduce the relative humidity in the lower layers, and the clouds dissipate.

Although mP air masses from the Atlantic may produce an occasional unwelcome "northeaster" during the winter, summertime incursions of this air mass often bring pleasant weather. Like the Pacific source region, the northwestern Atlantic is dominated by high pressure during the summer (see Figure 7–5). So the upper air is stable because of subsidence and the lower air is essentially stable because of the chilling effect of the cold water. As the circulation on the southern side of the anticyclone carries this stable and relatively dry mP air into New England, and occasionally as far south as Virginia, the region enjoys clear, cool weather and good visibility.

Maritime Tropical (mT) Air Masses

Maritime tropical air masses affecting North America most often originate over the warm waters of the Gulf of Mexico, the Caribbean Sea, or the adjacent western Atlantic Ocean (Figure 8–1). The tropical Pacific is also a source region for mT air. However, the land area affected by this latter source is small compared with the size of the region influenced by air masses produced in the Gulf of Mexico and adjacent waters.

As expected, mT air masses are warm to hot and they are humid. In addition, they are often unstable. It is through invasions of mT air masses that the subtropics export much heat and moisture to the less endowed areas to the north. Consequently, these air masses are important to the weather whenever present because they are capable of contributing significant precipitation.

Maritime tropical air masses from the Gulf-Caribbean-Atlantic source region greatly affect the weather of the United States east of the Rocky Mountains. Although the source region is dominated by the North Atlantic subtropical high, the air masses it produces are not stable but are neutral or unstable because the source region is located on the weak western edge of the anticyclone where pronounced subsidence is absent.

During winter when cP air dominates the central and eastern United States, mT air seldom enters this part of the country. When an invasion does occur, the lower portions of the air mass are chilled and stabilized as it moves northward. Its classification is changed to mTw. As a result, the formation of convective showers is precluded. Widespread precipitation does occur, however, when a northward-moving mT air mass is pulled into a traveling cyclone and forced to ascend. In fact, much of the wintertime precipitation over the eastern and central states results when mT air from the Gulf is lifted along fronts in traveling cyclones. When northward-moving mT air does not become part of a cyclonic storm, the chilling of the air mass by the cold ground on occasion produces dense advection fogs.

During the summer mT air masses from the Gulf, Caribbean, and adjacent Atlantic cover a much wider area of North America and are present for a greater percentage of the time than during the winter. As a result, they exert a strong and often dominating influence over the summer weather of the United States east of the Rocky Mountains. This influence is due to the general sea-to-land (monsoonal) airflow over the eastern portion of North America during the warm months, which brings more frequent incursions of mT air that penetrate much deeper into the continent than during the winter months. Consequently, these air masses are largely responsible for the hot and humid conditions that prevail over the eastern and central United States.

Initially summertime mT air from the Gulf is unstable. As it moves inland over the warmer land, it becomes an mTk air mass as daytime heating of the surface layers further increases the air's instability. Because the relative humidity is high, only modest lifting is necessary to bring about active convection, cumulus development, and thunderstorm or shower activity. This is, indeed, a common warm weather phenomenon associated with mT air.

It should also be noted here that air masses from the Gulf–Caribbean–Atlantic source region are the primary source of much, if not most, of the precipitation received in the eastern two-thirds of the United States. Pacific air masses contribute little to the water supply east of the Rockies because the western mountains effectively "drain" the moisture from the air by numerous episodes of orographic uplift. Figure 8–5, which shows the distribution of average annual precipitation for the eastern two-thirds of the United States by using **isohyets** (lines connecting places having equal rainfall), illustrates this situation nicely. The pattern of isohyets shows the greatest rainfall in the Gulf region and a decrease in precipitation with increasing distance from the mT source region.

Isohyet

Compared to mT air from the Gulf of Mexico, mT air masses from the Pacific source region have much less of an impact on North American weather. In winter only northwestern Mexico and the extreme southwestern United States are influenced by air from the tropical Pacific. Because the source region lies along the eastern side of the Pacific anticyclone, subsidence aloft produces upper-level stability. When the air mass moves northward, cooling at the surface also causes the lower layers to become more stable, often resulting in fog or drizzle. If lifted along a front or forced over mountains, moderate precipitation results.

For many years the summertime influence of air masses from the tropical Pacific source region on the weather of the southwestern United States and northern Mexico was thought to be minimal. It was widely believed that the source of moisture for the infrequent summer thunderstorms that occur in the region could be traced to occasional westward thrusts of mT air from the Gulf of Mexico, but the Gulf of Mexico is no longer believed to be the primary supplier of moisture for the area west of the Continental Divide. Rather, it has been demonstrated that the tropical North Pacific west of central Mexico is a more important source of moisture for this area. In summer the mT air moves northward from its Pacific source region up the Gulf of California and into the interior of the western United States. This movement, which is confined largely to July and August, is essentially monsoonal in character. That is, the inflow of moist air is a response to the thermally produced low over the heated landmass.

Continental Tropical (cT) Air Masses

North America narrows as it extends southward through Mexico; therefore the continent has no extensive source region for continental tropical air masses. Only in summer do northern interior Mexico and adjacent parts of the arid southwestern

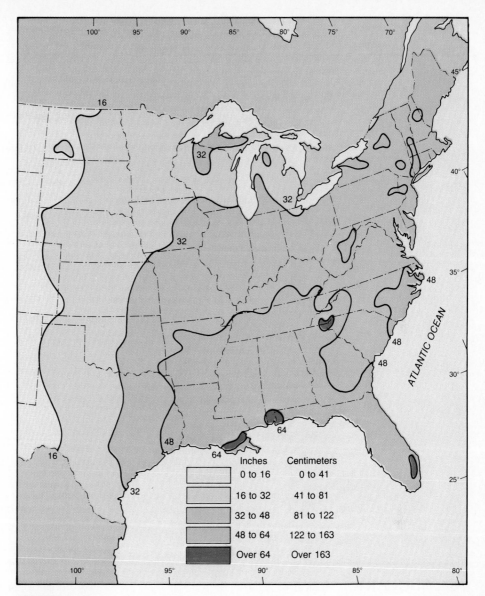

FIGURE 8–5
Average annual precipitation for the eastern two-thirds of the United States. Note the general decrease in yearly precipitation totals with increasing distance from the Gulf of Mexico, the source region for mT air masses. Isohyets are labeled in inches. (Courtesy of Environmental Data Service, NOAA)

United States produce hot, dry cT air. Because of the intense daytime heating at the surface, a steep lapse rate and turbulence to considerable heights are found. However, although the air is unstable, it generally remains clear because of extremely low humidity. Consequently, the prevailing weather is hot, with an almost complete lack of rainfall. Large daily temperature ranges are the rule. Although cT air masses are usually confined to the source region, occasionally they move into the southern Great Plains. If the cT air persists for long, drought may occur.

231

REVIEW

1. Define the terms air mass and air-mass weather.

2. What two criteria must be met for an area to be an air-mass source region?

3. Why are regions that have a cyclonic circulation generally not conducive to air-mass formation?

4. On what bases are air masses classified? Compare the temperature and moisture characteristics of the following air masses: cP, mP, mT, cT.

5. Why were mA and cE left out of the air-mass classification scheme?

6. What do the lowercase letters k and w indicate about an air mass? List the general weather conditions associated with k and w air masses.

7. How might vertical movements induced by a pressure system or topography act to modify an air mass?

8. What two air masses are most important to the weather of the United States east of the Rocky Mountains? Explain your choice.

9. What air mass influences the weather of the Pacific Coast more than any other?

10. Why do cA and cP air masses often sweep so far south into the United States?

11. Describe the modifications that occur as a cP air mass traverses one of the Great Lakes in the winter.

12. Why do mP air masses from the North Atlantic source region seldom affect the eastern United States?

13. What air mass and source region provide the greatest amount of moisture to the eastern and central United States?

14. For each statement below, indicate which air mass is most likely involved and from what source region it came:
    a. summer drought in the southern Great Plains
    b. wintertime advection fog in the Midwest
    c. heavy winter precipitation in the western mountains
    d. summertime convectional showers in the Midwest and East
    e. a "northeaster"

VOCABULARY REVIEW

air mass

air-mass weather

source region

polar (P) air mass

arctic (A) air mass

tropical (T) air mass

equatorial (E) air mass

maritime (m) air mass

continental (c) air mass

lake-effect snow

northeaster

isohyet

9

Weather Patterns

Middle-Latitude Cyclone

Wave Cyclone

Earlier we examined the basic elements of weather as well as the dynamics of atmospheric motions. We are now ready to consider these diverse phenomena in a unified framework that will serve as the basis for understanding day-to-day weather in the middle latitudes. For our purposes, the middle latitudes refer to the region between southern Florida and Alaska. The primary weather producer here is called the **middle-latitude cyclone** or **wave cyclone.** Middle-latitude cyclones are large low-pressure systems that generally travel from west to east across the United States (Figure 9–1). Lasting from a few days to more than a week, these weather systems have a counterclockwise circulation pattern with a net inflow toward their centers. Most middle-latitude cyclones also have a cold front and frequently a warm front extending from the central area of low pressure. Mass convergence in the area of low pressure and forceful lifting of air along the frontal zones initiates cloud development and frequently abundant precipitation.

POLAR FRONT THEORY

As early as the 1800s it was known that cyclones (lows) were the bearers of precipitation and severe weather. Thus the barometer was established as the main tool in "forecasting" day-to-day weather changes. These early methods of weather prediction largely ignored the importance of air-mass interactions in the formation of cyclones. Consequently, it was not possible to determine the conditions under which cyclone development was favorable. The first encompassing model of the wave cyclone was constructed by a group of Norwegian scientists during World War I. The Norwegians were then cut off from weather reports, especially those from the Atlantic. To counter this deficiency, a tight network of weather stations was established throughout the country. In the years that followed several Norwegian-trained meteorologists made great advances in broadening our understanding of the weather. Included in this early group were J. Bjerknes, V. Bjerknes, H. Solberg, and T. Bergeron. In 1918 J. Bjerknes published his theory of cyclone formation in an article entitled "On the Structure of Moving Cyclones." This theory, which became known as the **polar front theory,** provides us with a useful model of the wave cyclone. Although some changes in the original version were necessary because of subsequent discoveries, the main tenets of this theory remain an integral part of meteorological thought.

In the Norwegian model the wave cyclone develops in conjunction with a frontal zone known as the polar front. Recall that the polar front separates cold

Polar Front Theory

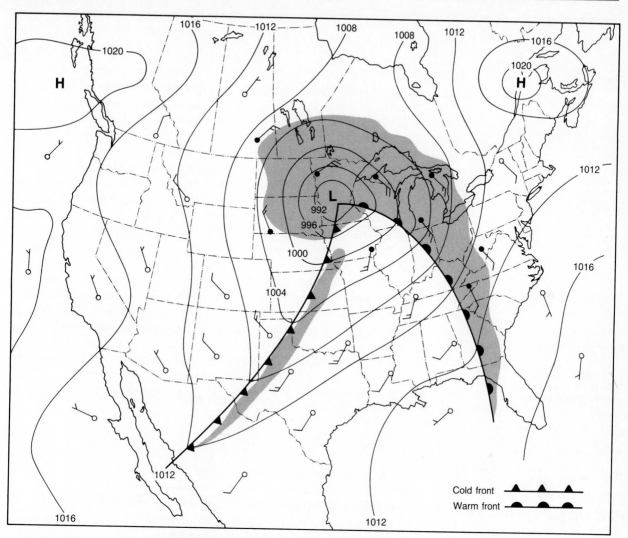

FIGURE 9–1
Simplified weather map showing the circulation of a middle-latitude or wave cyclone.
The dark areas indicate regions of probable precipitation.

polar air from warm subtropical air. During cool months the polar front is generally
well defined and forms an essentially continuous band around the earth that can
be recognized on upper-air charts. Near the surface, this frontal zone is often
broken into distinct segments separated by regions that experience a more gradual
temperature change. Other fronts capable of generating cyclones have also been
discovered. It was learned, for example, that fronts can form between continental
arctic and continental polar air masses as well as between continental polar and
maritime polar air masses. Because wave cyclones form in conjunction with fronts,
we first consider the nature of fronts and the weather associated with their movement and then apply what we have learned to the cyclone model.

FRONTS

Front

Fronts are defined as boundary surfaces separating air masses of different densities, one warmer and often higher in moisture content than the other. Thus, ideally, fronts can form between any two contrasting air masses. Considering the vast size of the air masses involved, these 15- to 200-kilometer-wide bands of discontinuity are relatively narrow. On the scale of a weather map, they are normally narrow enough to be satisfactorily represented by a broad line.

Above the ground, the frontal surface slopes at a low angle so that warmer air overlies cooler air as shown in Figure 9–2. In the ideal case, the air masses on both sides of the front move in the same direction and at the same speed. Under this condition the front acts as a barrier with which the air masses must move but cannot penetrate. Generally, however, the pressure field across a front is such that air on one side is moving faster in the direction perpendicular to the front than the air mass on the other side of the front. Thus one air mass actively advances into another and "clashes" with it. As a result, the Norwegian meteorologists visualized these zones of interaction as being analogous to battle lines and tagged them fronts. As one air mass moves into another, some mixing does occur along the frontal surface, but for the most part, the air masses retain their identity as one air mass is displaced upward over the other. No matter which air mass is advancing, it is always the warmer, lighter air that is forced aloft, whereas the cooler, heavier air acts as the wedge on which lifting takes

Overrunning

place. The term **overrunning** is generally applied to warm air gliding up a cold air mass.

Warm Fronts

Warm Front

When the surface position of a front moves so that warm air occupies territory formerly covered by cooler air, it is called a **warm front** (Figure 9–2). On a weather map the surface position of a warm front is denoted by a line with semicircles extending into the cooler air. East of the Rockies warm tropical air often enters the United States from the Gulf of Mexico and overruns receding cool air. As the cold wedge retreats, friction slows the advance of the surface position of the front more so than its position aloft so that the boundary separating these air masses acquires a very gradual slope. The average slope of a warm front is about 1:200, which means that if you went a distance of 200 kilometers ahead of the surface location of a warm front, you would find the frontal surface at a height of 1 kilometer overhead.

As warm air ascends the retreating wedge of cold air, it cools by adiabatic expansion to produce clouds and frequently precipitation. Typically the sequence of clouds shown in Figure 9–3(a) precedes a warm front. The first sign of the approach of a warm front is the appearance of cirrus clouds overhead. These high clouds form 1000 kilometers or more ahead of the surface front where the overrunning warm air has ascended high up the wedge of cold air. Another indication of the approach of a warm front is provided by aircraft contrails. On a clear

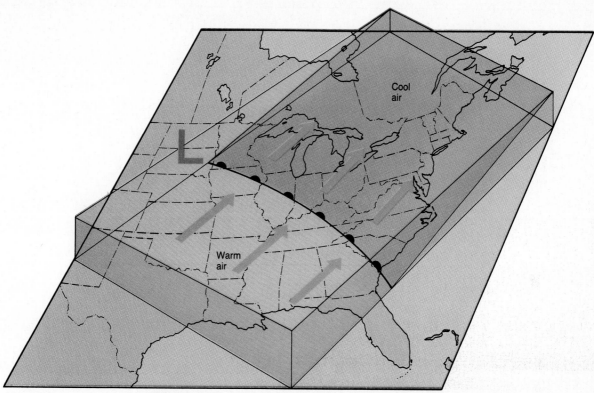

FIGURE 9–2
Warm front produced as warm air glides up over a cold air mass.

day, when these condensation trails persist for several hours, you can be fairly certain that relatively warm moist air is ascending overhead. As the front nears, cirrus clouds grade into cirrostratus that blend into denser sheets of altostratus. About 300 kilometers ahead of the front, thicker stratus and nimbostratus clouds appear and rain or snow commences. Because of their slow rate of advance and very gradual slope, warm fronts usually produce light-to-moderate precipitation over a large area for an extended period. Warm fronts, however, are occasionally associated with cumulonimbus clouds and thunderstorms (Figure 9–3b). Such a situation occurs when the overrunning air is inherently unstable and the front is rather sharp. When these conditions exist, cirrus clouds are generally followed by cirrocumulus clouds, giving us the familiar "mackerel sky" that warned sailors of an impending storm, as indicated by the following weather proverb:

Mackerel scales and mares' tails
Make lofty ships carry low sails.

At the other extreme, a warm front associated with a rather dry air mass could pass unnoticed by those of us at the surface.

As we can see from Figure 9–3, the precipitation associated with a warm front occurs ahead of the surface position of the front. Some of the rain that

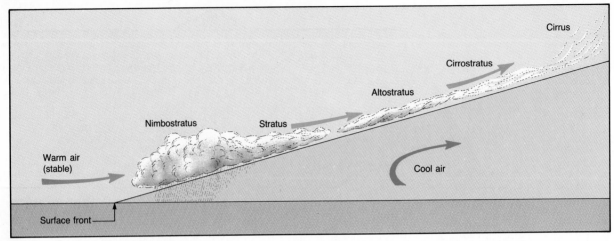

(a)

(b)

FIGURE 9–3
(a) Warm front with stable air and associated stratiform clouds. Precipitation is moderate and occurs within a few hundred kilometers of the surface front. (b) Warm front with unstable air and cumuliform clouds. Precipitation is heavy near the surface front.

falls through the cool air below the clouds evaporates. As a result, the air directly beneath the cloud base often becomes saturated and a stratus cloud deck develops. These clouds occasionally grow rapidly downward, which can cause problems for pilots of small aircraft that require visual landings. One minute pilots may encounter visibility that is adequate and the next find themselves in the middle of a cloud mass that has the landing strip "socked in." Occasionally during the winter a relatively warm air mass is forced over a body of subfreezing air. When this occurs it can lead to our most hazardous driving conditions. The icy roads are produced when raindrops become supercooled as they fall through the sub-freezing air. Upon colliding with the road surface they freeze to produce the icy layer called glaze.

With the passage of a warm front, a rather gradual increase in temperature occurs. As you would expect, the increase is most apparent when a large temperature contrast exists between adjacent air masses. Moreover, a wind shift from the east to the southwest is generally noticeable. The reason for this shift will be evident later. The moisture content and stability of the encroaching warm air mass largely determine the time period required for clear skies to return. During the summer cumulus and occasionally cumulonimbus clouds are embedded in the warm unstable air mass that follows the front. These clouds can produce precipitation, but it is usually randomly organized and restricted in extent.

Cold Fronts

Cold Front

When cold air actively advances into a region occupied by warmer air, the zone of discontinuity is called a **cold front** (Figure 9–4). As is the case with warm fronts, friction tends to slow the surface position of a cold front more so than its position aloft. Because of the relative positions of the adjacent air masses, however, the cold front steepens as it moves. On the average, cold fronts are about twice as steep as warm fronts, having a slope of perhaps 1:100. In addition, the average cold front advances at speeds of 35 kilometers per hour as compared with speeds

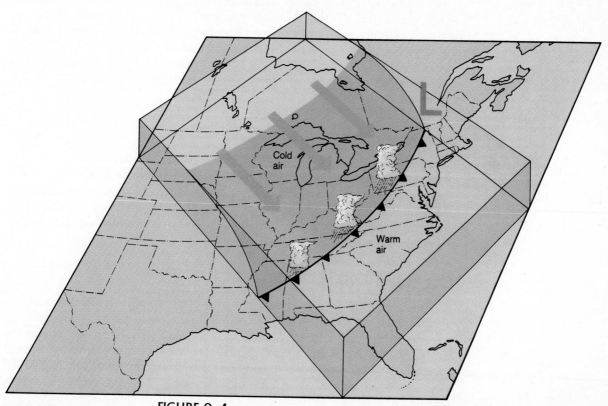

FIGURE 9–4
Fast-moving cold front and cumulonimbus clouds. Often thunderstorms occur if the warm air is unstable.

of 25 kilometers per hour for warm fronts. These two differences, rate of movement and steepness of slope, largely account for the more violent nature of cold-front weather compared with the weather generally accompanying a warm front. The displacement of air along a cold front is often rapid enough to permit the released latent heat to increase the air's buoyancy appreciably. The sudden downpours and vigorous gusts of wind associated with mature cumulonimbus clouds (Figure 9–4) frequently result. Because cold fronts produce roughly the same amount of lifting as a warm front but over a shorter distance, the intensity of precipitation is greater, but the duration is shorter.

The arrival of a cold front is sometimes preceded by altocumulus clouds. As the front approaches, generally from the west or northwest, towering clouds can often be seen in the distance. Near the front a dark band of ominous clouds foretells of the ensuing weather. A marked temperature drop and a wind shift from the south to west or northwest usually accompany frontal passage. The sometimes violent weather and the sharp temperature contrast along the cold front are indicated by the symbol used to depict them on a weather map. The symbol consists of a line with triangle-shaped points extending into the warmer air mass.

The weather behind a cold front is dominated by a subsiding and relatively cold air mass. Thus clearing conditions prevail after the front passes. Although general subsidence causes some adiabatic heating, it has a minor effect on surface temperatures. In winter the clear skies associated with these cold outbreaks further reduce surface temperatures because of more rapid radiation cooling at night. If the continental polar air mass, which most frequently accompanies a cold front, moves into a relatively warm and humid area, surface heating can produce shallow convection, which, in turn, may generate low cumulus or stratocumulus clouds behind the front.

Stationary Fronts

Stationary Front

Occasionally the flow on both sides of a front is almost parallel to the position of the front. The surface position of the front does not move and it is so named a **stationary front.** On a weather map stationary fronts are shown with triangular points on one side of the front and semicircles on the other (see Figure 8–2). At times some overrunning occurs along a stationary front; then precipitation of the gentle warm-front type is most likely.

Occluded Fronts

Occluded Front

Another common front is the **occluded front.** Here an active cold front overtakes a warm front, as shown in Figure 9–5. As the advancing cold air wedges the warm front upward, a new front emerges between advancing cold air and the air over which the warm front is gliding. The weather of an occluded front is generally complex. Most precipitation is associated with the warm air being forced aloft, but when conditions are suitable, the newly formed front is capable of initiating precipitation of its own.

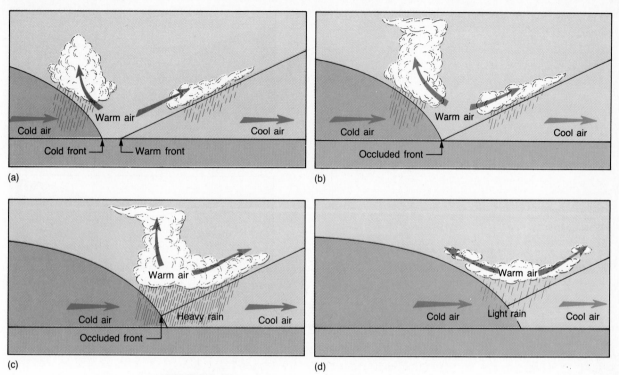

FIGURE 9–5
Stages in the formation and eventual dissipation of an occluded front.

Cold-Type Occluded Front

Warm-Type Occluded Front

In the occluded front shown in Figure 9–6(a) the air behind the cold front is colder than the air underlying the warm front it is overtaking. It is the most common occluded front to form east of the Rockies and is termed a **cold-type occluded front.** It is also possible for the air behind the active cold front to be warmer than the air underlying the warm front. These **warm-type occluded fronts** frequently occur along the Pacific Coast, where somewhat milder maritime polar air invades polar air that had its origin over the continent. Notice in Figure 9–6(a) and (b) that in the warm-type occluded front the warm front aloft, and hence the precipitation, often precedes the arrival of the surface front but that the situation is reversed for the cold-typed occluded front. It should also be noted that cold-type occluded fronts may frequently resemble cold fronts in terms of the type of weather generated.

A word of caution is in order concerning the weather associated with various fronts. Although the preceding discussion should be an aid in recognizing the weather patterns associated with fronts, remember that these descriptions are rather idealized. The weather generated along any individual front may or may not conform fully to our picture. Fronts, like all aspects of nature, never lend themselves to classification as nicely as we would like.

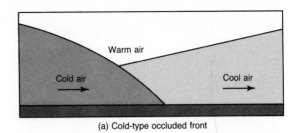

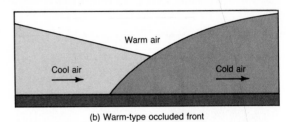

FIGURE 9–6
Occluded fronts of the (a) cold type and (b) warm type.

WAVE CYCLONE

The Norwegian model of the middle-latitude or wave cyclone was created primarily from near-surface observations. As data from the middle and upper troposphere and from satellite images became available, some modifications were necessary. Yet this model is still an accepted working tool in interpreting the weather. It provides a visual picture of the dynamic atmosphere as it generates a storm. If you keep this model in mind when you consider changes in the weather, the observed changes will no longer come as a surprise. You should begin to see some order in what had appeared to be disorder, and you might even occasionally "predict" the impending weather.

Life Cycle of a Wave Cyclone

According to the wave cyclone model, cyclones form along fronts and proceed through a somewhat predictable life cycle. This cycle can last for a few hours or for several days, depending on whether conditions for development are favorable. Figure 9–7 is a schematic representation of the stages in the development of a "typical" wave cyclone. As the figure shows, cyclones originate along a front where air masses of different densities (temperatures) are moving parallel to the front in opposite directions. In the classic model this would be continental polar air associated with the polar easterlies north of the front and maritime tropical air of the westerlies south of the front. The result of this opposing airflow is the development of cyclonic shear, which produces a net counterclockwise rotation. To better visualize this effect, place a pencil between the palms of your hands.

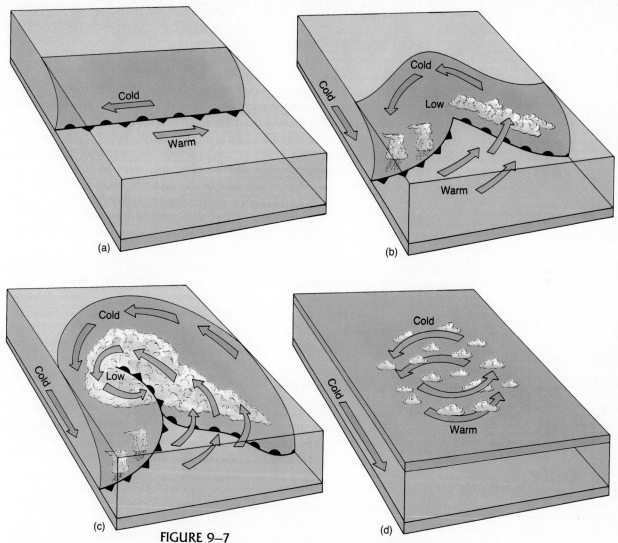

FIGURE 9–7
Stages in the life cycle of a middle-latitude cyclone as proposed by J. Bjerknes.

Now move your right hand ahead of your left hand and notice that your pencil rotates in a counterclockwise fashion. It is possible for cyclonic flow to develop in other ways that may also initiate a wave cyclone. In any event, under the correct conditions the frontal surface will take on a wave shape. These waves are analogous to the waves produced on the surface of a water body by moving air except that the scale is different. The waves generated between two contrasting air masses are usually several hundred kilometers long. Some waves tend to dampen out whereas others become unstable and grow in amplitude. The latter change in shape with time much like a gentle ocean swell does as it moves into shallow water and becomes a tall breaking wave (see Figure 9–7).

Once a small wave forms, warm air invades this weak spot along the front and extends itself poleward and the surrounding cold air moves equatorward. This change causes a readjustment in the pressure field that results in almost circular isobars, with the low pressure centered at the apex of the wave. The resulting cyclonic circulation is shown in Figure 9–1. Once the cyclonic circulation develops, we would expect general convergence to result in vertical lifting, especially where warm air is overrunning colder air. We can see in Figure 9–1 that the air in the warm sector is flowing from the southwest toward colder air flowing from the southeast. Because the warm air is moving faster than the cold air in a direction perpendicular to this front, we can conclude that warm air is invading a region formerly occupied by cold air; therefore this must be a warm front. Similar reasoning indicates that in the rear of the cyclonic disturbance cold air is underrunning the air of the warm sector, generating a cold front there. Generally the position of the cold front advances faster than the warm front and begins to close the warm sector, as shown in Figure 9–7. This process, called **occlusion,** results in the formation of an occluded front with the displaced warm sector located aloft. The cyclone enters maturity (maximum intensity) when it reaches this stage in its development. A steep pressure gradient and strong winds develop as lifting continues. Eventually all the warm sector is forced aloft and cold air surrounds the cyclone at low levels. Once the sloping discontinuity (front) between the air masses no longer exists, the pressure gradient weakens. At this point the cyclone has exhausted its source of energy and the storm comes to an end.

Occlusion

Idealized Weather of a Wave Cyclone

As stated earlier, the cyclone model provides a useful tool for examining the weather patterns of the middle latitudes. Figure 9–8 illustrates the distribution of clouds and thus the regions of possible precipitation associated with a mature wave cyclone. Compare this drawing to the satellite image of a cyclone shown in Figure 9–9.

Guided by the westerlies aloft, cyclones generally move eastward across the United States, so that we can expect the first signs of their arrival in the west. Frequently in the region of the Mississippi valley, however, cyclones begin a more northeasterly trajectory and occasionally move directly northward. Typically a midlatitude cyclone requires 2 to 4 days to pass over a given region. During that relatively short time period rather abrupt changes in atmospheric conditions can be experienced, particularly in the spring of the year when the largest temperature contrasts occur across the midlatitudes.

Using Figure 9–8 as a guide, let us consider these weather producers and the conditions we should expect as they pass an area in the spring of the year. To facilitate our discussion, profiles are provided along lines A–E and F–G. Imagine the change in weather as you move from right to left along profile A–E. At point A the sighting of high cirrus clouds would be the first sign of the approaching cyclone. These high clouds can precede the surface front by 1000 kilometers or more, and they are normally accompanied by falling pressure. As the warm front advances, a lowering and a thickening of the cloud deck are noticed. Within 12

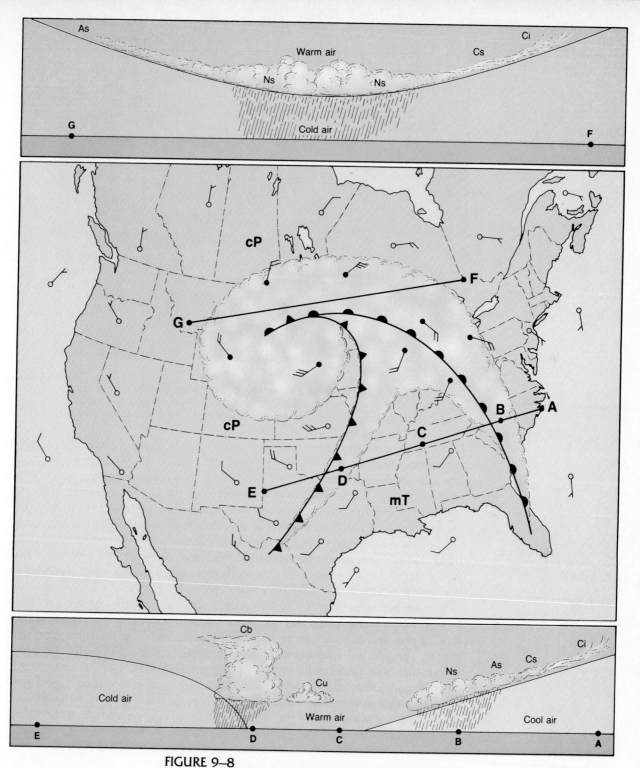

FIGURE 9–8
Cloud patterns usually associated with a mature middle-latitude cyclone. Upper and lower sections are vertical cross sections along F–G and A–E, respectively; the middle section represents a map view.

FIGURE 9–9
Satellite view of a mature cyclone situated over the eastern one-third of the United
States. It is easy to see why we often refer to the cloud pattern of a cyclone as
having a "toboggan" shape. Photograph taken on June 12, 1975. (Courtesy of NOAA)

to 24 hours after the first sighting of cirrus clouds, light precipitation usually
commences (point *B*). As the front nears, the rate of precipitation increases, a
rise in temperature is noticed, and winds begin to change from an easterly to a
southerly flow. With the passage of the warm front, the area is under the influence
of the maritime tropical air mass of the warm sector (point *C*). Generally the
region affected by this sector of the cyclone experiences warm temperatures,
southerly winds, and usually clear skies, although fair-weather cumulus, or altocumulus are not uncommon here. The rather pleasant weather of the warm sector
passes quickly in the spring of the year and is replaced by gusty winds and precipitation generated along the cold front. The approach of a rapidly advancing cold
front is marked by a wall of rolling black clouds (point *D*). Severe weather accompanied by heavy precipitation and occasionally hail or a tornado is common at this
time of year. In addition, intense squall-line thunderstorms of short duration frequently precede the cold front. On many occasions, the squall activity is more
severe than that associated with the cold front itself. The passage of the cold
front is easily detected by a wind shift, the southerly flow is replaced by winds
from the west to northwest, and there is a pronounced drop in temperature.
Also, rising pressure hints of the subsiding cool, dry air behind the front. Once
the front passes, the skies clear quickly as the cooler air invades the region (point
E). A day or two of almost cloudless deep blue skies are often experienced unless
another cyclone is edging into the region.

A very different set of weather conditions prevails in regions that encounter
the portion of the cyclone that contains the occluded front as shown along profile

F–G. Here temperatures remain cool during the passage of the storm; however, a continual drop in pressure and increasingly overcast conditions strongly hint at the approach of the low-pressure center. This sector of the cyclone most often generates snow or icing storms during the cool months. Moreover, the occluded front often moves more slowly than the other fronts; thus the wishbone-shaped frontal structure shown in Figure 9–8 rotates in a counterclockwise manner so that the occluded front appears to "bend over backward." This effect adds to the misery of the region influenced by the occluded front, for it remains over the area longer than the other fronts. Also, the storm reaches its greatest intensity during occlusion; consequently, the area affected by the developing occluded front can expect to receive the brunt of the storm's fury.

A few observations on the foregoing discussion can serve to illustrate the value of knowing the cyclone model. In particular, shifts in wind direction are useful in predicting the impending weather (see Figure 9–8). Notice that, with the passage of both the warm and cold fronts, the wind arrows changed position in a clockwise direction; for example, with the passage of the warm front, the wind changed from easterly to southerly. From nautical terminology the word **veering** is applied to a wind shift in a clockwise manner. Because clearing conditions occur with the passage of both fronts, veering winds are indicators that the weather will improve. On the other hand, the area located in the northern portion of the cyclone will experience winds that shift in a counterclockwise direction, as can be seen in Figure 9–8. Winds that shift direction in a counterclockwise direction are said to be backing. With the approach of a wave cyclone, **backing** winds indicate cool temperatures and continued foul weather.

Veering Wind Shift

Backing Wind Shift

The preceding discussion provides only a brief sketch of the weather normally experienced with the passage of a wave cyclone. After we examine the daily weather map in Chapter 11, we will use this aid to look more closely at cyclonic weather as we consider a "real" cyclone.

CYCLOGENESIS

Cyclogenesis

We saw in the Norwegian model that **cyclogenesis** (cyclone formation) occurs where a frontal surface is distorted into a wave-shaped discontinuity. Several surface factors are thought to produce a wave in a frontal zone. Topographic irregularities, such as mountains, temperature contrasts as between sea and land, or ocean current influences can disrupt the general zonal flow sufficiently to produce a wave along a front. It is also believed that the disturbance of one storm center may initiate cyclogenesis elsewhere. But more often than not, the initiation of a cyclonic system can be attributed to the flow aloft in the vicinity of the polar jet stream.

When the earliest studies of cyclones were made, little data were available on the nature of the airflow in the middle and upper troposphere. Since then a close relationship between surface disturbances and the flow aloft has been established. Whenever the flow aloft is relatively straight—that is, from west to east—

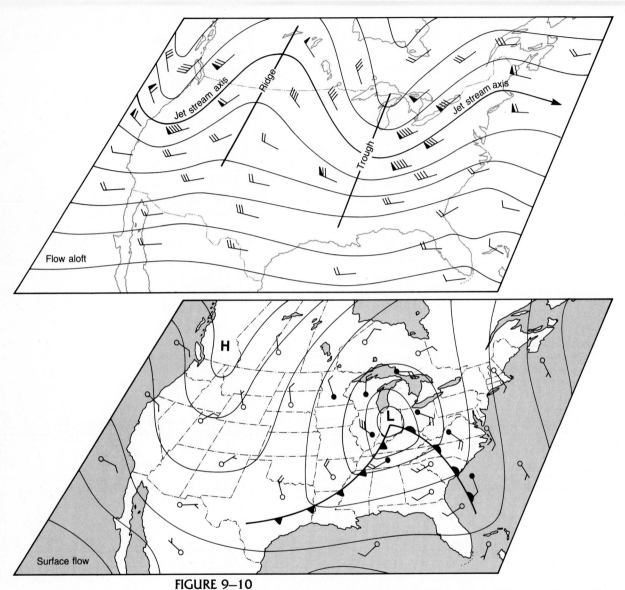

FIGURE 9–10
Drawing showing the relative positions of the wavy flow pattern aloft and surface cyclones and anticyclones.

little cyclonic activity occurs at the surface. However, when the upper air begins to meander widely in a north-to-south manner, thereby producing high-amplitude waves consisting of alternating troughs and ridges, surface cyclonic activity intensifies. Also, when surface cyclones form, almost invariably they are centered directly below the jet stream axis and slightly downstream from an upper-level trough (Figure 9–10).

Before discussing how cyclones are generated by the flow aloft, let us review the nature of cyclonic and anticyclonic winds. Recall that airflow about a surface low is inward, a fact that leads to mass convergence (coming together). The resulting accumulation of air must be accompanied by a corresponding increase in surface pressure. Consequently, we might expect a surface low-pressure system to "fill" rapidly and be eliminated, just as the vacuum in a coffee can is quickly equalized when we open it. This process, however, does not occur. On the contrary, cyclones often exist for a week or longer. In order for this to occur, surface convergence must be offset by a mass outflow at some level aloft (Figure 9–11). As long as divergence (spreading out) aloft is equal to, or greater than, the surface inflow, the low pressure, with its accompanying convergence, can be sustained.

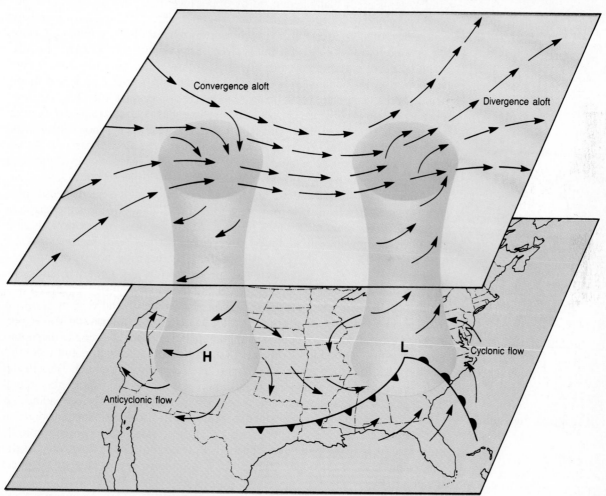

FIGURE 9–11
Idealized diagram depicting the support that divergence and convergence aloft provide to cyclonic and anticyclonic circulation at the surface.

Because cyclones are bearers of stormy weather, they have received far more attention than their counterparts, anticyclones. Yet a close relationship exists that makes it difficult to separate any discussion of these two pressure systems. The surface air that feeds a cyclone, for example, generally originates as air flowing out of an anticyclone. Consequently, cyclones and anticyclones are typically found adjacent to one another (Figure 9–10). Like the cyclone, an anticyclone depends on the flow aloft to maintain its circulation. In this instance, divergence at the surface is balanced by convergence aloft and general subsidence of the air column (Figure 9–11).

We have seen that airflow aloft plays an important role in maintaining cyclonic and anticyclonic circulation. In fact, more often than not, these rotating surface wind systems are actually generated by upper-level flow. In order for a middle-latitude cyclone to form, two important conditions must be met. First, a cyclonic airflow pattern must be established. In the northern hemisphere cyclonic circulation has a counterclockwise rotation and is directed toward the center of low pressure. Second, and of equal importance, the inward flow of air near the surface must be supported by outflow aloft. It is the role of upper-level divergence in the vicinity of the jet stream that is believed to be most important in cyclone development. Upper-air divergence creates an environment analogous to a partial vacuum, which initiates upward flow. The fall in surface pressure that accompanies the outflow aloft will induce inward flow at the surface. The Coriolis effect will then come into play to produce the curved flow pattern associated with cyclonic flow.

Divergence aloft does not involve the outward clockwise movement of air, as occurs about a surface anticyclone. Instead the airflow aloft is nearly geostrophic (see Figure 9–10). Its path is from west to east and generally takes the shape of sweeping curves. One mechanism responsible for the mass transport of air aloft is a phenomenon known as **speed divergence.** It has been known for some time that wind speeds along the axis of the jet stream are not constant. Some regions experience much higher wind speeds than others. On entering a zone of maximum wind velocity, air accelerates and therefore experiences divergence. On the other hand, when air leaves a zone of maximum wind velocity, a pileup (convergence) results. Analogous situations occur on a tollway, in the region between toll stations. On exiting one toll booth and entering the zone of maximum speed, we find automobiles diverging (spreading out). As automobiles slow to pay the next toll, they experience convergence (coming together). Thus in the zone upstream from a region of maximum speed, the air experiences speed divergence, whereas downstream mass convergence occurs. The spreading out of the airstream, called **directional divergence,** can also contribute to divergence aloft, just as convergence of an airstream can cause mass convergence. Typically the region that favors directional divergence is located slightly downwind of the axis of an upper-air trough. Conversely, convergence occurs downwind of an upper-air ridge.

Upper-level airflow is also important in developing cyclonic and anticyclonic **vorticity,** which is the tendency of the air to rotate in either a cyclonic or anticyclonic manner. Recall that the polar jet stream generally flows from west to east and frequently develops large meanders. To visualize how this wavy flow aloft

Speed Divergence

Directional Divergence

Vorticity

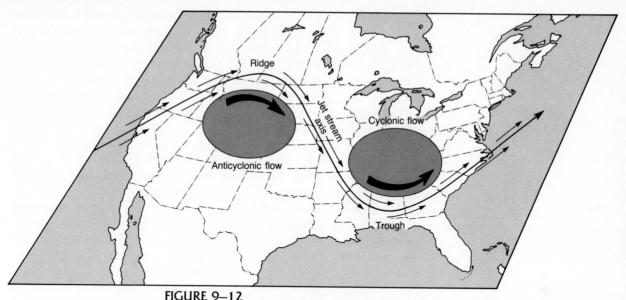

FIGURE 9–12
Vorticity provided by flow in the jet stream generates cyclonic flow near a trough and anticyclonic flow adjacent to a ridge.

causes rotation of an air mass, examine Figure 9–12. Notice that an air mass located south of the jet stream near a ridge of high pressure acquires an anticyclonic rotation, whereas an air mass located north of the jet stream adjacent to a trough of low pressure acquires a cyclonic rotation. In this manner, cyclonic vorticity provided by a trough in the jet stream is a major contributor to the strong cyclonic circulation associated with middle-latitude storms. In addition, recall that the region aloft that favors directional divergence is located slightly ahead of the axis of an upper-air trough. Consequently, surface cyclones generally form directly below a trough in the polar jet stream and continue to develop downstream from these upper-level waves that support their growth (Figure 9–10). On the other hand, the zone in the jet stream that generally experiences directional convergence and anticyclonic rotation is located near a ridge. The piling up of air in this region of the jet stream leads to subsidence and increased surface pressure. Hence this is a favorable site for the development of a surface anticyclone. Because of the significant role that the upper-level flow has on cyclogenesis, it should be evident that any attempt at weather prediction must strongly consider the airflow pattern aloft.

TRAVELING CYCLONES AND ANTICYCLONES

Not only are waves in the westerlies an important cause of cyclonic development, but the flow aloft is also important in determining how rapidly these pressure systems advance and the direction they will follow. Compared with the flow aloft

(500-millibar level), cyclones generally travel at somewhat less than half the speed. Normally they advance at a rate of from 20 to 50 kilometers per hour so that distances of roughly 240 to 1200 kilometers are traversed each day. The faster speeds correspond with the colder months when temperature gradients are greatest.

One of the most exacting tasks in weather forecasting is predicting the path of cyclonic storms. As stated earlier, the flow aloft tends to steer developing pressure systems. Let us examine an example of this steering effect by noting how changes in the upper flow correspond to changes in the direction of the path taken by a cyclone. Figure 9–13 illustrates the changing position of a wave cyclone over a period of a few days. Notice that on March 21 the 500-millibar contours are relatively flat and that for the following 2 days the cyclone moves in a rather straight easterly direction. By March 23 the 500-millibar contours make a sharp bend northward on the eastern side of a trough situated over Wyoming. Notice that on the next day the path of the cyclone makes a similar northward migration. Although an admittedly oversimplified example, it does illustrate the steering effect of upper-level flow. It should be remembered that we examined the effect of upper airflow on cyclonic movement after the fact. In order to make useful predictions of future positions of cyclones, an accurate appraisal of changes in the westerly flow aloft is required. For this reason, predicting the behavior of the wavy flow aloft forms an important part of modern weather forecasting.

As evidenced by the preceding example, wave cyclones have a tendency to migrate first in an easterly direction and then travel a more northeastward path (on some occasions, developing storms move southeastward prior to migrating northward). The northeastward trajectory of cyclones frequently causes them to merge with one of the two semipermanent low-pressure systems (the Aleutian and Icelandic lows) that are centered at roughly 60°N latitude. It is probably more accurate to state that the great number of traveling cyclones that merge in these regions show up in the mean picture as semipermanent low-pressure centers.

The cyclones that reach the West Coast originate in the Pacific along the polar front. These storms usually occur as members of a "family" of cyclones in various stages of development, as shown in Figure 9–14. Most of these systems move northeastward toward the Gulf of Alaska, where they merge with the Aleutian low. During the winter months these storms develop farther south in the Pacific and often reach the coast of the contiguous 48 states, occasionally moving as far south as southern California. These systems provide the rather short winter rainy season that affects California and the western states in general. Most Pacific storms do not cross the Rockies intact, but many redevelop on the lee side of these mountains. A favorite site for redevelopment is Colorado, but other common sites of formation exist as far south as Texas and as far north as Alberta. The cyclones that form in Canada tend to move southward toward the Great Lakes and then turn northeastward and move out into the Atlantic. Cyclones that redevelop over the Plains states generally migrate eastward until they reach the central United States, where a northeastward or even northward trajectory is followed. Most of these storms also traverse the Great Lakes region, making this the most storm-ridden region in the country. Not all cyclones that affect the United States originate

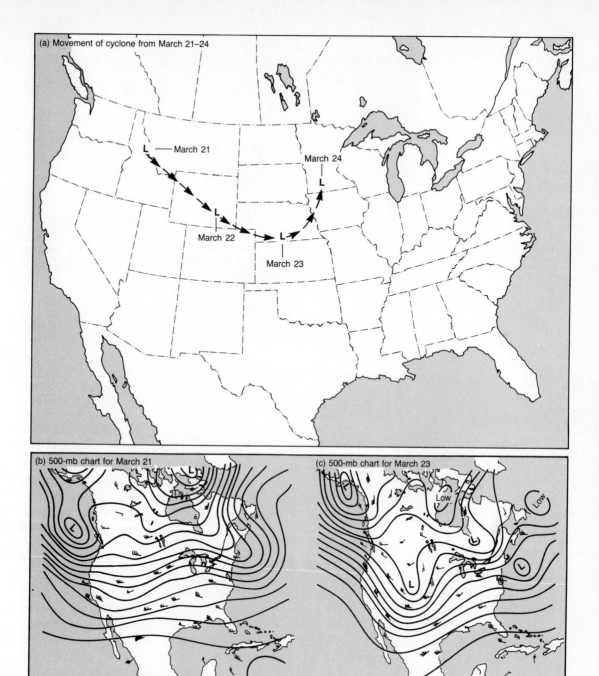

FIGURE 9–13

Steering of cyclonic storms. (a) Notice that the cyclone (low) moved almost in a straight southeastward direction from March 21 to March 23. On the morning of March 23 it abruptly turned northward. This change in direction corresponds to the change from rather flat contours (b) on the upper-air chart for March 21 to curved contours (c) on the chart for March 23.

FIGURE 9–14
Two well-developed cyclones are apparent in this satellite image from their characteristic toboggan-shaped cloud patterns. Located in the area of only scattered clouds between this so-called family of cyclones is an anticyclone. (Courtesy of NOAA)

in the Pacific. Some form over the Plains states and are associated with an influx of maritime tropical air from the Gulf of Mexico. Another area where cyclogenesis occurs is east of the southern Appalachians. These storms tend to move northward with the Gulf Stream and eventually merge with the Icelandic low.

Although the paths of cyclones tend to be northeastward, anticyclones more frequently move southeastward. However, some anticyclones are embedded between members of a family of cyclones and thus travel northeastward with the group. After the last member of a family of cyclones passes through the United States, the cold anticyclone situated behind the cold front breaks out and moves southward. This process produces the occasional cold waves that are experienced deep into the southern states during the winter. These anticyclones normally continue to move toward the subtropical high-pressure center situated over the Atlantic. Because anticyclones are associated with clear conditions and so are "weatherless," much less study has been conducted on their development and movement. However, because anticyclones occasionally become stagnant and remain over a region for several days, they are becoming ever more important to people trying to cope with air pollution. Recall that it is the stable and rather calm conditions associated with high pressure that contribute to pollution episodes. Large stagnant anticyclones are also important because they often block the eastward migration of cyclonic centers. This effect can keep one section of the country dry for a week or longer while another area is continually under the influence of a cyclonic storm.

THERMAL LOWS

Not all low-pressure systems move and not all of them are associated with fronts. As indicated earlier, the name cyclone merely implies a low-pressure system that may or may not produce severe weather. Two large-scale lows that exist without fronts are tropical cyclones (hurricanes) and thermal lows. Tropical cyclones are discussed in Chapter 10.

Thermal Low

Thermal lows are found in two major regions of the world: in the area near the Persian Gulf and in the American Southwest. Both regions are generally dominated by the circulation of the semipermanent subtropical highs that were considered earlier in the discussion of global circulation. The general subsidence that prevails in these systems provides the clear skies that, in turn, allow the intense surface heating associated with these areas around the globe. During the hottest summer months, however, these regions develop lows that dominate the surface circulation. These lows form because the intense surface heating causes expansion of the overlying air column; hence the name thermal lows. This effect generates a general outflow aloft that encourages surface inflow. The formation of this circulation is similar to the development of the sea breeze considered earlier. In the United States a thermal low center develops each summer over southwestern Arizona. It is believed that this low contributes to a low-level influx of relatively moist air from the Gulf of California during the hottest months of the year. It would then account for the general increase in precipitation that this region experiences in the hot summer months compared with the cooler spring season. For example, Phoenix, Arizona, receives only meager precipitation all year long, but on the average, it receives ten times more in July than it does in May.

REVIEW

1. If you were located 400 kilometers ahead of the surface position of a warm front, how high would the frontal surface be above you?

2. Why is cold-front weather usually more severe than warm-front weather?

3. Explain the basis for the following weather proverb:

   Rain long foretold, long last;
   Short notice, soon past.

4. Distinguish between cold-type and warm-type occluded fronts.

5. Although the formation of an occluded front represents the period of maximum intensity for a wave cyclone, it also marks the beginning of the end of the system. Explain why such is the case.

6. For each of the weather elements listed here, describe the changes that an observer experiences when a wave cyclone passes with its center north of the observer.
   - **a.** wind direction
   - **b.** pressure tendency
   - **c.** cloud type
   - **d.** cloud cover
   - **e.** precipitation
   - **f.** temperature

7. Describe the weather conditions that an observer would experience if the center of a wave cyclone passed to the south.

8. Distinguish between veering and backing winds.

9. Briefly explain how the flow aloft initiates and maintains cyclones at the surface.

10. Explain why predicting the behavior of upper-level flow is an important task in modern weather forecasting.

11. Why does a city like Phoenix, Arizona, experience a precipitation maximum in midsummer?

VOCABULARY REVIEW

middle-latitude cyclone

wave cyclone

polar front theory

front

overrunning

warm front

cold front

stationary front

occluded front

cold-type occluded front

warm-type occluded front

occlusion

veering wind shift

backing wind shift

cyclogenesis

speed divergence

directional divergence

vorticity

thermal low

10

Severe Weather

Occurrences of severe weather have a fascination that ordinary weather phenomena cannot provide. The lightning display generated by a severe thunderstorm can be a spectacular event that elicits both awe and fear (Figure 10–1, see page 273). Of course, hurricanes and tornadoes also attract a great deal of much-deserved attention. A single tornado outbreak or hurricane can cause billions of dollars in property damages as well as many deaths. Table 10–1 presents a 26-year record of fatalities in the United States caused by these important meteorological hazards. During this span, an estimated 5658 people lost their lives to lightning, tornadoes, and hurricanes. It may surprise many that the number of deaths attributed to lightning was greater than those caused by either tornadoes or hurricanes. However, this may not necessarily be the case in the future. As we shall learn later in this chapter, many of our coastal areas may be vulnerable to a hurricane disaster.

FIGURE 10–1
A time exposure of cloud-to-ground lightning associated with an intense thunderstorm. (Photo by D. Baumhefner, National Center for Atmospheric Research/National Science Foundation)

WHAT'S IN A NAME?

In Chapter 9 we examined the middle-latitude cyclones that play such an important role in causing day-to-day weather changes. Yet the use of the term "cyclone" is often confusing. To many people, the term implies only an intense storm, such as a hurricane or a tornado. When a hurricane unleashes its fury on India or Bangladesh, for example, it is usually reported in the media as a cyclone (the local term denoting a hurricane in that part of the world). Similarly, tornadoes are occasionally referred to as cyclones. This custom is particularly common in portions of the Great Plains of the United States. Recall that in the *Wizard of Oz* Dorothy's house was carried from her Kansas farm to the land of Oz by a cyclone. Indeed, the nickname for the athletic teams at Iowa State University is the *Cyclones*. Although hurricanes and tornadoes are, in fact, cyclones, the vast majority of cyclones are not hurricanes or tornadoes. The term "cyclone" simply refers to the circulation around any low-pressure center, no matter how large or intense it is.

Tornadoes and hurricanes are both smaller and more violent than middle-latitude cyclones. Whereas middle-latitude cyclones may have a diameter of 1600 kilometers or more, hurricanes average only 600 kilometers across, and tornadoes, with a diameter of just ¼ kilometer, are much too small to show up on a weather map.

The thunderstorm, a much more familiar weather event, hardly needs to be distinguished from tornadoes, hurricanes, and midlatitude cyclones. Unlike the flow of air about these latter storms, the circulation associated with thunderstorms is characterized by strong up-and-down movements. Winds in the vicinity of a thunderstorm do not follow the inward spiral of a cyclone, but they are typically variable and gusty.

Although thunderstorms form "on their own" away from cyclonic storms, they also form in conjunction with cyclones. For instance, thunderstorms are frequently spawned along the cold front of a midlatitude cyclone, where on rare occasions a tornado may descend from the thunderstorm's cumulonimbus tower. Hurricanes also generate widespread thunderstorm activity. Thus thunderstorms are related in some manner to all three types of cyclones mentioned here.

THUNDERSTORMS

Almost everyone has observed a small-scale phenomenon that is a result of the vertical movements of relatively warm, unstable air. Perhaps you have seen a dust devil form over an open field on a hot day and whirl its dusty load to great heights or seen a bird glide effortlessly skyward on an invisible thermal of hot air. These examples illustrate the dynamic thermal instability that occurs during the development of a **thunderstorm.** At any given time over the earth's surface,

Thunderstorm

nearly 2000 thunderstorms are in progress, mostly in tropical regions. Thunderstorm activity is associated with cumulonimbus clouds that generate heavy rainfall, thunder, lightning, and occasionally hail.

Stages in the Development of a Thunderstorm

All thunderstorms require warm, moist air, which, when lifted, will release sufficient latent heat to provide the buoyancy necessary to maintain its upward flight. Although this instability and associated buoyancy are triggered by a number of different processes, all thunderstorms have a similar life cycle.

Because instability is enhanced by high surface temperatures, thunderstorms are most common in the afternoon and early evening. Surface heating is generally not sufficient in itself, however, to cause the growth of towering cumulonimbus clouds. A solitary cell of rising hot air produced by surface heating alone could, at best, produce a small cumulus cloud. Mixing between the moist air of the infant cloud and the cool, dry air aloft causes evaporation that dissipates the cloud in 10 to 15 minutes. The development of a 12-kilometer or, on rare occasions, a 20-kilometer cumulonimbus tower requires a continuous supply of moist air. Each new surge of warm air rises higher than the last, adding to the height of the cloud (Figure 10–2). This phase in the development of a thunderstorm, called the **cumulus stage,** is dominated by updrafts (Figure 10–3). These updrafts must occasionally reach speeds of 160 kilometers per hour to accommodate the large hailstones they are capable of carrying upward.

Cumulus Stage

Once the cloud passes beyond the freezing level, the Bergeron process begins producing precipitation. Usually within an hour of its inception, the accumulation

FIGURE 10–2
Cumulonimbus clouds in various stages of development. (Courtesy of R. R. Braham, Cloud Physics Laboratory, The University of Chicago)

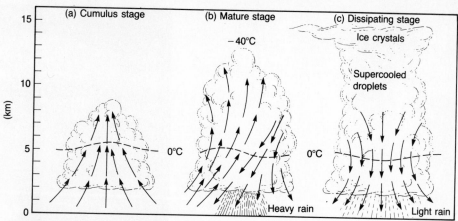

FIGURE 10–3

Stages in the development of a thunderstorm. During the cumulus stage strong updrafts act to build the storm. The mature stage is marked by heavy precipitation and cool downdrafts in part of the storm. When the warm updrafts disappear completely, precipitation becomes light and the cloud begins to evaporate.

Entrainment

of precipitation in the cloud is too great for the updrafts to support. The falling precipitation causes drag on the air and initiates a downdraft. The creation of the downdraft is further aided by the influx of cool, dry air surrounding the cloud, a process termed **entrainment.** This process intensifies the downdraft because the air added during entrainment is cool and therefore heavy; possibly of greater importance, it is dry. It thus causes some of the falling precipitation to evaporate (a cooling process), thereby cooling the air within the downdraft. As the resulting downdraft leaves the bottom of the cloud, precipitation is released, marking the beginning of the cloud's **mature stage.** At the surface the cool downdraft spreads laterally and can be felt before the actual precipitation reaches the ground. The sharp cool gusts at the surface are indicative of the downdrafts aloft. During the mature stage updrafts exist side by side with downdrafts and continue to enlarge the cloud. When the cloud grows to the top of the unstable region often located at the base of the warmer stratosphere, the updrafts spread laterally and produce the characteristic anvil top (Figure 10–4). Generally ice-laden cirrus clouds make up the top and are spread downwind by rapid winds aloft. The mature stage is the most active period of a thunderstorm. Gusty winds, lightning, heavy precipitation, and sometimes hail are experienced.

Dissipating Stage

Once a downdraft begins, the vacating air encourages more entrainment of the cool, dry air surrounding the cell. Eventually downdrafts dominate throughout the cloud and initiate the **dissipating stage** (Figure 10–3). The cooling effect of falling precipitation and the influx of colder air aloft mark the end of the thunderstorm activity. Without a supply of moisture, the cloud will soon evaporate. The life span of a single cumulonimbus cell within a thunderstorm complex is only an hour or two, but as the storm moves, fresh supplies of warm, water-laden air generate new cells to replace those that are dissipating.

Mature Stage

FIGURE 10–4
Satellite view near the Florida coast of several well-
developed cumulonimbus clouds exhibiting anvil tops.
(Courtesy of NASA)

To summarize, the stages in the development of a thunderstorm are as follows:

1. The cumulus stage in which updrafts dominate throughout the cloud and growth from a cumulus to a cumulonimbus cloud occurs
2. The mature stage characterized by violent weather when downdrafts are found side by side with updrafts
3. The dissipating stage dominated by downdrafts and entrainment causing evaporation of the structure

Thunderstorm Formation

The greatest number of thunderstorms occur in association with relatively short-lived cumulonimbus clouds that produce local precipitation. Occasionally, however, thunderstorms grow much larger and remain active for a few hours. By definition, these **severe thunderstorms** produce frequent lightning accompanied by locally damaging winds or hail that is 2 centimeters or more in diameter. Because of some differences in the circumstances of their formation, we can divide thunderstorms into two types: (1) isolated thunderstorms produced within a warm, humid air mass and (2) severe thunderstorms produced by forceful lifting along a cold front.

Severe Thunderstorm

Air-Mass Thunderstorms

Air-Mass Thunderstorm

In the United States **air-mass thunderstorms** generally occur in warm, moist (maritime tropical) air that originates over the Gulf of Mexico and moves northward.

Recall that air of this type contains most of its moisture in the lower portion and can be rendered unstable when lifted. Because this air becomes unstable most often in the spring and summer when it is warmed sufficiently from below, air-mass thunderstorms are most frequent then. They also have a strong preference for midafternoon when the surface temperatures are highest. Some do occur after sunset when the growth of immature cells becomes restimulated by cloud-top cooling. Because local differences in surface heating aid in the growth of air-mass thunderstorms, they generally occur as scattered, isolated cells instead of being organized in narrow bands like frontal thunderstorms.

Mountainous regions, such as the Rockies in the West and the Appalachians in the East, experience a greater number of air-mass thunderstorms than do the Plains states. The air near the mountain slope is heated more intensely than air at the same elevation over the adjacent lowlands. A general upslope movement then develops during the daytime and generates thunderstorm cells. These cells remain almost stationary above their source area, the slopes below.

Although the growth of thunderstorms is aided by high surface temperatures, most air-mass thunderstorms are not generated solely by surface heating, mountainous areas being the possible exception. Recall that the thunderstorms that form over the peninsula of Florida are believed to be the result of converging winds from the ocean on both sides. Similarly, the air-mass thunderstorms that occur over much of the eastern two-thirds of the United States occur as part of the general convergence associated with a passing midlatitude cyclone. Near the equator, they form in association with the convergence along the equatorial low.

A few places in the United States—western Kansas and Nebraska in particular—experience a thunderstorm maximum at night, perhaps because of general cooling and downflow in the adjacent mountains to the west. As the air flows out of the mountains during the night, it produces a general convergence over the plains that triggers these storms.

Severe Thunderstorms

When a weather forecast predicts severe thunderstorms, many people become concerned and uneasy. The reason for these feelings is easily understood because thunderstorms are responsible for *all* forms of what meteorologists call "severe local storms." This includes straight-line winds that damage crops and property, hail that ruins crops, tornadoes, heavy rains that lead to flash floods, and lightning-induced deaths and fires.

Most severe thunderstorms in the midlatitudes form along or ahead of cold fronts that accompany wave cyclones. As cold air advances into a region of warm air, the less dense warm air is displaced upward along the front. If the rising air is sufficiently moist, this mechanical lifting leads to condensation and triggers the release of large reserves of latent heat that renders the air unstable. In this case, vertical cloud growth and thunderstorm development begins.

The conditions normally required for severe thunderstorm formation include a warm, moist, conditionally unstable air mass, a strong cold front to provide the needed lift, and airflow aloft that favors the formation of strong updrafts. In the United States these conditions are most prevalent in the spring and early summer when warm, moist maritime tropical (mT) air from the Gulf of Mexico clashes with cold, dry continental polar (cP) air from Canada. At this time of year the temperature contrast between these two air masses is great and cold fronts normally move rapidly, adding to the vertical acceleration of the displaced air. Also, because of the sharp temperature contrast, a steep pressure gradient exists aloft that generates a strong jet stream parallel to the front. Near the cold front, divergence in the upper-level jet favors upward flow and cloud formation. Such upper-level support is critical; without it, severe thunderstorms would seldom develop.

The most violent weather phenomenon associated with a thunderstorm is a tornado. Although the typical tornado-producing thunderstorm is often 15 kilometers high and has a lifespan of 2 to 3 hours, it generally produces only one short-lived tornado. Because of the potentially destructive nature of tornadoes, meteorologists are interested in learning more about the thunderstorms that spawn them.

Atmospheric studies suggest that the existence of an inversion layer a few kilometers above the surface contributes to the formation of most tornado-producing thunderstorms. Recall that temperature inversions represent very stable atmospheric conditions that restrict vertical air motions. The presence of an inversion seems to aid the production of a few very large thunderstorms by inhibiting the formation of many smaller ones (Figure 10–5). This situation develops because the inversion prevents the mixing of warm, humid air in the lower troposphere with cold, dry air above. Consequently, surface heating continues to increase the temperature and moisture content of the layer of air trapped below the inversion. Eventually the inversion is locally eroded by strong mixing from below. The unstable air below "erupts" explosively at these sites, producing unusually large

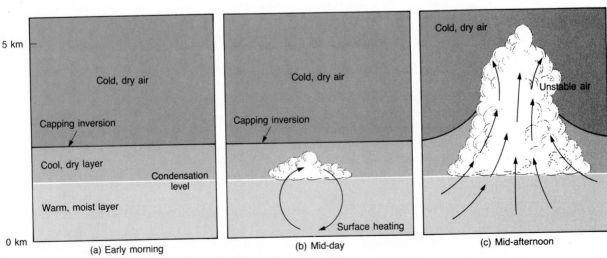

FIGURE 10–5
The formation of severe thunderstorms can be enhanced by the existence of a temperature inversion located a few kilometers above the surface.

FIGURE 10–6
The dark overcast of a mammatus sky, with its characteristic downward bulging
pouches, often precedes a squall line. (Courtesy of NOAA)

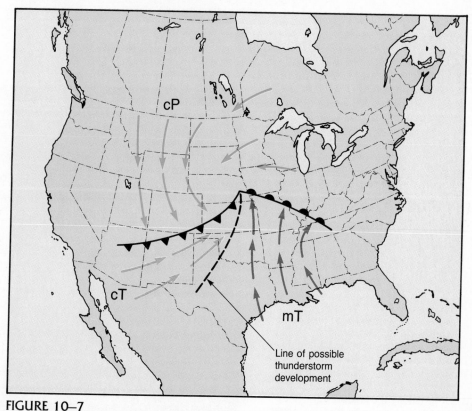

FIGURE 10–7
Squall-line thunderstorms frequently develop along a boundary separating warm, dry continental tropical air and warm, moist maritime tropical air.

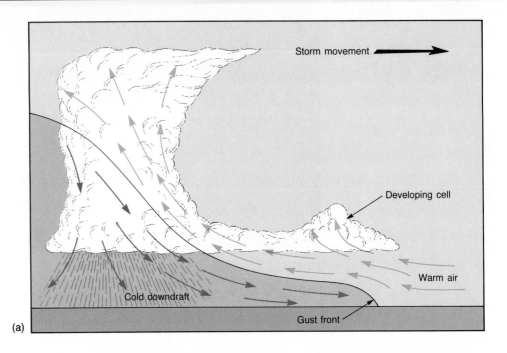

(a)

FIGURE 10–8
(a) A squall line helps to propagate itself by producing downdrafts that initiate cloud development by lifting the warm air ahead of the active thunderstorms. (b) Roll clouds like this one over Miles City, Montana, are produced along the leading edge of a squall line. (Photo by Phil Roskowski, National Center for Atmospheric Research)

(b)

cumulonimbus clouds. It is from such clouds with their concentrated, persistent updrafts that tornadoes are spawned.

Not all severe thunderstorms occur along cold fronts. In some instances, a line of thunderstorms may form as much as 300 kilometers ahead of the cold front along a narrow belt called a **squall line.** The approach of a squall line is often preceded by a *mammatus sky* consisting of dark cloud rolls that have downward pouches (Figure 10–6). A sharp veer in the wind direction, cooler temperatures, and gusty conditions capable of inflicting heavy destruction are also common. In addition, the most violent of all windstorms, the tornado, is an occasional companion.

Some squall-line thunderstorms form when continental tropical (cT) air from the southwestern United States is pulled into a middle-latitude cyclone as shown in Figure 10–7. The denser cT air acts much like a cold front to forcefully displace the lighter mT air upward.[1] On other occasions, squall lines are initiated by disturbances in the airflow aloft.

Once formed, a squall line helps propagate itself by aiding in the development of new cells downwind. As shown in Figure 10–8, the downdrafts from the thunderstorm cells of the squall disturbance produce an advancing wedge of cold air. The leading edge of this advancing cold air is called a **gust front.** Lifting of warm air along the gust front initiates the development of new cells ahead of the squall line. Thus the squall-line disturbance generally moves ahead of and parallel to the cold front at speeds often exceeding that of the cold front.

Squall Line (margin)

Gust Front (margin)

Thunder and Lightning

By international agreement a storm is classified as a thunderstorm only after thunder is heard. Because thunder is produced by lightning, lightning must also be present (Figure 10–9). **Lightning** is similar to the electrical shock you may have experienced on touching a metal object on a very dry day. Only the intensity is different.

During the development of a large cumulonimbus cloud a separation of charge occurs, which simply means that part of the cloud obtains an excess negative charge whereas another part acquires an excess positive charge. The object of lightning is to equalize these electrical differences by producing a negative flow of current from the region of excess negative charge to the region with excess positive charge or vice versa. Because air is a poor conductor of electricity (good insulator), the electrical potential (charge difference) must be very high before lightning will occur, on the order of 3000 volts per meter.

The origin of charge separation in clouds, although not fully understood, must hinge on rapid vertical movements within, for lightning occurs primarily in the violent mature stage of a cumulonimbus cloud. Because the formation of these tall clouds is chiefly a summertime phenomenon in the midlatitudes, it

Lightning (margin)

---

[1] Warm, dry air is more dense than warm, humid air, for the molecular weight of water vapor (H_2O) is only about 62 percent as great as the molecular weight of the mixture of gases that make up dry air.

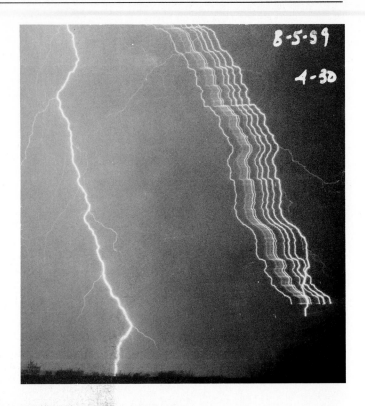

FIGURE 10–9
Photograph showing multiple lightning stroke of a
single flash as recorded by a moving-film camera.
(Courtesy of Marx Brook)

also explains why lightning is seldom observed there in the winter. Furthermore, lightning rarely occurs before the growing cloud penetrates the 5-kilometer level, where sufficient cooling begins to generate some ice crystals. Some cloud physicists believe that charge separation occurs during the formation of ice pellets. Experimentation shows that as droplets begin to freeze, positively charged ions are concentrated in the colder regions of the droplets, whereas negatively charged ions are concentrated in the warmer regions. Thus as the droplets freeze from the outside in, they develop a positively charged ice shell and a negatively charged interior. As the interior begins to freeze, it expands and shatters the outside shell. The small positively charged ice fragments are carried upward by turbulence, and the relatively heavy droplets eventually carry their negative charge toward the cloud base. As a result, the upper part of the cloud is left with a positive charge and the lower portion of the cloud maintains an overall negative charge with small positively charged pockets (Figure 10–10). As the cloud moves, the negatively charged cloud base alters the charge at the earth's surface directly below by repelling negatively charged particles. Thus the earth's surface beneath the cloud acquires a net positive charge. These charge differences build to millions and even hundreds of millions of volts before a lightning stroke acts to discharge the negative region of the cloud by striking the positive area of the earth below, or, more frequently, the positively charged portion of that cloud, or a nearby cloud.

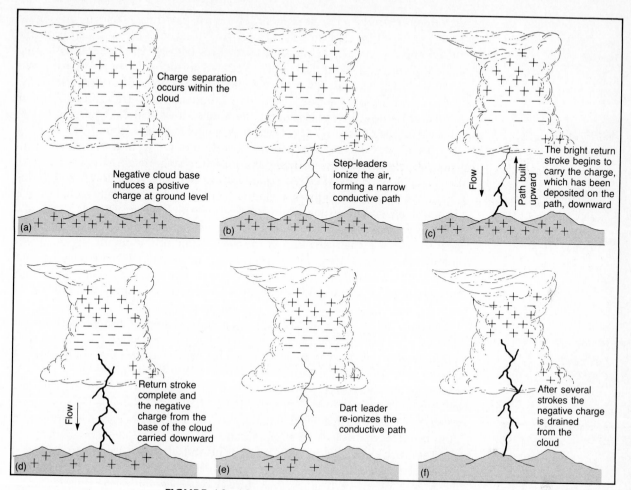

FIGURE 10–10
Discharge of a cloud via cloud-to-ground lightning. Examine this drawing carefully while reading the text.

Cloud-to-earth strokes are of most interest and have been studied in detail. Moving-film cameras have greatly aided in these studies. They show that the lightning we see as a single flash is really several very rapid strokes between the cloud and the earth (Figure 10–9). We will call the total discharge, which lasts only a few tenths of a second and appears as a bright streak, the **flash.** Individual components that make up each flash are called **strokes.** Each stroke is separated by roughly 50 milliseconds, and there are usually three to four strokes per flash. When a lightning flash appears to flicker, it is because your eyes discern the individual strokes that make up this discharge. Moreover, each stroke consists of a downward propagating leader that is immediately followed by a luminous return stroke.

Each stroke is believed to begin when the electrical field near the cloud

Flash

Stroke

Leader

base frees electrons in the air immediately below, thereby ionizing the air (Figure 10–10). Once ionized, the air becomes a conductive path having a radius of roughly 10 centimeters and a length of 50 meters. This path is called a **leader.** During this electrical breakdown the mobile electrons in the cloud base begin to flow down this channel. This flow increases the electrical potential at the head of the leader, which causes a further extension of the conductive path through further ionization. Because this initial path extends itself earthward in short, nearly invisible

Step Leader

bursts, it is called a **step leader.** Once this channel nears the ground, the electrical field at the surface ionizes the remaining section of the path. With the path completed, the electrons that were deposited along the channel begin to flow downward. The initial flow begins near the ground. As the electrons at the lower end of this conductive path move earthward, electrons positioned successively higher up the channel begin to migrate downward. Because the path of electron flow is continually being extended upward, the accompanying electric discharge has been appropri-

Return Stroke

ately named a **return stroke.** As the wave front of the return stroke moves upward, the negative charge that was deposited on the channel is effectively lowered to the earth. It is this intense return stroke that illuminates the conductive path and discharges the lowest kilometer or so of the cloud. During this phase tens of coulombs of negative charge are lowered to the ground.[2]

The first stroke is usually followed by additional strokes that apparently drain charges from higher areas within the cloud. Each subsequent stroke begins with

Dart Leader

a **dart leader** that once again ionizes the channel and carries the cloud potential toward the earth. The dart leader is continuous and less branched than the step leader. When the current between strokes has ceased for periods greater than one-tenth of a second, further strokes will be preceded by a stepped leader whose path is different from that of the initial stroke. The total time of each flash consisting of three or four strokes is about 0.2 second.

The electrical discharge of lightning heats the air (much like a wire is heated by an electric current) and causes it to expand explosively. This expansion produces

Thunder

the sound waves we hear as **thunder.** Because lightning and thunder occur simultaneously, it is possible to estimate the distance to the stroke. Lightning is seen instantaneously, but the rather slow sound waves, which travel approximately 330 meters per second, reach us a little later. If thunder is heard 5 seconds after the lightning is seen, the lightning occurred about 1650 meters away (approximately 1 mile).

The thunder that we hear as a rumble is produced along a rather long lightning path located at some distance from the observer. The sound that originates along the path nearest the observer arrives before the sound that originated farthest away. This factor lengthens the duration of the thunder. Reflection of the sound waves further delays their arrival and adds to this effect. When lightning occurs more than 20 kilometers away, thunder is rarely heard. This type of lightning, popularly called *heat lightning*, is no different from the lightning that we associate with thunder.

---

[2] A coulomb is a unit of electrical charge equal to the quantity of charge transferred in 1 second by a steady current of 1 ampere.

LIGHTNING: REFUTING THE MISCONCEPTIONS

H. Michael Mogil*

Although references to thunderbolts and lightning flashes are found in the literature from ancient to modern times, most persons are unaware of their effects on life and property. For example, according to the U.S. Department of Agriculture's Forest Service, lightning is the leading cause of forest fires in the western United States. During one 10-year period, lightning accounted for 45 percent of Alaskan forest fires and more than 80 percent of the acreage burned. Similar fires resulted in the destruction of over 2 million acres of forest in California and Alaska alone during July and August 1977. Economic and ecological losses are incalculable.

Beyond such economic losses, lightning is responsible for more deaths than any other *stormy* weather event. Lack of respect of this danger by the general public may be attributed to the fact that limited publicity is given to such incidents and that most reported incidents involve only one or two persons. Floods and tornadoes are more likely to make news headlines because they affect larger areas and often cause multiple deaths, injuries, and widespread damage.

Between January 1940 and December 1976, lightning (either directly or indirectly) was reported to have killed more than 7500 Americans and injured more than 20,000 others, giving annual averages of approximately 200 killed and 550 injured. Incomplete reporting has resulted in conservative numbers, and we estimate that the average annual death toll from lightning in the conterminous United States is probably double that number.

Lightning doesn't strike the same place twice.

Rubber tires from cars make occupants safe from lightning.

Common misconceptions about lightning such as these are shattered almost every day.

There appears to be no basis for the statement, "Lightning doesn't strike the same place twice." In fact, according to Uman, "much of what is known about lightning today has been discovered precisely because lightning does strike the same structure over and over again." For example, the Empire State Building in New York City is struck on the average about 23 times per year.

But lightning striking the same place more than once also affects people, as the following incidents illustrate.

In 1978, a boy was killed in Knightdale (Wake County, North Carolina) as he stood in the doorway of a garage. About 10 minutes earlier his father was hit outside the building working on some equipment.

In 1979, a home in Tennessee was destroyed by a lightning-caused fire. This was the third time lightning had struck the home since it was built in 1970.

There are *no absolutely safe* places from lightning, although enclosed automobiles and buildings are among the safest. And reliance on the misconception that "rubber tires make people in cars safe from lightning" could lead some to believe that they are safe from lightning because they are wearing rubber-soled shoes. But lightning, which travels many miles through insulating air, is not halted by a few inches or even a yard of insulating rubber. In the case of an enclosed automobile, the lightning usually travels along the metal skin of the vehicle and then jumps to the ground either through the air, along the wet tire surface, or by traveling through the tire (causing it to fail). The latter may become more of a problem as steel-belted tires find increasing use.

Another misconception that deserves attention is the one that suggests safety because the storm has

*H. Michael Mogil served as an emergency warnings meteorologist at the National Weather Service Headquarters in Silver Spring, Maryland, from 1975 to 1980. He has also spent time at the National Weather Service Forecast Offices in Fort Worth, Texas, and San Francisco, California. This essay is drawn largely from an article by Mr. Mogil entitled "Lightning in 1978," which appeared in *Weatherwise*, 32, no. 1 (1979), 17–20. Some portions of this essay were taken from two other *Weatherwise* articles coauthored by Mr. Mogil: "Lightning—A Preliminary Reassessment" (with Marjorie Rush and Mary Kutka), 30, no. 5 (1977), 192–199; and "Update on Lightning" (with James L. Campbell), 33, no. 1 (1980), 36–37.

passed (or because the storm is still far enough away so there is time to get to a safe place). Recent unpublished studies undertaken as a part of the Florida Area Cumulus Experiment and photographic evidence suggest that lightning can often occur outside of precipitation, or even outside of clouds. Unfortunately, a lightning strike during a little league game at Ames, Iowa, on June 15, 1978, supports these findings. In this incident, the game, which had been delayed because of thunderstorms and rain, was resumed as the storm moved north. A "bolt from the blue" injured five youngsters.

Although lightning tends to strike the highest object, it can still impact people nearby, even though the people may not be the highest objects or the people may be in a relatively safe place (e.g., inside a house). The lightning injuries to people talking on telephones and the deaths and injuries to people standing beneath or near trees attest to this.

TORNADOES

Tornadoes are local storms of short duration that must be ranked high among nature's most destructive forces (Figure 10–11). Their sporadic occurrence and violent winds cause many deaths each year (see Table 10–1). The nearly total destruction in some stricken areas has led many to liken their passage to bombing raids during war.

FIGURE 10–11
Hubbard, Ohio, was hard hit by a tornado that moved across the area on the evening of May 31, 1985, killing 18 people. (Photo by Paul Tople. Reprinted with permission of the *Akron Beacon Journal*)

Table 10–1 Estimated Total Annual Fatalities from Three Important Meteorological Hazards, 1960–85

Year	Lightning	Tornado	Hurricane
1960	97	46	65
1961	113	51	46
1962	120	28	4
1963	210	31	11
1964	108	73	49
1965	125	296	75
1966	76	98	54
1967	73	114	18
1968	103	131	9
1969	93	66	256
1970	111	72	11
1971	113	156	8
1972	91	27	121
1973	105	87	5
1974	102	361	1
1975	91	60	21
1976	72	44	9
1977	98	43	0
1978	88	53	35
1979	63	84	22
1980	76	28	2
1981	67	24	0
1982	77	64	3
1983	77	34	22
1984	67	122	4
1985	74	94	30
Total	2490	2287	881
Average per year	95.77	87.96	33.88

SOURCE: National Oceanic and Atmospheric Administration.

Tornado

Tornadoes, sometimes called twisters or cyclones, are violent windstorms that take the form of a rotating column of air or *vortex* that extends downward from a cumulonimbus cloud. Pressures within some tornadoes have been estimated to be as much as 10 percent lower than immediately outside the storm. Drawn by the much lower pressure in the center of the vortex, air near the ground rushes into the tornado from all directions. As the air streams inward, it is spiraled upward around the core until it eventually merges with the airflow of the parent thunderstorm deep in the cumulonimbus tower.

Because of the rapid drop in pressure, air sucked into the storm expands and cools adiabatically. If the air cools below its dew point, the resulting condensation creates a pale and ominous-appearing cloud that may darken as it moves across the ground picking up dust and debris (Figure 10–12). Occasionally when the inward spiraling air is relatively dry, no condensation funnel forms because the drop in pressure is not sufficient to cause the necessary adiabatic cooling. In

FIGURE 10–12
A tornado is a violently rotating column of air in contact with the ground. The air column is visible when it contains condensation or when it contains dust and debris. Often the appearance is the result of both. When the column of air is aloft and does not produce damage, the visible portion is properly called a *funnel cloud*. (Photograph by Tom Carter, courtesy of the *Peoria Journal Star*)

such cases, the vortex is only made visible by the material that it vacuums from the surface and carries aloft.

Because of the tremendous pressure gradient associated with a strong tornado, maximum winds are believed to approach 480 kilometers (300 miles) per hour. But this figure is simply an estimate that is based on the analysis of motion pictures and damage to engineered structures. Reliable wind-speed measurements are lacking. Similarly, the changes that occur in atmospheric pressure with the passage of a tornado are also estimates and are based on a few storms that happened to pass a nearby weather station. Thus meteorologists have had to base their tornado models on a relatively scant observational foundation. Although additional data might be gathered if shelters and instruments capable of withstanding the fury of a tornado were placed in an area, such an effort would probably not be worthwhile. Tornadoes are highly localized and randomly distributed, and the probability of placing a set of instruments at the right site would thus be infinitesimally small. For example, the probability of a tornado striking a given point in the region most frequently subject to tornadoes is about once in 250 years. The development of Doppler radar, however, has increased our ability to study tornado-producing thunderstorms. As we shall see, the development of this new technology is allowing meteorologists to gather important new data from a safe distance.

The Development and Occurrence of Tornadoes

Tornadoes form in association with severe thunderstorms that produce high winds, heavy (sometimes torrential) rainfall, and often damaging hail. Although hail may or may not precede a tornado, the portion of the thunderstorm adjacent to large hail is often the area where strong tornadoes are most likely to occur. Fortunately, less than 1 percent of all thunderstorms produce tornadoes. Although weather scientists are still not sure what triggers tornado formation, it has become apparent that they are the product of the interaction between the strong updrafts in the thunderstorm and the winds in the troposphere. In spite of recent advances in modeling the many variables that eventually produce a strong tornado, our knowledge still must be considered limited. Nevertheless, the general atmospheric conditions that are most likely to develop into tornado activity are known.

Severe thunderstorms—and hence tornadoes—are most often spawned along the cold front or squall line of a middle-latitude cyclone. Throughout the spring, air masses associated with midlatitude cyclones are most likely to have greatly contrasting conditions. Continental polar air from the Canadian Arctic may still be very cold and dry, whereas maritime tropical air from the Gulf of Mexico is warm, humid, and unstable. The greater the contrast, the more intense the storm. Because these two contrasting air masses are most likely to meet in the central United States, it is not surprising that this region generates more tornadoes than any other area of the country or, in fact, the world. Figure 10–13, which depicts the average annual tornado incidence in the United States over a 27-year period, readily substantiates this fact.

An average of about 770 tornadoes are reported each year in the United States. Still, the actual numbers that occur from one year to the next vary greatly. During the 1953–1986 period, for example, yearly totals ranged from a low of

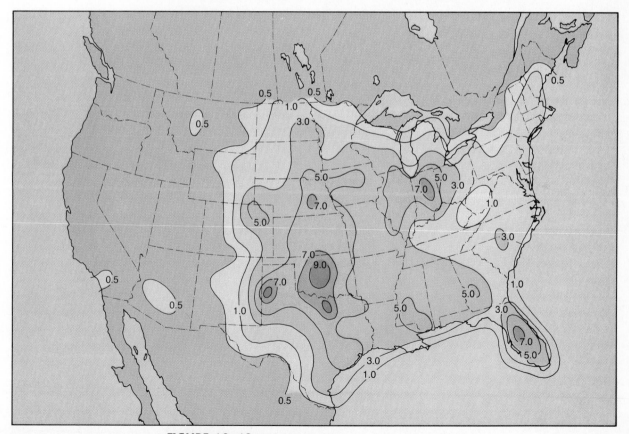

FIGURE 10–13
Average annual tornado incidence per 10,000 square miles (26,000 square kilometers) for a 27-year period. (Courtesy of NOAA)

421 in 1953 to a high of 1102 in 1973. Tornadoes occur during every month of the year. April through June is the period of greatest tornado frequency in the United States, and December and January are the months of lowest activity. Of the more than 25,000 confirmed tornadoes reported over the contiguous 48 states during the 1953–1986 period, an average of almost five per day occurred during May. At the other extreme, a tornado was reported only every other day in January. Typically, 54 percent of all tornadoes take place during the spring. Fall and winter, on the other hand, together account for only 19 percent (Figure 10–14). In February, when the incidence of tornadoes begins to increase, the center of maximum frequency lies over the central Gulf states. During March this center moves eastward to the southeastern Atlantic states, where tornado frequency reaches a peak in April. During May and June the center of maximum frequency moves through the southern Plains and then to the northern Plains and Great Lakes area. This drift is due to the increasing penetration of warm, moist air while contrasting cool, dry air still surges in from the north and northwest. Thus when the Gulf states are substantially under the influence of warm air after May, there is no cold-air intrusion to speak of, and tornado frequency drops. Such is the case across the country after June. Winter cooling permits fewer and fewer encounters between warm and cold air masses, and tornado frequency returns to its lowest level by January.

The average tornado has a diameter of between 150 and 600 meters, travels across the landscape at approximately 45 kilometers per hour, and cuts a path

FIGURE 10–14
Average number of tornadoes and tornado days each month in the United States for a 27-year period. (Courtesy of NOAA)

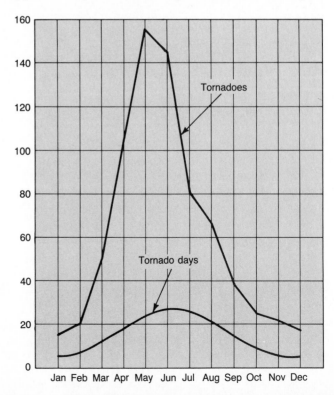

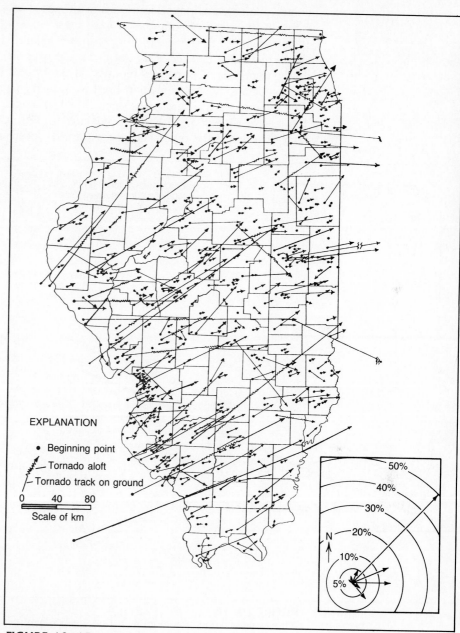

FIGURE 10–15

Paths of Illinois tornadoes (1916–1969). Since most tornadoes occur slightly ahead of a cold front, in the zone of southwest winds, they tend to move toward the northeast. Tornadoes in Illinois verify this. Over 80 percent exhibited directions of movement toward the northeast through east. (After John W. Wilson and Stanley A. Changnon, Jr., *Illinois Tornadoes*, Illinois State Water Survey Circular 103, 1971, pp. 10, 24)

about 26 kilometers long. Because many tornadoes occur slightly ahead of a cold front, in the zone of southwest winds, most move toward the northeast. The Illinois example demonstrates this fact nicely (Figure 10–15). Figure 10–15 also shows that many tornadoes do not fit the description of the "average" tornado. Of the hundreds of tornadoes reported in the United States each year, over half are comparatively weak and short lived. Most such small tornadoes have lifetimes of 3 minutes or less and paths that seldom exceed 1 kilometer in length and 100 meters wide. Typical wind speeds are on the order of 150 kilometers per hour or less. On the other end of the tornado spectrum are the infrequent and

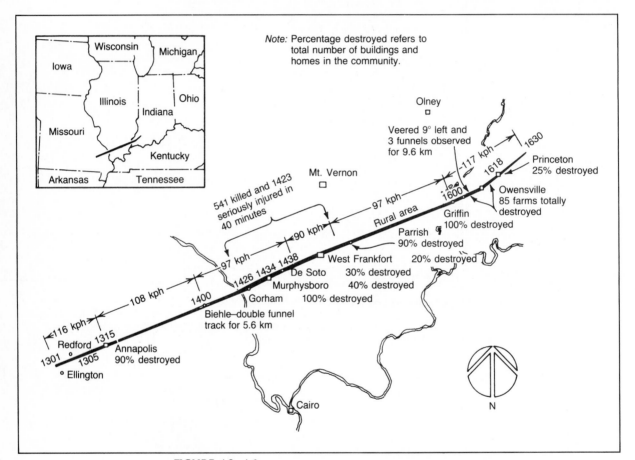

FIGURE 10–16

"One tornado among the more than 13,000 which have occurred in the United States since 1915 easily ranks above all others as the single most devastating storm of this type. Shortly after its occurrence on 18 March 1925, the famed Tri-State tornado was recognized as the worst on record, and it still ranks as the nation's greatest tornado disaster. The tornado remained on the ground for 219 miles. The resulting losses included 695 dead, 2027 injured, and damages equal to $43 million in 1970 dollars. This represents the greatest death toll ever inflicted by a tornado and one of the largest damage totals." (Map and description from J. W. Wilson and S. A. Changnon, Jr., *Illinois Tornadoes*, Illinois State Water Survey Circular 103, 1971, p. 32)

often long-lived violent tornadoes. Although large tornadoes constitute only a small percentage of the total reported, their effects are often devastating. Such tornadoes may exist for periods in excess of 3 hours and produce an essentially continuous damage path more than 150 kilometers long and perhaps a kilometer or more wide. Maximum winds range upward to 450 kilometers per hour (Figure 10–16).

Tornado Destruction

Tornadoes have accomplished many seemingly impossible tasks, such as driving a piece of straw through a thick wooden plank and uprooting huge trees (Figure 10–17). In 1931 a tornado actually carried an 83-ton railroad coach and its 117 passengers 24 meters through the air and dropped them in a ditch.

Fujita Intensity Scale (F-scale)

Most tornado losses are associated with a few storms that strike urban areas or devastate entire small communities. The amount of destruction wrought by such storms depends largely on the strength of the winds. One commonly used guide to tornado intensity was developed by T. Theodore Fujita at the University of Chicago and is appropriately called the **Fujita Intensity Scale** or simply the **F-scale** (Table 10–2). Because tornado winds cannot be measured directly, a rating on the F-scale is determined by assessing the worst damage produced by a storm.

The drop in atmospheric pressure associated with the passage of a tornado plays a minor role in the damage process. Most structures have sufficient venting to allow for the sudden drop in pressure. Opening a window, once thought to be a way to minimize damage by allowing inside and outside atmospheric pressure to equalize, is no longer recommended. In fact, if a tornado gets close enough to a structure for the pressure drop to be experienced, the strong winds probably will have already caused significant damage.

FIGURE 10–17
The force of the wind during a tornado in Clarendon, Texas, in 1970 was enough to drive a wooden stick through a 4-centimeter metal pipe. (Courtesy of NOAA)

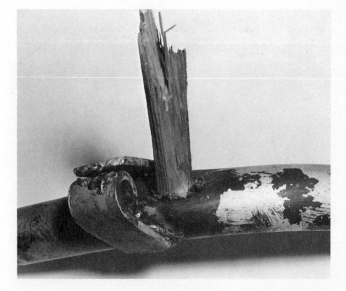

TABLE 10–2 Fujita Intensity Scale

Scale	km/hr	Expected Damage
F0	<116	Light damage
F1	116–180	Moderate damage
F2	181–253	Considerable damage
F3	254–332	Severe damage
F4	333–419	Devastating damage
F5	>419	Incredible damage

Although the greatest part of tornado damage is caused by violent winds, most tornado injuries and deaths result from flying debris. On the average, tornadoes cause more deaths each year than any other weather event except lightning. The average annual death toll from tornadoes during the 26-year period depicted in Table 10–1 is about 88 people. By examining this table, it is clear that the actual number of deaths each year can depart significantly from the average. On April 3–4, 1974, for example, an outbreak of 148 tornadoes brought death and destruction to a 13-state region east of the Mississippi River. More than 300 people died and nearly 5500 people were injured in this worst tornado disaster in half a century. In one statistical study that examined a 29-year period, there were 689 tornadoes that caused loss of life. This figure represented slightly less than 4 percent of the total 19,312 reported storms. Although the percentage of tornadoes that result in death is small, every tornado is potentially lethal. If you examine Figure 10–18, which compares tornado fatalities with storm intensities, the results

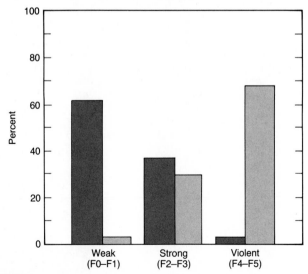

FIGURE 10–18
Percentage of tornadoes in each intensity category (black) and percentage of fatalities associated with each category. (From Joseph T. Schaefer et al., "Tornadoes—When, Where and How Often," *Weatherwise*, 33, No. 2 [1980], 57)

are quite interesting. It is clear from this graph that the majority (63 percent) of all tornadoes are weak and that the number of storms decreases as tornado intensity increases. The distribution of tornado fatalities, however, is just the opposite. Although only 2 percent of tornadoes are classified as violent, they account for nearly 70 percent of the deaths.

If there is some question about the causes of tornadoes, there certainly is none about the destructive effects of these violent storms. A severe tornado leaves the affected area stunned and disorganized and may require a response of the magnitude demanded in war.

Predicting Tornadoes

Because severe thunderstorms and tornadoes are small and short-lived phenomena, they are among the most difficult weather features to forecast precisely. Two researchers at the National Severe Storms Laboratory describe the difficulty of tornado forecasting this way:

> They are less specific than we would like, because our observations are too sparse to describe atmospheric variability on the scale producing the tornado or thunderstorm phenomena. A severe thunderstorm may extend 10 to 25 miles, and exist for about 6 hours, while the distance between primary surface weather stations is about 100 miles and between upper-air stations more than 200 miles. Observations are made hourly at the surface stations (more often under special conditions) but usually at only 12-hour intervals at the upper-air stations. Even if our knowledge were otherwise adequate to the task, the current weather observing system limits us to indicating the probability of thunderstorms and accompanying tornadoes in regions much larger than the storms.[3]

Tornado Watch

Such forecasts, called **tornado watches,** generally refer to situations that are expected to begin from 1 to 7 hours after the forecast is issued in areas covering about 65,000 square kilometers. Between 35 and 40 percent of these predictions are correct; that is, one or more tornadoes occurred somewhere in the region. The incorrect predictions are about evenly divided between cases when no tornadoes are sighted and cases when tornadoes occur outside but near the forecast areas.

Tornado Warning

Whereas a tornado watch is designed to alert people to the possibility of tornadoes, a **tornado warning** is presently issued when a tornado has actually been sighted in an area or is indicated by radar. When severe weather threatens, radar screens are monitored for very intense echoes, which, in turn, are associated with heavy precipitation and the greater likelihood of hail, strong winds, and tornadoes. In addition, the echo from a tornadic storm within about 100 kilometers of the radar sometimes displays a hook-shaped appendage, as shown in Figure 10–19.

---

[3] Robert Davies-Jones and Edwin Kessler, *Weather and Climate Modification*, edited by Wilmot N. Hess (New York: John Wiley & Sons, Inc., 1974), p. 566.

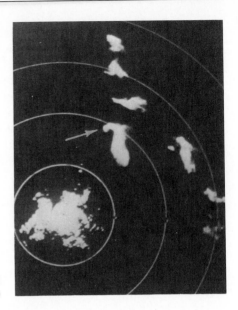

FIGURE 10–19
Sometimes tornadoes appear as hook-shaped echoes on a radar screen. (Courtesy of NOAA)

If the direction and approximate speed of the storm are known, an estimate of its most probable path can be made. Because tornadoes often move erratically, the warning area is fan shaped downwind from the point where the tornado has been spotted. Since the late 1960s warnings have been given for most major tornadoes. It is believed that such warnings substantially reduce the number of deaths and serious injuries that might otherwise occur. However, additional deaths and serious injuries could be averted if more people would take adequate safety measures after a tornado warning is issued. This point was dramatically illustrated in a study made following the tornado that struck Wichita Falls, Texas, on April 10, 1979. The study revealed that

> twenty-six (60 percent) of the 43 traumatic deaths and 30 (51 percent) of the 59 serious injuries occurred in people who, despite ample warning, went to their cars to drive out of the storm's path. These people had a risk of serious or fatal injury of 23 per 1000. People who remained indoors and in stationary homes were at a relatively low risk (3 per 1000) if they took simple precautions; people in mobile homes were at greatest risk (85 per 1000).[4]

As noted earlier, the probability of one place being struck by a tornado, even in the area of greatest frequency, is slight. Nevertheless, although the probabilities may be small, tornadoes have provided many mathematical exceptions. For example, the small town of Codell, Kansas, was hit 3 years in a row—1916, 1917, and 1918—and each time on the same date, May 20! Needless to say, tornado watches and warnings should never be taken lightly.

---

[4] Roger I. Glass et al., "Injuries from the Wichita Falls Tornado: Implications for Prevention," *Science*, 207, no. 4432 (1980), 735.

Doppler Radar

The tornado warning system presently used throughout the United States relies heavily on visual sightings by a few trained observers as well as by members of the general public. Unfortunately, such a system is prone to incomplete coverage and mistakes. The errors are most likely to occur at night when tornadoes may go unnoticed or harmless clouds may be mistaken for funnel clouds. So there may be a lack of adequate warning on the one hand or unnecessary warnings on the other.

Unfortunately, these problems cannot be solved by using conventional weather radar. The radar units now in routine use work by transmitting short pulses of electromagnetic energy. A small fraction of the waves that are sent out is scattered by a storm and returned to the radar. The strength of the returning signal indicates rainfall intensity and the time difference between the transmission and return of the signal indicates the distance to the storm. But in order to identify tornadoes and severe thunderstorms, we must be able to detect the characteristic circulation patterns associated with them. Conventional radar cannot do so except in the rare case when spiral rain bands occur in association with a tornado and give rise to the hook-shaped echo mentioned earlier.

Doppler Radar

Many difficulties that currently limit the accuracy of tornado warnings may be reduced or eliminated in the future if an advancement in radar technology, called **Doppler radar,** is put into general use. Doppler radar not only performs the same tasks as conventional radar but also has the ability to detect motion directly. It does so by comparing the frequency of the reflected signal to that of the original pulse. The movement of precipitation toward the radar increases the frequency of reflected pulses, whereas motion away from the radar decreases the frequency. These frequency changes are then interpreted in terms of speed toward or away from the Doppler radar unit. It is this same principle that allows police radar to determine the speed of moving cars. An analogous change in frequency that you may be familiar with occurs when a train passes with its whistle blowing. As the train approaches, the sound seems to have a higher than normal pitch; when it is moving away, the pitch seems lower than normal.

Mesocyclone

Within a radius of about 230 kilometers, a Doppler radar unit can detect the initial formation and subsequent development of a **mesocyclone,** an intense rotating wind system within a thunderstorm that precedes tornado development. Almost all (96 percent) mesocyclones produce damaging hail, severe winds, or tornadoes. Those that produce tornadoes (about 50 percent) can often be distinguished by their stronger wind speeds and their sharper gradients of wind speeds.

Carefully planned tests conducted over several years showed that Doppler radar provided an average warning time before tornado touchdown of 21 minutes. By comparison, the average warning time provided by visual observations is less than 2 minutes. Furthermore, an additional advantage was a decreased false-alarm rate. Predictions based on Doppler radar were correct 75 percent of the time, a considerable improvement over present standards.

The benefits of Doppler radar are many. As a research tool, it is not only providing data on the formation of tornadoes but is also helping meteorologists

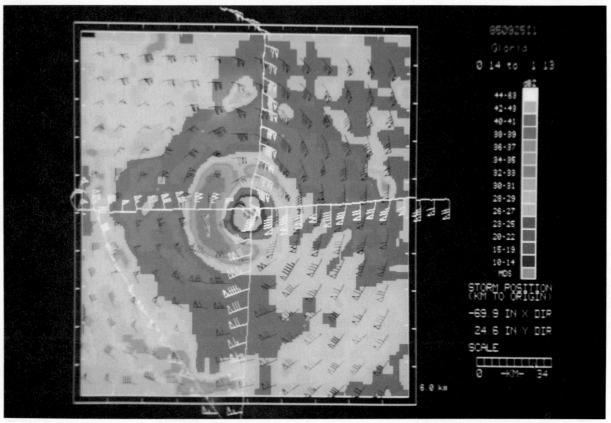

FIGURE 10–20
Airborne Doppler radar observations of Hurricane Gloria, September 25, 1985. Radar reflectivity (rainfall intensity) is depicted by color variations. The strength and direction of winds is shown with white barbs. (Courtesy of Frank Marks, Environmental Research Laboratories, Hurricane Research Division, NOAA)

gain new insights into thunderstorm development, the structure and dynamics of hurricanes, and air turbulence hazards that plague aircraft (Figure 10–20). As a practical tool for tornado detection, it has significant advantages over a system that uses observers and conventional radar. Recognizing these advantages, the National Weather Service plans to replace its aging system of weather radars with more advanced systems that will use Doppler principles of direct wind speed and direction measurements. The new radars are part of the 10-year *Next Generation Weather Radar* (*NEXRAD*) program aimed at improving the forecasting of severe storms, tornadoes, and flash floods.

HURRICANES

Hurricane

The whirling tropical cyclones that occasionally have wind speeds exceeding 300 kilometers per hour are known in the United States as **hurricanes**—the greatest storms on earth. Hurricanes are among the most destructive of natural disasters. When a hurricane reaches land, it is capable of annihilating coastal areas and killing tens of thousands of people. On the positive side, however, these storms provide essential rainfall over many areas they cross. Consequently, a resort owner in Florida may dread the coming of hurricane season, but a farmer in Japan may welcome its arrival.

Most hurricanes form between the latitudes of 5 and 20 degrees over all the tropical oceans except the South Atlantic and the eastern South Pacific. These intense tropical storms are known in various parts of the world by different names. In the western Pacific they are called *typhoons*, and in the Indian Ocean, including the Bay of Bengal and Arabian Sea, they are simply called *cyclones*. In the following discussion these storms will be referred to as hurricanes.

Profile of a Hurricane

Although numerous tropical disturbances develop each year, only a few reach hurricane status, which by international agreement requires wind speeds in excess of 115 kilometers per hour and a rotary circulation. Mature hurricanes average about 600 kilometers across, although they can range in diameter from 100 kilometers up to about 1500 kilometers. From the outer edge of the hurricane to the center the barometric pressure has sometimes dropped 60 millibars, from 1010 to 950 millibars. The lowest pressure ever recorded in the United States was 892.31 millibars, which was measured during a hurricane in September 1935. A steep pressure gradient like that shown in Figure 10–21 generates the rapid, inward spiraling winds of a hurricane. As the air moves closer to the center of the storm, its velocity increases.

Law of Conservation of Angular Momentum

In order to understand better why wind speeds increase near the storm center, we must mention the **law of conservation of angular momentum.** This law states that the product of the velocity of an object around a center of rotation (axis) and the distance squared of the object from the axis is constant. Therefore when a parcel of air moves toward the center of a storm (the center of rotation), the product of its distance squared and velocity must remain unchanged. Consequently, as air moves in toward the center of a storm, the rotational velocity must increase. A common example of the conservation of angular momentum occurs when a figure skater starts whirling on the ice with both arms extended. Her arms are traveling in a circular path about an axis (her body). When the skater pulls her arms inward, she decreases the radius of the circular path of her arms. As a result, her arms go faster and the rest of her body must follow, thereby increasing her rate of spinning.

Eye Wall

As the inward rush of warm, moist surface air approaches the core of the storm, it turns upward and ascends in a ring of cumulonimbus towers (Figure 10–22). This doughnut-shaped wall of intense convective activity surrounding the center of the storm is called the **eye wall.** It is here that the greatest wind speeds and heaviest rainfall occur.

Eye

At the very center of the storm is the **eye** of the hurricane. This well-known feature is a zone where precipitation ceases and winds subside. It offers a brief but deceptive break from the extreme weather in the enormous curving wall

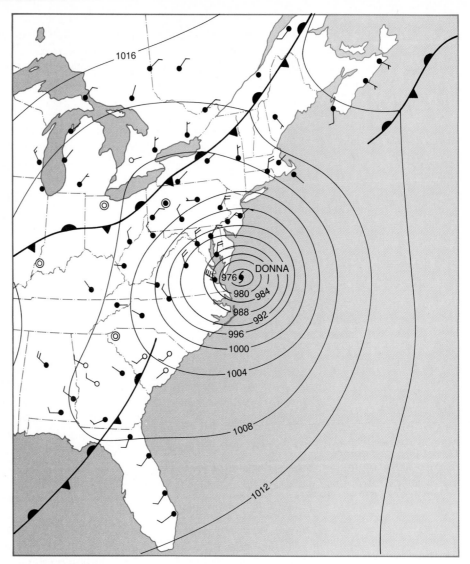

FIGURE 10–21
Hurricane Donna on the surface weather map, 1200 Greenwich mean time, September 12, 1960.

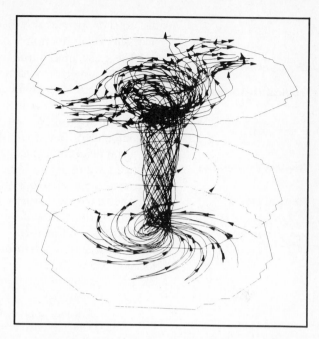

FIGURE 10–22
Circulation in a well-developed hurricane. Cyclonic winds are experienced at the surface. At the core of the storm the air is whirled in great spirals upward. Outward flow dominates near the top of the storm and seems to be aided by upper-level airflow.

clouds that surround it. The air within the eye gradually descends and heats by compression, making it the warmest part of the storm. Although many people believe that the eye is characterized by clear blue skies, such is usually not the case because the subsidence in the eye is seldom strong enough to produce cloudless conditions. Although the sky appears much brighter in this region, scattered clouds at various levels are common.

Hurricane Formation and Decay

A hurricane can be described as a heat engine that is fueled by the latent heat liberated when huge quantities of water vapor condense. The amount of energy produced by a typical hurricane in just a single day is truly immense—roughly equivalent to the entire electrical energy production of the United States in a year. The release of latent heat warms the air and provides buoyancy for its upward flight. The result is to reduce the pressure near the surface, which, in turn, encourages a more rapid inflow of air. To get this engine started, a large quantity of warm, moist air is required, and a continuous supply is needed to keep it going.

Hurricanes develop most often in the late summer when ocean waters have reached temperatures of 27°C or higher and are thus capable of providing the necessary heat and moisture to the air. This ocean-water temperature requirement is thought to account for the fact that hurricanes do not form over the relatively cool waters of the South Atlantic and the eastern South Pacific. For the same reason, few hurricanes form poleward of 20 degrees of latitude. Although water temperatures are sufficiently high, hurricanes do not form within 5 degrees of the equator, presumably because the Coriolis effect is too weak to initiate the necessary rotary motion.

The exact mechanism of formation is not completely understood, but it is known that smaller tropical cyclones initiate the process. The U.S. National Weather Service labels such incipient storms **tropical disturbances.** Although many tropical disturbances occur each year, only a few develop into full-fledged hurricanes. Recall that tropical cyclones are called hurricanes only when their winds exceed 115 kilometers per hour. By international agreement, lesser tropical cyclones are given different names based on the strength of their winds. When a cyclone's strongest winds do not exceed 61 kilometers per hour, it is called a **tropical depression;** when winds are between 61 and 115 kilometers per hour, the cyclone is termed a **tropical storm.** Each year between 80 and 100 tropical storms develop over the earth. Of them, usually half or more eventually reach hurricane status.

Tropical Disturbance

Tropical Depression

Tropical Storm

Hurricanes diminish in intensity whenever they (1) move over ocean waters that cannot supply warm moist tropical air, (2) move onto land, or (3) reach a location where the large-scale flow aloft is unfavorable. Richard Anthes describes the possible fate of hurricanes in the first category as follows:

> Many hurricanes approaching the North American or Asian continents from the southeast are turned toward the northeast, away from the continents, by the steering effect of an upper-level trough. This recurvature carries the storms toward higher latitudes where the ocean temperatures are cooler and an encounter with cool, dry polar air masses is more likely. Often the tropical cyclone and a polar front interact, with cold air entering the tropical cyclone from the west. As the release of latent heat is diminished, the upper-level divergence weakens, mean temperatures in the core fall and the surface pressure rises.[5]

Whenever a hurricane moves onto land, it loses its punch rapidly. The most important reason for this rapid demise is the fact that the storm's source of warm, moist air is cut off. When an adequate supply of water vapor does not exist, condensation and the release of latent heat must diminish. In addition, because the land is usually cooler than the ocean, the low-level air is chilled rather than warmed. Moreover, the increased surface roughness over land results in a rapid reduction in surface wind speeds. This factor causes the winds to move more directly into the center of the low, thus helping to eliminate the large pressure differences.

Hurricane Destruction

North Atlantic hurricanes develop in the trade winds, which generally move these storms from east to west at about 25 kilometers per hour. Then almost without exception, hurricanes curve poleward and are deflected into the westerlies, which increase their forward motion up to a maximum of 100 kilometers per hour.

---

[5] *Tropical Cyclones: Their Evolution, Structure, and Effects*, Meteorological Monographs, 19, no. 41 (1982), 61. Boston: American Meteorological Society.

FIGURE 10–23
The aftermath of the Galveston Hurricane of 1900. Entire blocks were swept clean, while mountains of debris accumulated around the few remaining buildings. (Courtesy of the Library of Congress)

Some move toward the mainland, but their irregular paths make prediction of their movement difficult.

A location only a few hundred kilometers from a hurricane—just one day's striking distance away—may experience clear skies and virtually no wind. Before the age of weather satellites, such a situation made it difficult to warn people of impending storms.

The worst natural disaster in U.S. history came as a result of a hurricane that struck an unprepared Galveston, Texas, on September 8, 1900. The strength of the storm, together with the lack of adequate warning, caught the population by surprise and cost the lives of 6000 people in the city and at least 2000 more elsewhere (Figure 10–23). Fortunately, hurricanes are no longer the unheralded killers they once were. Since the launching of the first meteorological satellite in 1960, meteorologists have been able to identify and track tropical storms even before they become hurricanes. Once a storm develops cyclonic flow, and the spiraling bands of clouds so typical of a hurricane, it receives continual monitoring (Figure 10–24).

(a) August 5, 1980

(b) August 7, 1980

(c) August 8, 1980

(d) August 10, 1980

FIGURE 10–24

Satellite images showing the movement of Hurricane Allen, one of the strongest storms of the century. The eye of the storm is most obvious on the images for August 7 and 8. Notice how the storm diminishes in intensity when it begins to move over land on August 10. (Courtesy of the National Environmental Satellite Service)

Although the amount of damage caused by a hurricane depends on several factors, including the size and population density of the area affected and the near-shore bottom configuration, certainly the most significant factor is the strength of the storm itself. By studying the storms of the past, a scale has been established that is used to rank the relative intensity of hurricanes (Table 10–3). A 5 represents the worst storm possible, and a 1 is least severe. The famous Galveston hurricane just mentioned, with winds in excess of 209 kilometers per hour and a pressure of 931 millibars, would be placed in category 4. Storms that fall into category 5 are rare. Only three have hit the United States this century: Camille hit Mississippi in 1969, a Labor Day hurricane struck the Florida Keys in 1935, and Allen hit the south Texas coast in 1980.

Damage caused by hurricanes can be divided into three categories: (1) wind damage, (2) storm surge, and (3) inland freshwater flooding. Although wind damage

TABLE 10–3 Saffir–Simpson Hurricane Scale

Scale Number (category)	Central Pressure (millibars)	Winds (km/hr)	Storm Surge (meters)	Damage
1	≥980	119–153	1.2–1.5	Minimal
2	965–979	154–177	1.6–2.4	Moderate
3	945–964	178–209	2.5–3.6	Extensive
4	920–944	210–250	3.7–5.4	Extreme
5	<920	> 250	>5.4	Catastrophic

is perhaps the most obvious of the categories, it is not directly responsible for the greatest amount of destruction. It does not mean, however, that wind damage cannot be significant. For some structures, the force of the wind is sufficient to cause total destruction. Mobile homes are particularly vulnerable. In addition, the strong winds can create a dangerous barrage of flying debris.

Storm Surge

Without question, the most devastating damage is caused by the storm surge. It not only accounts for a large share of coastal property losses, but it is also responsible for 90 percent of all hurricane-caused deaths. A **storm surge** is a dome of water 65 to 80 kilometers long that sweeps across the coast near the point where the eye makes landfall. If all wave activity were smoothed out, the storm surge is the height of the water above normal tide level (Figure 10–25). In addition, tremendous wave activity is superimposed on the surge. We can easily imagine the damage that this surge of water could inflict on low-lying coastal areas (Figure 10–26). In the delta region of Bangladesh, for example, most of the land is less than 2 meters above sea level. When a storm surge superimposed on normal high tide inundated that area on November 13, 1970, the official death toll was 200,000; unofficial estimates ran to 500,000. It was one of the worst natural disasters of modern times.

The torrential rains that accompany most hurricanes represent a third significant threat—flooding. Whereas the effects of storm surge and strong winds are concentrated in coastal areas, heavy rains may affect places hundreds of kilometers from the coast for several days after the storm has lost its hurricane-force winds. A well-known example of such destruction is Agnes. Although it was only a category 1 storm on the Saffir-Simpson scale, it was one of the costliest hurricanes of the century, creating more than $2 billion in damage and taking 122 lives. The greatest destruction was attributed to flooding in the northeastern portion of the United States, especially in Pennsylvania, where record rainfalls occurred. Another good example is Camille. Although this storm is best known for its exceptional storm surge and the devastation it brought to coastal areas, the greatest number of deaths associated with this storm occurred in the Blue Ridge Mountains of Virginia two days after Camille's landfall. Here many places received more than 25 centimeters of rain and severe flooding took more than 150 lives.

In the United States early warning systems have greatly reduced the number of deaths caused by hurricanes. At the same time, however, there has been an

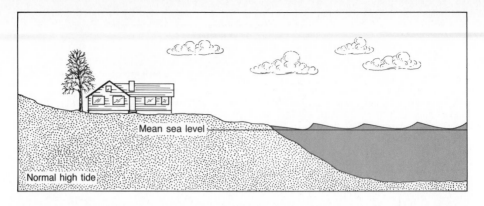

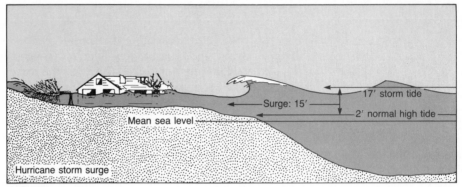

FIGURE 10–25
Superimposed upon normal high tide, a storm surge can devastate a coastal area. (Source: NOAA)

astronomical rise in the amount of property damage. The primary reason, of course, has been the rapid population growth in coastal areas. For example, during the 1960–1980 period, Florida, one of the states most vulnerable to hurricanes, doubled its coastal population. At the same time, Texas, another highly vulnerable state, experienced a 64 percent population increase in its 17 coastal counties. The National Weather Service is concerned that such population increases could set the stage for a major hurricane disaster because the evacuation of large numbers of people might require greater warning times than are presently available.

Hurricane Modification

Methods of halting hurricane destruction cannot rely on tracking alone. If possible, methods that reduce the intensity of the storm must also be designed. This is the goal of the hurricane modification experiments that have been going on since 1961. Recall that the energy for hurricanes comes from the release of latent heat in the ring of rising air, called the eye wall. The object of hurricane modification has been to seed portions of the storm beyond this zone of maximum wind velocity. Theoretically, at least, doing so would cause supercooled droplets to freeze, thus releasing latent heat that would then stimulate cloud growth beyond the eye wall. In this way, the inward spiraling flow of moist air would be intercepted and caused to ascend at a greater distance from the storm center. The net effect

(*continued on page 296*)

(a)

(b)

FIGURE 10–26

The 25-foot storm surge of Hurricane Camille at Pass Christian, Mississippi, devastated the coast. (a) The Richelieu Apartments before the hurricane. This substantial looking three-story building was directly across the highway from the beach. (b) The same apartments after the hurricane. (Photos by Chauncey T. Hinman)

IS THE UNITED STATES HEADED FOR HURRICANE DISASTER?

A Statement of Concern by the American Meteorological Society as adopted by the Council on January 12, 1986[*]

We are more vulnerable to hurricanes in the United States now than we have ever been in our history. Millions of people have been attracted to our beautiful coasts, exposing a large population to the threat from hurricanes and the storm surge associated with them. Unfortunately, the rate of improvement in achieving greater forecast lead times is not keeping pace with the time requirements for evacuating an ever-increasing coastal population. This situation demands increased emphasis on both meteorological research to improve predictions and on emergency plans for evacuation of populations threatened by storms.

During the past decade, a number of comprehensive evacuation studies by emergency management authorities have been completed and the results are alarming. It would take 20 to 30 hours to evacuate some of the most hurricane-prone locations, including the Galveston Bay area, the Tampa Bay area, the southwest Florida coast, the Florida Keys, and southeast Florida coast. Evacuation times for Corpus Christi, Texas, Beaumont, Texas, and Charleston, South Carolina, approach 15 hours. We do not even know the time required to evacuate our most vulnerable city, New Orleans, nor the coastal islands of New Jersey, which are a mecca for hundreds of thousands of summertime vacationers. Studies of these areas have yet to be completed, but it is estimated that their evacuation times are similar to that for the Galveston Bay area. Coastal residents find it difficult to believe that evacuation times are as great as indicated by comprehensive studies. According to a survey conducted a few years ago of 300 coastal residents, the median estimated time to evacuate the Sarasota, Florida, coastal area was one and a half hours. In reality, detailed

analysis reveals that it would take 16 hours to move threatened residents to safety.

Traditionally, the goal of the National Hurricane Center has been to provide 12 daylight hours of warning time for evacuation. Obviously, the time required to evacuate many coastal areas far exceeds the lead time of warnings. If we attempt to provide warning lead times that are compatible with evacuation studies, substantial unnecessary evacuations will occur because of the inherent uncertainty in long-range hurricane prediction.

Consider the problem that arises when an intense hurricane threatens an area that requires 24 hours to evacuate. If a perfect forecast could be made, the length of the coastal region that requires evacuation would normally be less than 100 miles. However, the mean error in predicting the position of a hurricane 24 hours in advance is slightly more than 100 miles. To provide for this uncertainty in the evacuation decision, approximately 300 miles of coast would have to be evacuated. Consequently, two-thirds of those who evacuated would have done so unnecessarily. Stated another way—for every three times evacuation is advised, only once will it prove necessary. This raises the issue of credibility. How many times will people evacuate unnecessarily before they lose faith in the warnings and ignore the advice to evacuate?

Hurricanes occasionally behave in unexpected ways that preclude the issuance of warnings in sufficient time to complete evacuation. Hurricane Alicia in 1983 had the potential to produce a catastrophe. It formed as a weak storm in the Gulf of Mexico in an environment that did not appear favorable for strengthening. Local officials in Galveston decided against complete evacuation of coastal areas. When Alicia strengthened significantly in the 18 hours before landfall, it was too late to totally evacuate the threatened area. Large loss of life was averted by the presence of a 15-foot seawall that was built to protect the city following the record hurricane disaster of 1900 that claimed 6000 lives. Few coastal locations have such massive seawalls. If a similar situation occurs in an unprotected area, resulting casualties could number in the hundreds or thousands.

Unanticipated acceleration of a hurricane causes similar warning problems. In 1938, a major hurricane was just east of the Bahamas. It accelerated northwestward and crossed Long Island, New York, only 30 hours later, at a speed approaching 60 miles per hour. Hurricane Hazel struck North Carolina at 50 miles per hour in 1954. Little warning preceded the arrival of these storms, and even today it might not be possible to provide more than 6 hours warning in similar circumstances.

One emergency option that should be studied is the concept of vertical evacuation (moving to the upper floors of substantial multi-story buildings). Even though it was introduced nearly 15 years ago, vertical evacuation has not been thoroughly evaluated. Consequently, not one community has chosen to adopt this as part of its procedures.

On the average, two hurricanes strike the United States every year. During the past few decades, however, the number of severe hurricanes that has affected the United States has been below average. Fewer hurricanes crossed our coastlines during the 1970s than during any other decade of the century. More than 35 years have passed since the Florida peninsula was hit by a devastating storm. Ironically, this fortunate anomaly only adds to the current warning problem. Recent studies show that more than 80 percent of the 40 million residents of counties bordering the Atlantic Ocean and Gulf of Mexico have never experienced a major hurricane.

Our vulnerability was particularly evident in 1985 when six hurricanes and two tropical storms struck the United States coastline. Although the sustained winds of these storms at landfall were not as strong as those of the most intense hurricanes, there were 37 deaths and property damage exceeded $4 billion. Stronger winds at the time of landfall could only increase these figures.

Technology has improved our capability to observe hurricanes and to display this information, for example, on television; forecasting hurricane motion and intensity still requires much research in order to achieve the skill necessary to adequately support the warning and evacuation components of the problem.

There is an urgent need in this nation to renew our efforts in dealing with the hurricane problem, including

1. Increasing research and development efforts focused directly on the hurricane prediction problem.

2. Completion of comprehensive evacuation studies for the entire coastline to determine the magnitude of the United States hurricane problem.

3. Development, by coastal communities, of emergency procedures to be initiated should a meteorological surprise occur which does not permit adequate lead time for evacuation.

4. Encouragement of governments at the local, state, and federal levels to adopt realistic growth-management regulations that properly consider the hurricane evacuation problem.

5. Development of objective decision-making techniques that will allow government officials to knowledgeably consider the uncertainties in the hurricane warnings when formulating "action" decisions.

6. Development of hurricane education programs that inform people of potential dangers, dispel misconceptions, and permit the public to derive maximum benefit from forecasts and warnings.

This statement is a plea for the protection of the lives and property of United States citizens. If we do not move forward quickly in seeking solutions to the hurricane problem, we will pay a severe price. The price may be thousands of lives.

would be an outward displacement of the eye wall and a reduction in maximum wind speeds (Figure 10–27). Recall from the discussion of the conservation of angular momentum that because the distance between an inward moving parcel of air and the center of the storm would increase, the rotational velocity would need to decrease.

Thus far experimental results are inconclusive and have not yet verified the hypothesis on which the attempted modification was based. It is safe to say that hurricane modification is still in its developmental stages and that many questions are still unresolved.

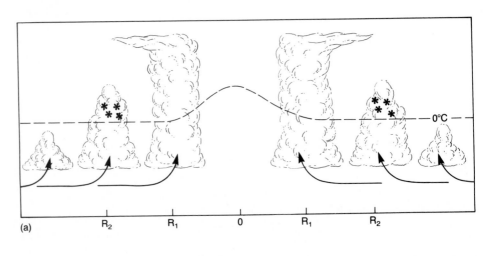

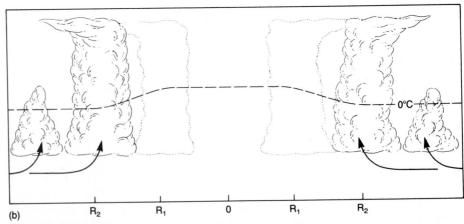

FIGURE 10–27
Schematic diagram showing hypothetical effect of seeding cumulus clouds in a hurricane. Seeding supercooled water (represented by dots) at radius R_2 in (a) results in growth of clouds at R_2 and a decay of old eye wall clouds at R_1 in (b). (From Richard A. Anthes, *Tropical Cyclones: Their Evolution, Structure and Effects*, Meteorological Monographs, Vol. 19, no. 41, p. 121, Boston: American Meteorological Society, 1982)

REVIEW

1. If you hear that a cyclone is approaching, should you immediately seek shelter?

2. What is the only fact of which you can be certain when you hear that a cyclone is approaching?

3. Compare the wind speeds and the sizes of middle-latitude cyclones, tornadoes, and hurricanes.

4. How do thunderstorms differ from the storms mentioned in question 3? In what ways are thunderstorms related to each one?

5. What is the primary requirement for the formation of thunderstorms?

6. Where would you expect thunderstorms to be most common on earth? In the United States?

7. At what time of day do most thunderstorms occur? Why?

8. Briefly describe the formation of a squall line.

9. How is thunder produced? How far away is a lightning stroke if the thunder is heard 15 seconds after the lightning is seen?

10. What is heat lightning?

11. Why do tornadoes have such high wind speeds?

12. Why are reliable measurements of pressure and winds in a tornado not available?

13. What general atmospheric conditions are most conducive to the formation of tornadoes?

14. When is the "tornado season"? Can you explain why it occurs when it does? Why does the area of greatest tornado frequency migrate?

15. Distinguish between a tornado watch and a tornado warning. Why is it that tornadoes cannot be predicted accurately?

16. What advantages does Doppler radar have over conventional radar?

17. Why do some areas of the world welcome the arrival of the hurricane season?

18. Using Figure 10–21, determine the pressure gradient (in millibars per 100 kilometers) from the 1004-millibar isobar to the center of the hurricane, a distance of approximately 400 kilometers.

19. When a parcel of air approaches the center of a hurricane, how does its speed change? What law explains this change?

20. Which of these statements about the eye of a hurricane are true and which are false?
    a. It is typically the warmest part of the storm.
    b. It is usually characterized by clear, blue skies.
    c. It is in the eye that winds are strongest.

21. Tropical storms that form near the equator do not acquire a rotary motion as cyclones of higher latitudes do. Why?

22. Which has the stronger winds, a tropical storm or a tropical depression?

23. Why does the intensity of a hurricane diminish rapidly when it moves onto land?

24. Hurricane damage can be divided into three broad categories. Name them. Which one of the categories is responsible for the greatest percentage of hurricane-related deaths?

25. A hurricane has slower wind speeds than a tornado, but a hurricane inflicts more total damage. How might this be explained?

26. The number of deaths in the United States attributable to hurricanes has continually declined over the last 50 years, but the number of tornado deaths has increased. Write an explanation to account for this situation.

27. When a storm warning is issued in some parts of the United States, some people board up their windows. Are they expecting a hurricane or a tornado? Explain.

VOCABULARY REVIEW

thunderstorm

cumulus stage

entrainment

mature stage

dissipating stage

severe thunderstorm

air-mass thunderstorm

squall line

gust front

lightning

flash

stroke

leader

step leader

return stroke

dart leader

thunder

tornado

Fujita intensity scale (F-scale)

tornado watch

tornado warning

Doppler radar

mesocyclone

hurricane

law of conservation of angular momentum

eye wall

eye

tropical disturbance

tropical depression

tropical storm

storm surge

11

Weather Analysis

Modern society's ever-increasing demand for more accurate weather forecasts is evident to most of us. The spectrum of needs for weather predictions ranges from the general public's desire to know if the weekend's weather will permit an outing at the beach to NASA's desire for clear skies on a particular launch date. Such diverse industries as airlines and fruit growers depend heavily on accurate weather forecasts. In addition, the designs of buildings, smokestacks, and many industrial facilities rely on a sound knowledge of the atmosphere. We are no longer satisfied with short-range predictions but instead are demanding accurate long-range predictions. Such a question as "Will the Northeast experience an unseasonably cold winter?" has become common. All these demands place a greater burden on the National Weather Service for better and longer-range forecasts.

Weather Forecasting

To produce even a short-range forecast is an enormous task involving numerous steps, including collecting, transmitting, and compiling weather data on a global scale. These data must then be analyzed so that an accurate assessment of the current conditions can be made. From current weather patterns a number of methods are used to determine the future state of the atmosphere, a task generally called **weather forecasting.** Although our goal here is to provide insight into the job of weather forecasting, many procedures used in modern weather prediction are beyond the scope of the text. Consequently, we can provide only a brief overview of this important aspect of the weather business.

SYNOPTIC WEATHER CHARTS

Weather Analysis

Before weather can be accurately predicted, the forecaster must have a firm grasp of current atmospheric conditions. This enormous task, called **weather analysis,** involves collecting, transmitting, and compiling millions of pieces of observational data. Because of the ever-changing nature of the atmosphere, this job must be accomplished as rapidly as possible. In this regard, modern high-speed computers have greatly aided the weather analyst. In addition to collecting an overwhelming quantity of data, the analyst must display it in a form that can be easily comprehended by the forecaster. This step is accomplished by placing the information on a

Synoptic Weather Chart

number of **synoptic weather charts** (Figure 11–1). They are called synoptic, which means coincident in time, because they display the weather conditions of the atmosphere at a given moment. These weather charts can be thought of as

FIGURE 11–1
Meteorologist inspecting a forecast chart prepared by
a computer at the National Meteorological Center,
Washington, D.C. (Courtesy of NOAA)

symbolic representations of the state of the atmosphere. Thus, to the trained
eye, a weather chart is a picture of the atmosphere that depicts its conditions,
including motion.

A vast network of weather stations is required to produce a weather chart
that will encompass enough to be useful for short-range forecasts. On a global
scale, the **World Meteorological Organization,** which consists of over 130
nations, is responsible for gathering the needed data and for producing some
general prognostic charts. About 10,000 surface stations located on both land
and on ships at sea report the atmospheric conditions four times each day at
000, 0600, 1200, and 1800 Greenwich mean time. In addition, satellite images
and radiosonde data are used to determine conditions aloft. This network is becoming more complete each year; nevertheless, even today vast sections of the globe,
especially the large portions of the oceans outside shipping lanes, are inadequately
monitored.

Once collected, the information is transmitted to three World Meteorological
Centers, located in Melbourne, Australia; Moscow, USSR; and Washington, D.C.
From these world centers the compiled data are sent to the National Meteorological
Center of each participating country. In the United States the National Meteorological
Center is located in Washington, D.C. From here the information is further dissemi-
nated to numerous regional and local meteorological centers where it is used to
provide more detailed forecasts for those respective areas.[1]

**World Meteorological
Organization (WMO)**

[1] See the essay "The National Weather Service," written by its director, Richard E. Hallgren, p. 314.

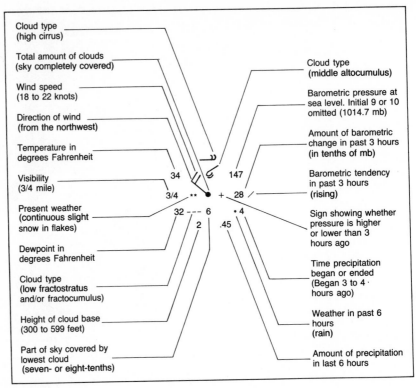

FIGURE 11–2
A specimen station model showing the location of general data. (Abridged from International Code)

Actual production of surface weather charts involves first plotting the data from selected observing stations. By international agreement, data are plotted by using the symbols illustrated in Figure 11–2. Normally data that are plotted include temperature, dew point, pressure and its tendency, cloud cover (height, type, and amount), wind speed and direction, and both current and past weather. They are always plotted in the same position around the station symbol for easy reading. Using Figure 11–2, for example, we see that the temperature is plotted in the upper-left corner of the sample model, and it will always appear in that location. The only exception to this arrangement is the wind arrow that is oriented with the direction of airflow. A more complete weather station model and a key for decoding weather symbols are found in Appendix B. Before you read the next section, which uses weather maps to trace the movement of a wave cyclone, it would be advantageous to become familiar with the station model if you have not already done so.

Once data have been plotted, isobars and fronts are added to the weather charts (Figure 11–3). Isobars are usually plotted on surface maps at intervals of 4 millibars in order to keep the map from becoming too cluttered. The positions

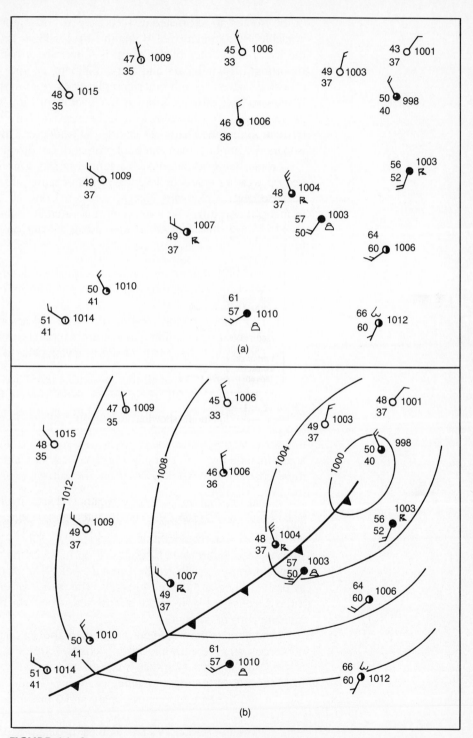

FIGURE 11–3
Simplified weather chart including (a) stations giving data for temperature, dew point, wind direction, sky cover, and barometric pressure and (b) isobars and cold front.

of the isobars are estimated as accurately as possible from the pressure readings available. Note in Figure 11–3 that the 1012-millibar isobar, which is located near the lower-left corner of the chart, is found about halfway between the stations reporting 1010 millibars and 1014 millibars, as you would expect. Frequently observational errors and other complications require that the analyst smooth the isobars so that they conform to the overall picture. Many irregularities in the pressure field are caused by local influences and have little bearing on the larger circulation that the charts are attempting to depict. Once the construction of the isobars is complete, the centers of high and low pressure are designated.

Fronts, being boundaries separating contrasting air masses, can often be identified on weather charts by locating places of rather abrupt changes in conditions. Because several elements change across a front, all are examined so that the frontal position might be located most accurately. Some of the most easily recognized changes that can aid in identifying a front on a surface chart are listed here:

1. Marked temperature contrast over a short distance
2. Wind arrows veering (turning in a clockwise direction) by as much as 90 degrees
3. Humidity variations commonly occurring across a front that can be detected by examining dew-point temperatures
4. Clouds and precipitation patterns giving clues to the positions of fronts

Notice in Figure 11–3 that all the conditions listed are easily detected across the frontal zone, but not all fronts are as easily detected as those on our sample map, for in some cases, surface contrasts may be rather subdued. When this happens, charts of the upper air, where flow is less complex, become an invaluable tool for detecting fronts. Other useful aids for determining the position of frontal zones include the most recently available weather charts. Generally, by examining these charts, it is possible to estimate the current position of a preexisting front from the wind field.

In order to describe the atmosphere as completely as possible, surface charts and charts for several levels of the atmosphere are produced. Upper-air charts are drawn on a regular basis for 850-, 700-, 500-, 300-, 200-, and 100-millibar surfaces (see Figure 6–10). Recall that on these charts height contours instead of isobars are used to depict the pressure field. The charts also contain isotherms as dashed lines at 5-degree intervals. This series of upper-air charts provides a three-dimensional view of the atmosphere from which the forecaster can better evaluate current conditions.

In addition to the aforementioned charts, the National Weather Service provides numerous other special-information charts. One of the most widely distributed is the Daily Weather Map that is constructed for 7:00 A.M. EST (1200 GMT). This map consists of a surface weather chart, a 500-millibar height contour chart, a highest and lowest temperatures chart, and a precipitation chart.[2] We can learn a

---

[2] The Daily Weather Maps are for sale by the Public Documents Department, U.S. Government Printing Office, Washington, DC 20402, at a reasonable annual subscription rate. These maps are mailed weekly and contain a series of charts for the preceding week.

great deal about the weather from careful examination of these weather charts. When understood, they provide a pictorial view of weather systems as they migrate across the United States. The following section uses the daily maps as an aid in examining the weather over the United States during a 3-day period. This discussion is intended not only to illustrate the information provided by weather charts but also to provide additional reinforcement of the wave cyclone model discussed earlier.

WEATHER OF A WAVE CYCLONE

March 1975 followed a rather uneventful February, but the circulation of the westerlies changed with the onset of spring. We have selected a cyclonic storm that was generated during this particular month to illustrate the nature of a real wave cyclone. Our "sample" storm was one of three cyclones to migrate across the United States during the latter half of March. This cyclone reached our West Coast on March 21 at a position over the Pacific a few hundred kilometers northwest of Seattle, Washington. Like many Pacific storms, this one rejuvenated over the western United States and moved eastward into the Plains states. By the morning of March 23 it was centered over the Kansas-Nebraska border, as shown in Figure 11–4. At this time the central pressure had reached 985 millibars, and its well-developed cyclonic circulation exhibited a warm front and a cold front. During the next 24 hours the forward motion of the storm's center became sluggish, it curved slowly northward to a position in northern Iowa, and the pressure continued to deepen to 982 millibars (Figure 11–5). Although the storm center advanced slowly, the associated fronts moved vigorously toward the east and somewhat northward. The northern sector of the cold front overtook the warm front and generated an occluded front, which by the morning of March 24 was oriented in nearly an east-west direction, as can be seen in Figure 11–5. This period in the storm's history marked one of the worst blizzards ever to hit the north-central states. While the winter storm was brewing in the north, the cold front marched from northwestern Texas to the Atlantic Ocean. During its 2-day trek to the ocean this violent cold front, with an associated squall line, generated numerous severe thunderstorms and spawned 19 tornadoes. By March 25 the low pressure had diminished in intensity (1000 millibars) and had split into two centers (Figure 11–6). Although the remnant low from this system, which was situated over the Great Lakes, generated some weather for the remainder of March 25, by the following day it had completely dissipated.

After a brief view, let us now revisit the passage of this cyclone in more detail by using the weather charts for March 23 through March 25 (Figures 11–4 to 11–6) as our guide. The weather map for March 23 clearly depicts a developing cyclone. The warm sector of this system, as exemplified by Fort Worth, Texas, is under the influence of a warm, humid air mass having a temperature of 70°F and a relative humidity of approximately 80 percent. (*Note*: The relatively humidity is determined from air temperature and dew point by using relative humidity and dew-point tables.) Notice in the warm sector that winds are from the south

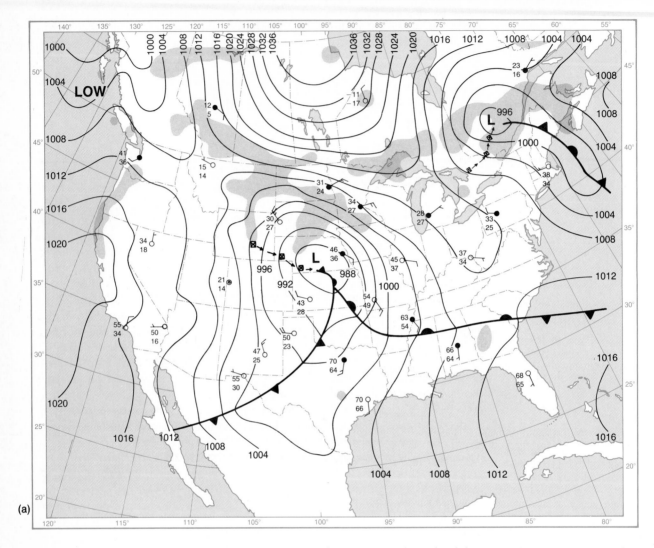

(a)

and are overrunning cooler air situated north of the warm front. In contrast, the air behind the cold front is 20 to 40°F cooler than the air of the warm sector and is flowing in a southeasterly direction, as depicted by the data for Roswell, New Mexico. This cold surge consists of air with below-freezing temperatures in the area just north of the Canadian border.

Prior to March 23 this system had generated some precipitation in the Northwest and as far south as California. On the morning of March 23, when the first map was constructed, however, little activity was occurring along the fronts, but as the day progressed, the storm intensified and changed dramatically. The map for March 24 illustrates the typical mature cyclone that developed during the next 24 hours. Careful examination of a few past weather symbols provides some insight into the fury of this storm. In particular, look at stations located ahead of

(b)

FIGURE 11–4
(a) Surface weather map for March 23, 1975. (b) Satellite photograph showing the cloud patterns on that day. (Courtesy of NOAA)

the cold front and north of the occluded front. Also, notice on the map of March 24 the closely spaced isobars that indicate the strength of this system as it affected the circulation of the entire eastern two-thirds of the United States. A quick glance at the wind arrows reveals a strong counterclockwise flow converging on the low. The activity in the cold sector of the storm just north of the occluded front produced one of the worst March blizzards since the Great Blizzard of March 1951. In the Duluth-Superior area winds were measured up to 81 miles per hour. Unofficial estimates of wind speeds in excess of 100 miles per hour were made on the aerial bridge connecting these cities. Winds blew 12 inches of snow into 10- to 15-foot drifts, and some roads were closed for 3 days (Figure 11–7). One large supper club in Superior was destroyed by fire because drifts prevented firefighting equipment from reaching the site of the blaze.

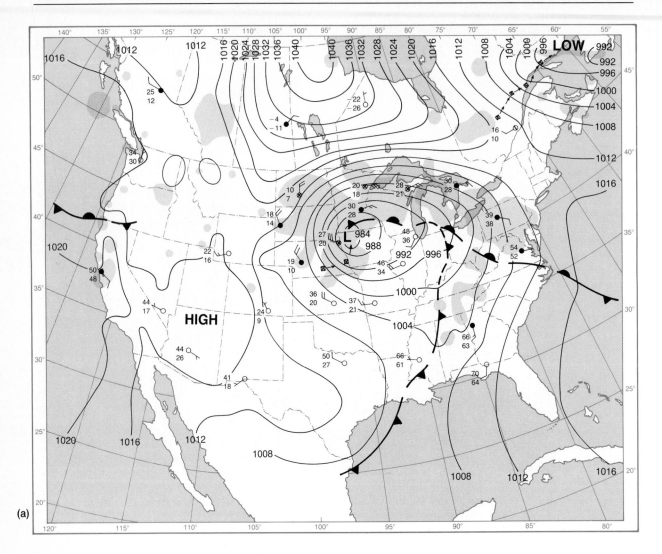

(a)

At the other extreme was the weather produced by the cold front as it closed in on the warm, humid air flowing into the warm sector. By the late afternoon of March 23 the cold front had generated a hailstorm in parts of eastern Texas. As the cold front and squall line moved eastward, they affected all of the southeastern United States except southern Florida. Throughout this region numerous thunderstorms were spawned. Although high winds, hail, and lightning caused extensive damage, the 19 tornadoes generated by the storm caused even greater death and destruction. The path of the front can be easily traced from the reports of storm damage. By the evening of March 23 hail and wind damage were reported as far east as Mississippi and Tennessee. Early on the morning of March 24 golf-ball-sized hail was reported in downtown Selma, Alabama. About 6:30 A.M. that day the "Governor's Tornado" struck Atlanta, Georgia. Here the storm displayed

(b)

FIGURE 11–5

(a) Surface weather map for March 24, 1975. (b) Satellite photograph showing the cloud patterns on that day. (Courtesy of NOAA)

its worst temper. Damage was estimated at over $50 million, three lives were lost, and 152 persons were injured. The 12-mile path of the "Governor's Tornado" cut through an affluent residential area of town that included the governor's mansion (hence the name). The official report on this tornado notes that no mobile homes lay in the tornado's path. Why do you suppose that fact was worth noting in a report on tornado destruction? The last damage along the cold front was reported at 4:00 A.M. on March 25 in northeastern Florida. Here hail and a small tornado caused minor damage. Thus a day and a half and some 1200 kilometers after the cold front became active in Texas, it left the United States and entered the Atlantic.

By the morning of March 24 we can see that cold polar air had penetrated deep into the United States behind the cold front (Figure 11–5). Forth Worth, Texas, which just the day before was situated in the warm sector, now experienced

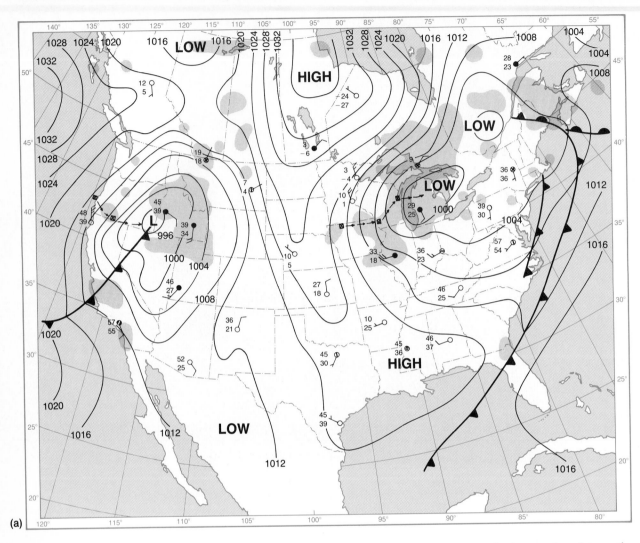

(a)

cool northwesterly winds. Below-freezing temperatures had moved as far south as northern Oklahoma. Notice, however, that by March 25 Forth Worth was again experiencing a southerly flow. We can conclude that it is a result of the decaying cyclone that no longer dominated the circulation in the region. We can also safely assume that a warming trend was experienced in Forth Worth over the next day or so. Also notice on the map for March 25 that a high was situated over southwestern Mississippi. The clear skies and calm conditions associated with the center of a subsiding air mass are well illustrated here.

The observant reader might have already noticed another cyclone moving in from the Pacific on March 25 as the other exited the country. This storm developed in a similar manner, but it was centered somewhat farther north. As you should have already guessed, another blizzard struck the northern Plains states and a

(b)

FIGURE 11–6
(a) Surface weather map for March 25, 1975. (b) Satellite photograph showing the cloud patterns on that day. (Courtesy of NOAA)

few tornadoes slipped through Texas, Arkansas, and Kentucky, while precipitation dominated the weather pattern in the central and eastern United States.

Having examined the general weather associated with the passage of this cyclone from March 23 through March 25, let us look at the weather experienced at a single location during this same time period. For this purpose, we have selected Peoria, a city in central Illinois located just north of the Springfield station shown on the weather charts. Before you read the description of Peoria's weather, try to answer the following questions by using Table 11–1, which provides weather observations at 3-hour intervals during this time period. Also use the three weather charts provided and recall the general wind and temperature changes expected with the passage of fronts. If you need to refresh your memory concerning the idealized weather associated with a wave cyclone, refer to Chapter 9.

FIGURE 11–7
Main street, Bowdle, South Dakota following a late spring blizzard. (Photo by Larry Gauer, *The Bowdle Pioneer*)

1. What type of clouds were probably present in Peoria during the early morning hours of March 23?
2. At approximately what time did the warm front pass through Peoria?
3. List two lines of evidence indicating that a warm front did pass through Peoria.
4. How did the wind and temperature changes during the early morning hours of March 23 indicate the approach of a warm front?
5. Explain the slight temperature increases experienced between 6:00 P.M. and 9:00 P.M. on March 23.
6. By what time had the cold front passed through Peoria?
7. List some changes that indicate the passage of the cold front.
8. Since the cold front had already gone through Peoria by noon on March 24, how do you account for the snow shower that occurred during the next 24 hours?
9. Did the thunderstorm in Peoria occur with the passage of the warm front, the cold front, or the occluded front?
10. Basing your answer on the apparent clearing skies late on March 25, would you expect the low temperature on March 26 to be lower or higher than on March 25? Explain.

Read the following discussion to obtain a more complete description of the weather you just reviewed.

We begin our weather observations in Peoria just after midnight on March

TABLE 11–1 Weather Data for Peoria, March 23 to 25

	Temperature (°F)	Wind Direction	Cloud Coverage (tenths)	Visibility (miles)	Weather and Precipitation
March 23					
00:00	43	ENE	5	15	
3:00 A.M.	43	ENE	8	15	
6:00 A.M.	42	E	5	12	
9:00 A.M.	50	ESE	10	10	
12:00 P.M.	61	SE	10	12	
3:00 P.M.	64	SE	10	10	Thunderstorm with rain showers
6:00 P.M.	64	SE	10	6	Haze
9:00 P.M.	65	S	10	10	Thunderstorm with rain
March 24					
00:00	57	WSW	10	15	
3:00 A.M.	47	SW	2	15	
6:00 A.M.	42	SW	6	15	
9:00 A.M.	39	SW	8	15	
12:00 P.M.	37	SSW	10	15	Snow showers
3:00 P.M.	33	SW	10	10	Snow showers
6:00 P.M.	30	WSW	10	10	Snow showers
9:00 P.M.	26	WSW	10	6	Snow showers
March 25					
00:00	26	WSW	10	15	
3:00 A.M.	25	WSW	10	8	Snow shower
6:00 A.M.	24	WSW	10	1	Snow
9:00 A.M.	26	W	10	3	Snow
12:00 P.M.	25	W	10	12	Snow
3:00 P.M.	27	WNW	10	12	Snow
6:00 P.M.	25	NW	9	12	
9:00 P.M.	24	NW	4	15	

23. The sky contains cirrus and cirrocumulus clouds and cool winds from the ENE dominate. As the early morning hours pass, we observe a slight wind shift toward the southeast and a very small drop in temperature. (Recall that veering winds are a sign of an approaching front.) Three hours after sunrise altocumulus clouds are replaced by stratus and nimbostratus clouds that darken the sky. Yet the warm front passes without incident. The 20°F increase in temperature and the wind shift from an easterly to a southerly flow, which occurred between 6:00 A.M. and noon on March 23, marked the passage of the warm front as Peoria entered the warm sector. The pleasant 60°F temperatures were welcomed in Peoria, which experienced a day 14°F above normal. But the mild weather was short-lived because the part of the cyclone that passed Peoria was near the apex of the storm, where the cold front is generally close behind the warm front. By

that afternoon the cold front had generated numerous thunderstorms from cumulonimbus clouds embedded in the warm sector just ahead of the cold front. Strong winds, ½-inch hail, and one tornado caused some local damage in the Peoria area. The temperature remained unseasonably warm during the thunderstorm activity. Later, during the evening of March 23, the passage of the cold front was marked by a wind shift, rapidly clearing skies, and a temperature drop of nearly 20°F, all in a period of less than 4 hours. Throughout March 24 southwest winds brought cold air around the back side of the intensifying storm (see Figure 11–6). Although the surface fronts had passed, this intense cyclone, with its occluded front aloft, was generating snowfall over a wide area. It is not unusual for an occluding cyclone to slow its movement like this one did. For about 24 hours, snow flurries dominated Peoria's weather picture. By noon on March 25 the storm had lost its punch and the pressure began to climb. The skies began to clear and the winds became northwesterly as the once tightly wound storm weakened. The mean temperature on March 25 was 16°F below normal, and the ground was covered with snow.

The effect of a spring cyclone on the weather of a midlatitude location should be apparent from this example. Within a period of 3 days, Peoria's temperatures changed from unseasonably warm to unseasonably cold. Thunderstorms with hail were followed by snow showers. It should also be evident that the north–south temperature gradient, which is most pronounced in the spring, generates these intense storms. Recall that it is the role of these storms to transfer heat from the tropics poleward. But because of the earth's rotation, this latitudinal heat exchange is complex, for the Coriolis force gives the winds a zonal (west-to-east) orientation. If the earth rotated more slowly, or not at all, a more leisurely north-south flow would exist and would possibly reduce the latitudinal temperature gradient. Thus the tropics would be cooler and the poles warmer and the midlatitudes would not experience such intense storms.

THE NATIONAL WEATHER SERVICE

Richard E. Hallgren[*]

The atmosphere serves as a link connecting all of humanity. In our national life, weather's effect on man is all pervasive. There isn't a part of the nation's economy that doesn't feel the impact of violent weather. And in much less striking ways, weather affects the lives of everyone every day.

The basic function of today's National Weather Service (NWS), as it has been throughout more than a century of weather service, is to provide scientific and technological assistance in the general field of the atmospheric sciences.

The Weather Service's primary responsibility and principal role has been and will continue to be to save lives, reduce injuries, and minimize property loss from the extreme weather events that plague our nation each year. The United States has a greater assortment of severe weather than any other nation in the world. Hurricanes, tornadoes, floods, flash floods, thunderstorms, and severe winter weather kill an inordinate number of our citizens, injure thousands more,

[*] Dr. Hallgren is the director of the National Weather Service. He has published numerous scientific and technical papers on meteorological systems, cloud physics, and atmospheric electricity. He is a Fellow and past president of the American Meteorological Society.

and cause damage estimated at well over $2 billion every year.

Organizationally, the NWS is a main component of the National Oceanic and Atmospheric Administration (NOAA). NOAA, created in 1970, reflects the acute linkage between global atmospheric and oceanic processes.

The NWS has an extensive operating program and the heart of its operations is at the more than 400 field offices and observatories in the 50 states and overseas. In one year about 3.5 million weather observations are taken at NWS facilities. Data from this network and from more than 150 other nations of the world flow to NWS national centers, forecast offices, weather service offices, and river forecast centers.

The forecast and warning program is carried out through three organizational levels.

The *National Meteorological Center* (*NMC*) located just outside Washington, D.C., is the nerve center of the forecast operation. It is responsible for the preparation of the synoptic-scale guidance material and long-range forecasts. The Center provides a single source for global analyses and prognoses. The *National Severe Storms Forecast Center* (*NSSFC*) in Kansas City, Missouri, maintains a constant watch for severe weather potential around the country and issues severe thunderstorm and tornado watches. Hurricane watches and warnings for the Atlantic, Caribbean, and Gulf of Mexico are issued by the *National Hurricane Center* (*NHC*) in Miami, Florida. Hurricane services for the eastern Pacific are provided by the San Francisco Forecast Office, and the Honolulu Forecast Office covers hurricanes in the central Pacific.

Thirteen river forecast centers around the country provide forecasts and warnings concerning floods and river stage levels for more than 2000 points along major U.S. river systems.

The Honolulu-based tsunami warning system provides early detection and timely warning of tsunami (sometimes referred to—erroneously—as "tidal waves"). Tsunami are caused by undersea earthquakes and volcanic eruptions throughout the Pacific basin.

The second echelon, the 52 Weather Service Forecast Offices (WSFOs), are the backbone of the field forecasting operation. These offices issue forecasts and warnings for entire states or for large portions of the states and assigned zones. State forecasts are issued twice daily for a 48-hour period and are updated as often as necessary. Extended forecasts—looking ahead 5 days—are issued daily for state-wide areas. The WSFOs also provide the main forecast support for marine and aviation programs as well as guidance for the agricultural and fire weather programs.

Each year NWS meteorologists provide specialized meteorological services to thousands of domestic and foreign air carriers and to more than 800,000 general aviation users.

Specialized weather forecasts are provided by meteorologists in agricultural areas, to help cut back on our $1.6 billion annual loss of weather-damaged crops.

Marine weather and oceanographic services are provided by special units—for small privately owned craft which operate near shore, for mammoth ocean-going vessels, and for the nation's fishing fleet.

NOAA offices across the land provide pollution-potential forecasts and other data to cities with air pollution problems.

Over 200 Weather Service Offices (WSOs) represent the third forecast level. They issue forecasts which are local adaptations of the zone forecasts. These offices also have the all-important county warning responsibility. Together with the WSFOs, they use the guidance material from the centers along with information from weather radar, environmental satellites, surface- and upper-air observations, and reports from volunteer storm spotters to issue and disseminate severe weather and flood warnings to every county, parish, and major metropolitan area in the nation.

The coming years will see some significant changes in the National Weather Service—both in the services offered, and the way in which they are provided. Because of national priorities and the scientific and technical opportunities, the NWS will give special emphasis to four broad areas:

1. Warnings of severe weather and flooding
2. Food and fiber production
3. Water resources management
4. Energy production, distribution, and use

A number of technological possibilities for detecting severe weather, for communicating, integrating and displaying the data, and for disseminating weather forecasts and warnings can now be clearly foreseen. The geostationary satellite, Doppler radar, and a whole range of ground-based remote-sensing systems will permit far better detection of severe weather. This new capacity to observe the small-scale atmospheric circulation will dramatically improve severe weather and flash-flood warnings. Low cost mini- and micro-computers now make it possible for forecasters to assimilate the information and make critical decisions quickly. Finally, the ability to disseminate the warnings through radio and TV systems gives the NWS the last link for a much improved severe weather and flash-flood warning system. A major goal of the Weather Service will be to put this technology to work for the safety and well-being of the public.

As our population grows, the demands for food, fiber, water, and energy increase dramatically. The best weather information possible is needed to manage these critical resources efficiently and effectively. For example, in food production the Weather Service must continue to help the farmer, but key information for national decisions on the production of food must also be provided. In water resources it is essential that the NWS provide the best prediction possible of the water flow in the rivers. And as the nation turns to the use of solar, wind, and ocean thermal energy, weather information becomes more and more vital. It will be a factor in determining the rate of production from these sources and have a day-to-day impact on the distribution systems.

The total Weather Service needs of this nation cannot and should not be met by the NWS alone. Private meteorologists will continue to play an important role, particularly in providing individualized services to major companies or specific site operations. The NWS must also coordinate many of its programs with other federal agencies and state and local governments. This coordination must be strengthened to ensure an effective overall Weather Service program in our nation.

The future is challenging, exciting, and bright for National Weather Service personnel and will be rewarding to the people we serve.

WEATHER FORECASTING

In most instances, accurate weather prediction is the ultimate goal of atmospheric research. The primary exceptions are the relatively rare attempts at modifying the weather. The key to improving forecasts lies in the meteorologist's ability to predict the future state of the atmosphere precisely. Robert Ryan, a broadcast meteorologist, summarized the difficulties involved in weather forecasting with the following remarks:

> Imagine a system on a rotating sphere that is 8000 miles wide, consists of different materials, different gases that have different properties (one of the most important of which, water, exists in different concentrations), heated by a nuclear reactor 93 million miles away. Then, just to make life interesting, this sphere is oriented such that, as it revolves around the nuclear reactor, it is heated differently at different locations at different times of the year. Then, someone is asked to watch the mixture of gases, a fluid only 20 miles deep, that covers an area of 250 million square miles, and to predict the state of that fluid at one point on the sphere two days from now. This is the problem weather forecasters face.[3]

[3] Robert T. Ryan, "The weather is changing . . . or Meteorologists and Broadcasters, the twain meet," *Bulletin of the American Meteorological Society*, 63, no. 3 (March 1982), 308.

Because of the complex and highly quantitative nature of modern weather forecasting, we can only highlight the approaches used here, which include, but are not limited to, the traditional synoptic approach, statistical methods, and numerical weather prediction. The object of each method is not only to project the location and possible intensification of existing pressure systems but also to determine the formation of new storm centers.

Synoptic Weather Forecasting

Synoptic weather forecasting was the primary method used in making weather predictions until the late 1950s. As the name implies, synoptic weather charts are the basis of these forecasts. From the careful study of weather charts over many years a set of empirical rules was established to aid the forecaster in estimating the rate and direction of weather system movements. When the forecaster knows the type of weather being generated along a front and is able to predict its motion, for example, a rather accurate forecast for the affected area can be made. But because cyclonic systems change so quickly, these forecasts are generally accurate only on a short-range basis of a few hours or perhaps a day.

Early attempts to predict cyclone development from synoptic charts relied heavily on the analysis of surface fronts. Since the discovery of the relationship between the flow aloft and surface weather, however, these efforts have been supplanted by the use of upper-air data. Recall, for instance, that wave cyclones can develop without the prior existence of surface fronts. Later it was shown that other methods besides synoptic analysis can more accurately predict the future state of the atmosphere. This statement is particularly true for forecasts made for periods longer than 1 or 2 days. Nevertheless, empirical rules applied to synoptic charts are still used by local forecasters in their attempt to pinpoint the occurrence of specific events, such as the arrival time of a storm.

Numerical Weather Prediction (NWP)

Modern weather forecasting relies heavily on **numerical weather prediction** (NWP). The word "numerical" is misleading, for all types of weather forecasting are based on some quantitative data and therefore could fit under this heading. Numerical weather prediction is based on the fact that the gases of the atmosphere obey a number of known physical principles. Ideally these physical laws can be used to predict the future state of the atmosphere, given the current conditions. This situation is analogous to predicting future positions of the moon based on physical laws and the knowledge of its current position. Still, the large number of variables that must be included when considering the dynamic atmosphere makes this task extremely difficult. To simplify the problem, numerical models were developed that omit some variables by assuming that certain aspects of the atmosphere do not change with time. Although these models do not fully represent the "real" atmosphere, their usefulness in prediction has been well established.

Most modern approaches to weather forecasting strive to predict the flow pattern aloft. Meteorologists use this information to project favorable sites for cyclogenesis. Even the most simplified models, however, require such a vast number of calculations that they could only be used after the advent of high-speed computers (Figure 11–8). Figure 11–9 summarizes the improvements that have been made since the mid-1950s in forecasting pressure and wind patterns in the middle levels (500-millibar level) of the atmosphere over North America. Note that terms like *barotropic* and *geostrophic* are names for the numerical models that were

FIGURE 11–8
High-speed computers used to formulate prognostic charts (Courtesy of NOAA)

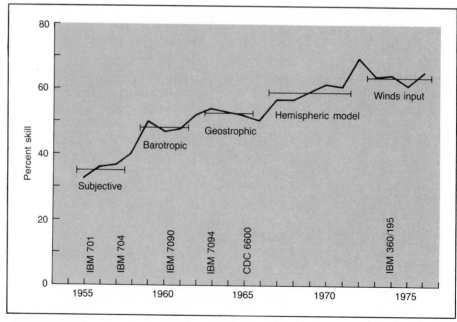

FIGURE 11–9
Average skill in predicting the middle- (500-millibar-) level atmospheric flow pattern over
North America 36 hours in advance. See the text for details. Here the skill score is designed
to eliminate the success that would be expected by random chance. (From F. Shuman,
"Numerical Weather Prediction." *Bulletin of the America Meteorological Society,* Vol. 59,
No. 1, Jan. 1978)

used during different periods. The dates along the bottom of the graph indicate when faster computers were available. It is apparent that forecasts of airflow aloft have improved significantly over the years. Still, meteorologists have not made as much progress in predicting the temperature variations and occurrences of precipitation that are expected to accompany changes in atmospheric circulation. As faster computers are developed and improvements in atmospheric models are made, we can expect even greater advances in the area of weather prediction. Only a few decades ago it was not uncommon to hear people say "It should be nice tomorrow; the weather report calls for rain." This attitude no longer prevails, thanks partly to advances in NWP.

Although the accuracy of numerical weather predictions greatly exceeds results obtained from more traditional methods, the prognostic charts obtained by these techniques are rather general. So the detailed aspects of the weather must still be determined by applying traditional methods to these charts. Furthermore, numerical forecasts are limited by deficiencies in observational data. Stated another way, a NWP can be no more complete or more accurate than the data that go into making it.

Statistical Methods (in forecasting)

Statistical methods are often used in conjunction with and to supplement the NWP. Statistical procedures involve the study of past weather data in order to uncover those aspects of the weather that are good predictors of future events. Once these relationships are established, current data can then be used to project future conditions. This procedure can be used to predict the overall weather, but it is most often used to determine one aspect of the weather at a time—for example, to project the maximum temperature for the day at a given location. Statistical data relating temperature to wind speed and direction, cloud cover, humidity, and to the season of the year are first compiled. These data are displayed on charts that provide a reasonable estimate of the maximum temperature for the day from these aspects of the current conditions.

Long-Range Forecasting

Long-range forecasting is another area in which statistical studies have proven valuable. Currently the National Weather Service prepares weekly, monthly, and seasonal weather outlooks. These are not weather forecasts in the usual sense; they are estimates of the rainfall and temperatures that should be expected during these periods. These projections only indicate whether the region will experience near-normal conditions. Detailed forecasts for more than a few days are currently beyond the capability of the National Weather Service.

The general monthly extended forecasts are produced by first constructing a mean 700-millibar contour chart for the coming month. This process requires taking into account the statistical records for that season of the year and altering them based on the known effects of such items as ocean temperatures and snow cover. Once this chart is compiled, the relationships between the flow aloft and the development and movement of surface weather patterns are considered in making a prediction for each segment of the United States.

Analog Method

Another statistical approach to weather forecasting is called the **analog method.** The idea here is to locate conditions in the weather records that are as nearly analogous to current conditions as possible. Once they are found, the sequence of weather events should parallel those of the past situation. Although it seems a straightforward method of prediction, it is not without its drawbacks. No two periods of weather are identical in all respects, and there are simply too many variables to match. Even when two periods seem well matched, the sequence of weather that follows may differ greatly in each case. The main problem with this method may well be the lack of complete-enough information, which also, of course, limits the usefulness of numerical weather predictions.

In summary, weather forecasts are produced by using several methods. The National Weather Service primarily uses numerical weather prediction methods to generate large-scale prognostic charts. These charts are then disseminated to regional and local forecast centers that use traditional synoptic and statistical methods to generate more specific forecasts.

WEATHER FORECASTING AND UPPER-LEVEL FLOW

In Chapter 9 we demonstrated that a strong correlation exists between cyclonic disturbances and the wavy flow in the westerlies aloft. By comparing surface charts and charts showing flow aloft, we can also see that upper-air patterns are somewhat simpler than the surface circulation. Because of these and other facts, modern weather forecasting depends heavily on predicting the weather from changes in the upper-level flow. Once the future state of the flow pattern aloft is determined, this information is used to make predictions concerning changes in the surface pressure systems.

Although much is still unknown about the wavy flow of the westerlies, some of its most basic features are understood with some degree of certainty. Among the most obvious features of the flow aloft are the changes that occur seasonally. The change in wind speed is reflected on upper-air charts by more closely spaced contour lines in the cool season. The seasonal fluctuation of wind speeds is a consequence of the seasonal variation of the temperature gradient. The steep temperature gradient across the middle latitudes in the winter months corresponds to stronger flow aloft. In addition, the position of the polar jet stream fluctuates seasonally. Its mean position migrates southward with the approach of winter and northward as summer nears. By midwinter, the jet core may penetrate as far south as central Florida. Because the paths of cyclonic systems are guided by the flow aloft, we can expect the southern tier of states to encounter most of their severe storms in the winter. During hot summer months the storm track is situated across the northern states and some cyclones never leave Canada. The northerly storm track associated with summer also applies to Pacific storms, which move toward Alaska during the warm months, thus producing a rather long dry season for much of our West Coast. In addition to the seasonal changes in storm tracks, the number of cyclones generated is also seasonal, with the largest number occurring in the cooler months when temperature gradients are greatest. This fact is in agreement with the role of cyclonic storms in the distribution of heat across the midlatitudes.

Even in the cool season the westerly flow goes through an irregular cyclic change. There may be periods of a week or more when the flow is nearly west

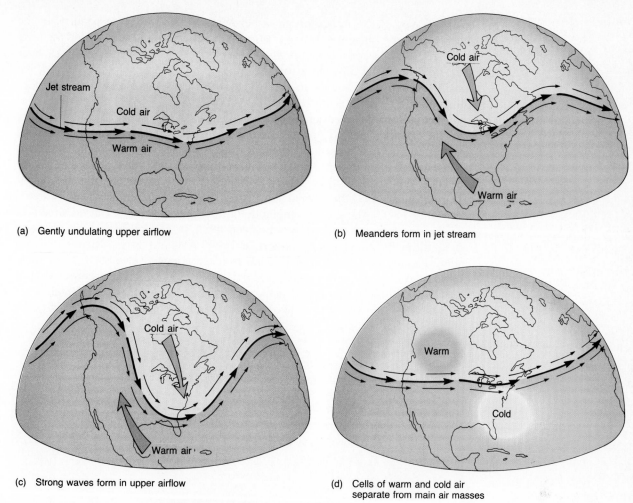

(a) Gently undulating upper airflow

(b) Meanders form in jet stream

(c) Strong waves form in upper airflow

(d) Cells of warm and cold air separate from main air masses

FIGURE 11–10
Cyclic changes that occur in the upper-level airflow of the westerlies. The flow, which has the jet stream as its axis, starts out nearly straight and then develops meanders that are eventually cut off. (After J. Namais, NOAA)

to east, as shown in Figure 11–10(a). Under these conditions relatively mild temperatures occur and few disturbances are experienced in the region south of the jet stream. Then, without warning, the upper flow begins to meander wildly and produces large-amplitude waves and a general north-to-south flow (Figure 11–10b). This change allows for an influx of cold air southward that intensifies the temperature gradient and the flow aloft. During these periods cyclonic activity dominates the weather picture. For a week or more, the cyclonic storms redistribute large quantities of heat across the midlatitudes by moving cold air southward and warm air northward. This redistribution eventually results in a weakened temperature gradient and a return to a flatter flow aloft and less intense weather at the surface. These cycles, consisting of alternating periods of calm and stormy weather, can last from 1 to 6 weeks.

Because of the rather irregular behavior of the circulation patterns of the flow aloft, long-range weather prediction still remains essentially beyond the forecaster's reach. Nonetheless, numerous attempts are being made to predict changes in the upper-level flow on a long-term basis more accurately. It is hoped that such research will allow us to answer questions like: Will next winter be colder than normal? Will California experience a drought next year?

Let us consider an example of the influence of the flow aloft on the weather for an extended period of time. As our example, we will examine an atypical winter. In a normal January an upper-air ridge is situated over the Rocky Mountains and a trough extends across the eastern two-thirds of the United States (see Figure 11–11). This "typical" flow pattern is believed to be caused by the mountains. During January 1977 the normal flow pattern was greatly accentuated, as illustrated in Figure 11–11. The greater amplitude of the upper-level flow caused an almost continuous influx of cold air into the deep south, producing record low temperatures throughout much of the eastern and central United States. Because of dwindling natural gas supplies, many industries experienced layoffs. Much of Ohio was hit so hard that 4-day weeks and massive shutdowns were ordered. Most of the East was in the deep freeze, but the westernmost states were under the influence of a strong ridge of high pressure. Generally mild temperatures and clear skies

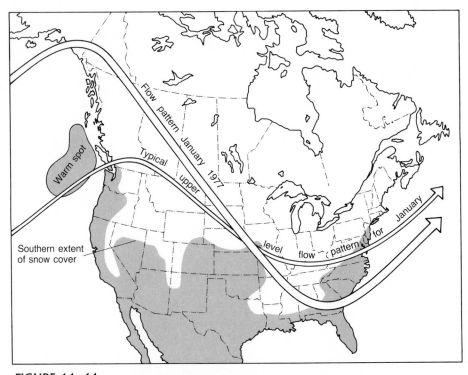

FIGURE 11–11
The unusually high amplitude experienced in the flow pattern of the prevailing westerlies during the winter of 1977 brought warmth to Alaska, drought to the west, and frigid temperatures to the central and eastern United States.

dominated their weather picture. But it was no blessing because this ridge of high pressure blocked the movement of Pacific storms that usually provide much needed winter precipitation. The shortage of moisture was especially serious in California, where January is the middle of its 3-month rainy season. Throughout most of the western states the winter rain and snow that supply water for summer irrigation was far below normal. This dilemma was compounded by the fact that the previous year's precipitation had also been far below normal and many reservoirs were almost empty. Although much of the country was concerned about economic disaster caused either by a lack of moisture or frigid temperatures, the highly accentuated flow pattern channeled unseasonably warm air into Alaska. Even Fairbanks, which generally experiences temperatures as low as 40°F below zero, had a rather mild January with numerous days having above-freezing temperatures.[4]

Although the effects of the upper-level flow on the weather are well documented, as in the preceding example, the causes of these fluctuations still elude meteorologists for the most part. Numerous attempts have been made to relate temperature variations to such diverse phenomena as sunspot cycles and volcanic activity. One attempt to relate the flow pattern experienced in 1977 to ocean temperatures has been given considerable attention. Jerome Namias of the Scripps Institution suggested prior to the winter of 1977 that above-average ocean temperatures in the eastern Pacific may cause a greater than average amplitude in the wavy pattern of the westerlies. It was also suggested that once snow is distributed farther south, the increases in albedo will further support the southward migration of the jet stream axis in this region. In other words, cold temperatures tend to perpetuate themselves by determining the position of the flow aloft.

FORECAST ACCURACY

At one time or another most people have asked, Why are the weather forecasts given on radio and television so often in error? As contradictory as it might sound, the answer lies to some degree in our desire for more accurate predictions. When weather forecasting was in its infancy, the public was pleased when even an occasional forecast came true. Today we expect more accurate predictions; consequently, we concentrate more on the relatively few incorrect forecasts and take for granted the more accurate predictions. Also, as weather forecasting became more accurate, attempts were made to make the forecast even more specific (Figure 11–12). Instead of just predicting the likelihood of precipitation, a modern forecast often gives the expected times of occurrence and the probable amount. Once I overheard a colleague complain that the forecast called for 3 to 6 inches of snow and we received only an inch. At one time this forecast would have been considered correct, but today it is viewed, at least by some, as inaccurate.

In general, we are safe in saying that modern weather forecasts are relatively

---

[4] For an excellent review of the winter weather of 1976–1977, see Thomas Y. Canby, "The Year the Weather Went Wild," *National Geographic*, 152, no. 6 (1977), 798–892.

FIGURE 11–12
Forecasters at the National Weather Service provide nearly two million predictions per year to the public and commercial interests. This meteorologist is operating a modern, computerized communications system designed to speed weather data handling. (Courtesy of NOAA)

accurate for the immediate future (from 6 to 12 hours) and are generally good for a day or two. After a forecast becomes a few days old, its accuracy decreases rapidly, however. Even today specific forecasts are only made for periods of 3 to 5 days in advance, and they are constantly being revised.

Persistence Forecast

When making short-range predictions of a few hours, it is difficult to improve on **persistence forecasts.** These forecasts assume that the weather occurring upstream will persist and move on and will affect the areas in its path in much the same way. Persistence forecasts are used by local forecasters in determining such events as the time of the arrival of a thunderstorm that is moving toward their region. Persistence forecasts do not account for changes that might occur in the intensity or in the path of a weather system, and they do not predict the formation or dissipation of cyclones. Because of these limitations and the rapidity with which weather systems change, persistence forecasts break down after 12 hours, or a day at most.

For periods of up to 5 days, forecasts made by numerical weather prediction methods and supplemented by traditional techniques are difficult to beat. They consider both the formation and the movement of pressure systems. Yet beyond 5 days, specific forecasts made by the more sophisticated methods prove no more accurate than projections made from past climatic data. The reasons for the limited range of modern forecasting techniques are many. As stated earlier, the network of observing stations is rather incomplete. Not only are large areas of the earth's land-sea surface inadequately monitored, but data gathering in the middle and upper troposphere is meager at best, except in a few regions. Moreover, the

laws governing the atmosphere are not completely understood and the current models of the atmosphere are not as complete as possible. Nevertheless, NWP has greatly improved the forecaster's ability to project changes in the upper-level flow. When the flow aloft can be tied more fully to surface conditions, weather forecasting should improve greatly.

You may have asked, Just how accurately is the weather currently being forecast? This question is more difficult to answer than you might imagine. One of the major problems is establishing when a forecast is correct. For example, if a forecast predicts a minimum temperature of 10°C and the temperature falls to 9°C, only 1 degree off, is that forecast incorrect? When a forecast calls for snow in Wisconsin and only the northern two-thirds of the state receives snow, is that forecast incorrect or is it two-thirds correct? The problems of assessing forecast accuracy are many, as can be seen from these examples.

The only aspect of the weather that is predicted as a percentage probability is rainfall. Here statistical data are used to indicate the number of times that precipitation occurred under similar conditions. Although the occurrence of precipitation can be predicted with about 80 percent accuracy, predictions on the amount and time of occurrence of precipitation are still fairly unreliable. Temperatures and wind directions can probably be most accurately predicted.

In the preceding discussion we highlighted the present state of weather forecasting. A policy statement by the American Meteorological Society summarizes the predictive skill of weather forecasts as follows:

For the period 0 to 12 hours: Weather forecasts of considerable skill and utility are attained. The spatial and temporal detail that can be included in the forecast decreases as the period increases. The behavior of small, short-lived, severe local storms is predictable for periods of the order of several minutes to an hour. The behavior of large features, such as squall lines, fronts, and organized areas of heavy precipitation, is predictable for periods of up to 6–12 hours. Weather changes associated with large cyclonic storms are well forecast for the entire range. Strong downslope winds, gorge winds, and other features induced by irregular terrain and surface inhomogeneities can be predicted for periods up to several hours ahead or longer.

For the period 12 to 48 hours: Skillful predictions of the development and movement of large extratropical weather systems, and of the associated day-to-day variations in temperature, precipitation, cloudiness, and air quality, can be made throughout this period. Useful predictions of tropical storm movement can also be made, although rapid changes in destructive potential are not well predicted. In addition, general areas within which severe storms and thunderstorms are likely to occur can often be specified up to 24 hours in advance. However, as indicated above, the exact times of occurrence and locations of individual local storms cannot be predicted at this range.

For the period two to five days: Large-scale circulation events such as major storms and cold waves usually can be anticipated several days in advance. Daily temperature forecasts decrease in skill, relative to climatology, from good at three days to fair by five days. Forecasts of precipitation occurrence show fair skill at three days and marginal skill at 5 days.

For the period 5 to 10 days: Mean temperatures for the period can be predicted with some skill. Daily maximum and minimum temperatures can be forecast with modest skill for the first two days of the period. Forecasts of average precipitation amount over the period exhibit marginal skill relative to climatology.

Monthly and seasonal forecasts: Slight skill exists in forecasting average temperatures for the month or season. No verifiable skill exists in forecasting day-to-day weather changes a month or a season in advance.[5]

SATELLITES IN WEATHER FORECASTING

The field of meteorology entered the space age on April 1, 1960, when the first weather satellite, TIROS 1, was launched. (TIROS stands for Television and Infra-Red Observation Satellite.) The early versions of this generation of weather satellites rotated for stability; consequently, for nearly 75 percent of the time their cameras were pointed away from the earth. Nevertheless, in its short life span of only 79 days, TIROS 1 radioed back thousands of pictures to the earth. A year before the ninth and last TIROS was launched in 1965 the first of the second-generation Nimbus (Latin for cloud) satellites was orbiting the earth (Figure 11–13). Later Nimbus satellites were equipped with infrared "cameras" capable of detecting cloud coverage at night. Another series of satellites was placed into polar orbits

FIGURE 11–13
Nimbus satellite. (Courtesy of NOAA)

[5] From the *Bulletin of the American Meteorological Society*, 64, no. 12 (December 1983), 1386.

Polar Satellite

Geostationary Satellite

so that they circled the earth in a north-to-south direction. The **polar satellites** orbit the earth at rather low altitudes (a few hundred kilometers) and require only 100 minutes per orbit. By properly orienting the orbits, these satellites drift about 15 degrees westward over the earth's surface during each orbit. Thus they are able to obtain photo coverage of the entire earth twice each day and coverage of a large region in only a few hours. By 1966 **geostationary satellites** were placed over the equator. These satellites, as their name implies, remain fixed over a point because their rate of travel corresponds to the earth's rate of rotation. In order to keep a satellite positioned over a given site, however, the satellite must orbit at a great distance from the earth's surface (about 35,000 kilometers). At this altitude the speed required to keep a satellite in orbit will also keep it moving with the rotating earth. But at this distance some detail is lost on the images.

So far weather satellites are not the modern weather forecasters some had hoped. Nevertheless, they greatly add to our knowledge of weather patterns. Most important, they helped to fill gaps in observational data, especially over some parts of the oceans. For example, examine the clarity with which the clouds outline the fronts of the wave cyclone shown in Figure 9–9. These wishbone-shaped whirls can be easily traced by satellites as they migrate across even the most remote portions of our planet. Of similar significance are the frequent discoveries of developing hurricanes that are made before they enter our surface observational network (Figure 11–14). In addition, infrared images from satellites have proven

FIGURE 11–14
Lined up in perfect order between Hawaii and Central America are three Pacific Storms. On the left is Hurricane Kate; in the center is a tropical depression that never reached hurricane force; and on the right Hurricane Liza which within hours smashed into La Paz and caused millions of dollars in damage. (Courtesy of NOAA)

FIGURE 11–15
Infrared image of the same cloud pattern as shown in Figure 11–16. On this image some of the clouds between the equator and 10°N latitude appear much whiter than others. These are the thicker, vertically developed clouds that have cold tops. Another band of rain clouds can be easily seen running from Wyoming to New Mexico. (Courtesy of NOAA)

very useful in determining regions of possible precipitation within a cyclone. Compare the infrared image in Figure 11–15 with the photograph taken in the visible wavelengths in Figure 11–16. Notice that in the photograph taken in the visible wavelengths it is not possible to distinguish one cloud type from another; that is, it is not possible to determine which clouds may be producing rain or snow and which clouds may be yielding only drizzle or no precipitation at all. All the clouds in the visible-wavelength photograph appear white. The infrared image, on the other hand, does indicate which clouds are the most probable precipitation producers, for here warm objects appear dark and cold objects appear white. Because high cloud tops are colder than low cloud tops, the infrared image allows us to distinguish between the higher thicker (cold) clouds that may be producing precipitation and the lower thinner (warm) clouds, that can, at best, only produce light drizzle.

Although weather satellites are still more weather observers than weather forecasters, a number of ingenious developments have made them more than simply TV cameras pointed at the earth from space. It is expected that satellites will be able directly or indirectly to determine wind speeds, humidity, and tempera-

FIGURE 11–16
Satellite view of the cloud distribution on June 4, 1976. Compare this with the
infrared photograph in Figure 11–15. (Courtesy of NOAA)

tures at various altitudes. Geostationary satellites are presently used to estimate
wind speeds from cloud movements. Further, satellites are being equipped with
instruments designed to measure temperatures at various elevations. The principle
used to perform this task involves the fact that different wavelengths of energy
are absorbed and radiated from different layers of the atmosphere. On a clear
night, for example, the atmosphere does not absorb radiation from the earth's
surface with a wave-length of 10 micrometers (recall that this is the atmospheric
window). Thus if we measure the radiation of 10 micrometers, we will have a
method of determining the earth's surface temperature. Because other wavelengths
indicate the altitudes within the atmosphere from which they originate, they too
can be used to measure temperatures. This method of obtaining temperature
readings is simpler in principle than in practice. Nevertheless, with refinements
it is hoped that this method may supplement conventional radiosonde data. Satellites
are also useful in tracking and transmitting information gathered from instruments
attached to balloons.

REVIEW

1. Compare the tasks of the weather analyst with those of a weather forecaster.

2. List the three types of weather forecasting procedures used by the National Weather Service and discuss the basis of each procedure.

3. Describe the statistical approach called the analog method. What are its drawbacks?

4. What features of the wavy flow of the westerlies exhibit seasonal variations?

5. What is a persistence forecast?

FIGURE 11–17

TABLE 11–2 Weather Data for March 5, 1964

Location	Temperature (°F)	Pressure (mb)	Wind Direction	Sky Cover (tenths)
Wilmington, NC	57	1006	SW	7
Philadelphia, PA	59	1001	S	10
Hartford, CT	47	1007	SE	Sky obscured
International Falls, MN	−12	1008	NE	0
Pittsburgh, PA	52	995	WSW	10
Duluth, MN	−1	1006	N	0
Sioux City, IA	11	1010	NW	0
Springfield, MO	35	1011	WNW	2
Chicago, IL	34	985	NW	10
Madison, WI	23	995	NW	10
Nashville, TN	40	1008	SW	10
Louisville, KY	40	1002	SW	5
Indianapolis, IN	35	994	W	10
Atlanta, GA	49	1010	SW	7
Huntington, WV	52	998	SW	6
Toronto, Canada	44	985	W	4
Seven Islands, Canada	24	1008	E	10
Albany, NY	50	998	SE	7
Savanna, GA	63	1012	SW	10
Jacksonville, FL	66	1013	WSW	10
Norfolk, VA	67	1005	S	10
Cleveland, OH	49	988	SW	4
Little Rock, AR	37	1014	WSW	0
Cincinnati, OH	41	997	WSW	10
Detroit, MI	44	984	SW	10
Montreal, Canada	42	993	E	10
Quebec, Canada	34	999	NE	Sky obscured

6. Why is the name "numerical weather forecasting" misleading?

7. The map in Figure 11–17 has the positions of several weather stations plotted on it. Using the weather data for March 5, 1964, which are given in Table 11–2, complete the following:

   a. On a copy of Figure 11–17, plot the temperature, wind direction, pressure, and sky coverage by using the international symbols given in Appendix B.

   b. Using Figure 11–3 as a guide, complete this weather map by adding isobars at 4-millibar intervals, the cold front and the warm front, and the symbol for low pressure.

   c. From your knowledge of the weather associated with a middle-latitude cyclone in the spring of the year, describe the weather conditions experienced on this date at each of the following locations:

   1. Philadelphia, Pennsylvania
   2. Seven Islands, Canada
   3. Sault Sainte Marie, Michigan
   4. Sioux City, Iowa

VOCABULARY REVIEW

weather forecasting

weather analysis

synoptic weather chart

World Meteorological Organization

synoptic weather forecasting

numerical weather prediction (NWP)

statistical methods (in forecasting)

long-range

anal

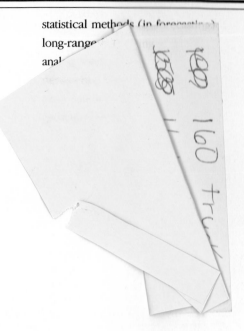

12

Optical Phenomena of the Atmosphere

One of the most spectacular and intriguing of natural phenomena must surely be the rainbow. Its splash of colors has been the focus of poets and artists alike, not to mention every amateur photographer within reach of a camera. In addition to rainbows, many other optical phenomena, such as halos, coronas, and mirages, are common events in our atmosphere. In this chapter we consider how the most familiar of these natural phenomena result. By learning how these spectacles occur, and by knowing when and where to look for them, you should become better able to identify each type. It is hoped that these fascinating displays will be more frequently witnessed as a result of this study.

NATURE OF LIGHT

The light that forms the array of colors that constitute the rainbow, or the deep blue color of the sky, originates as white (visible) light from the sun. It is the interaction of white sunlight with our atmosphere that creates the numerous optical phenomena that take place in the sky.

In Chapter 2 we considered some properties of light and how they contribute to occurrences like the blue sky and the red color of sunset. Here we examine other properties of light and describe how light interacts with the gases of the atmosphere, as well as with ice crystals and water droplets, to generate still other optical phenomena. We consider four basic properties of light: reflection, refraction, diffraction, and interference. The sections that follow consider reflection and refraction; diffraction and interference are considered later, in the section on coronas.

Reflection

Light traveling through the emptiness of outer space travels at a uniform speed and in a straight line. When light rays encounter a transparent material, such as a piece of glass, however, some rays bounce off the surface of the glass, whereas others are transmitted at a slower velocity through the glass. The rays that bounce back from the surface of the glass are said to be reflected. It is reflected light that allows you to see yourself in a mirror. The image that you see in a mirror originates as light that first reflected off you toward the mirror and then was bounced from the silvered surface of the mirror back to your eyes. When light

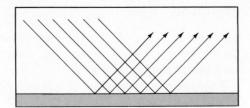

FIGURE 12–1
Reflection of light by a smooth surface.

Reflection, Law of

rays are reflected, they always bounce off the reflecting surface at the same angle at which they meet that surface (Figure 12–1). This principle is called the **law of reflection.** It states that the angle of incidence (incoming ray) is equal to the angle of reflection (outgoing ray).

Although the angle of incidence always equals the angle of reflection, not all objects are perfectly smooth. Consequently, when light encounters a rough surface, the rays will strike the surface at different angles, which tends to scatter the light rays (Figure 12–2). Even something as smooth appearing as a page in this book is rough enough to disperse the light in all directions, making it possible to see the print from any direction. In contrast, if this page were perfectly smooth and all the light was approaching from a specific direction, you would have to move your head to a position exactly opposite the light in order to read the print.

Internal Reflection

A type of reflection that is important to our discussion is **internal reflection.** Internal reflection occurs when light that is traveling through a transparent material, such as water, reaches the opposite surface and is reflected back into the transparent material. You can easily demonstrate this phenomenon by using a glass of water. Hold the glass of water directly overhead and look up through the water. You should be able to see clearly through the water, for very little internal reflection results when light strikes perpendicular to a surface. Keeping the glass overhead, move it sideways so that you look up at it at an angle. Notice how the underside of the surface of the water takes on the appearance of a silvered mirror. What you are observing is the total internal reflection that occurs when the light strikes the surface at an angle greater than 48 degrees from the vertical. Internal reflection is an important factor in the formation of optical phenomena, such as rainbows. In this instance, sunlight entering the raindrops strikes the opposite surface and is reflected back toward an observer.

FIGURE 12–2
Incident light striking a rough surface is diffused.

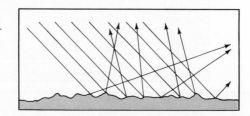

Refraction

When light strikes a transparent material such as water, the rays that are not reflected are transmitted through the water and are subjected to another well-known effect called refraction.

Refraction is the bending of light as it passes obliquely from one transparent medium to another. You have undoubtedly noticed this phenomenon as you have stood in a swimming pool and observed your apparently bent limbs. Obviously it had to be the light that played such a trick on your eyes.

Refraction is caused because the velocity of light varies, depending on the material that transmits it. In a vacuum, radiation travels at 3.0×10^{10} centimeters per second; and when it travels through air, its speed is slowed only slightly. However, in such substances as water, ice, or glass, its speed is slowed considerably.[1]

When light enters a material perpendicular to its surface, only the velocity is affected; yet whenever it encounters a transparent medium at some other angle, bending of the rays results (Figure 12–3). Why light is refracted (bent) can best be demonstrated by an analogy. Imagine how an automobile responds should the driver fall asleep as the car nears a curve. As the auto leaves the highway, the right front wheel will encounter the dirt shoulder before the left wheel does. Because of the soft nature of the dirt shoulder, the right wheel will slow while

[1] The speed of light is a constant; however, as it passes through various substances, this energy is delayed because of interaction with the electrons. The speed of light as it travels between these intervening electrons is the same as in a vacuum.

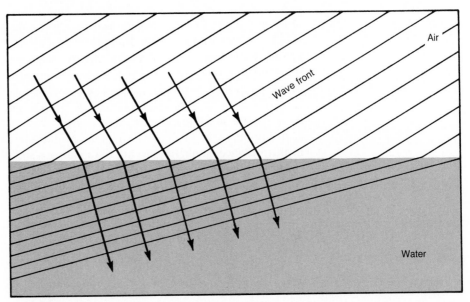

Refraction

FIGURE 12–3
Refraction (bending) occurs as light waves pass from one material to another. The light will change direction because one part of the wave front slows before the other part.

the left wheel, which is still on the pavement, will continue at the same rate of speed. The result will be a sudden turn of the auto toward the right, which will, it is hoped, waken the driver. Now if we can imagine that the path of the auto represents the path of light rays, while the pavement and softer shoulder represents air and water, respectively, we can see how light bends as it goes from air to water. As the light enters the water and is slowed, its path is diverted toward a line extending perpendicularly from the water's surface (Figure 12–3). Should the light pass from water into air, the bending will be in the opposite direction— that is, away from the perpendicular.

Recall that light bends because of a change in velocity. So it follows that the greater the difference in the velocity at which the light travels through the materials involved, the greater the bending. Because light travels only slightly slower in air than in a vacuum, the amount of refraction in air is small; thus air is said to have a small index of refraction. Water, on the other hand, has a much larger refractive index; so light will bend quite noticeably as it passes from air to water. The angle at which the light intersects the surface also affects the angle of refraction.

The bending of light caused by refraction is responsible for a number of common optical illusions. These optical happenings result because our brain perceives bent light as if it has traveled to our eyes along a straight path. Try to imagine "looking down" a bent light ray to view an object located around a corner. If you could see the object, your brain would place that object out from the corner in "plain sight." On occasion, you do see things that are "around the corner." One example is our view of the setting sun. Several minutes after the sun has actually slipped below the horizon, it still appears to us as a full disk. We will soon provide an explanation for this occurrence.

An illustration of how refraction causes optical illusions is found in Figure 12–4. Here we see how the refraction of light produces the apparent bending of

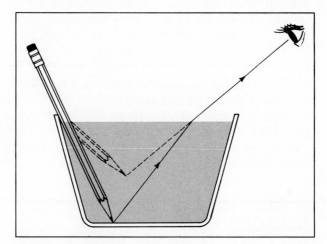

FIGURE 12–4
The pencil appears bent because the eye perceives light as if it were traveling along the dashed line rather than along the solid line which represents the path of the refracted light.

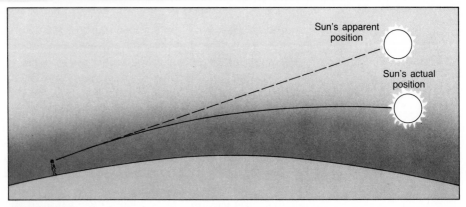

FIGURE 12–5
Light is refracted as it passes through the atmosphere resulting in an apparent displacement of the position of the sun.

a pencil that is immersed in water. The solid lines in this sketch show the actual path taken by the light, whereas the dashed lines indicate how we perceive those same light rays. As we look down at the pencil, the point appears closer to the surface than it actually is. Again, this situation occurs because our brain perceives that light as coming along the straight path indicated by the dashed line rather than along the actual bent path. Because all the light coming from the submerged portion of the pencil is bent similarly, this portion of the pencil appears nearer to the surface. Therefore where the pencil enters the water, it appears to be bent upward toward the surface of the water.

In addition to the abrupt bending of light as it passes obliquely from one transparent substance to another, light will also gradually bend as it traverses a material of varying density. As the density of a material changes, so does the velocity of light. Within the earth's atmosphere, for example, the density of air usually increases earthward. The result of this gradual density change are an equally gradual slowing and bending of light rays. These rays acquire a direction of curvature that has the same orientation as the earth's curvature. This bending is responsible for the apparent displacement of the position of the stars, moon, and sun. When the sun (or other celestial bodies) is near the horizon, this effect is particularly great, which explains why we can see the sun for a few minutes after it has set below the horizon. Figure 12–5 illustrates this situation. Recall that it is our inability to perceive light as bending that places the apparent position of the sun above the horizon.

MIRAGES

Mirage

One of the most interesting optical events common to our atmosphere is the **mirage.** Although this phenomenon is most often associated with desert regions, it can actually be experienced anywhere. One type of mirage occurs on very hot days when the air near the ground is much less dense than the air aloft. As noted, a change in the density of air is accompanied by a gradual bending of the light rays. When light is traveling through air that is less dense near the surface, the rays will develop a curvature in a direction opposite to the earth's curvature. As we can see in Figure 12–6, this direction of bending will cause the light reflected from a distant object to approach the observer from below eye level. Consequently, because the brain perceives the light as following a straight path, the object appears below its original position and is often inverted, as is the palm tree in Figure 12–6. The palm tree will appear inverted when the rays that originate near the top of the tree are bent more than those that originate near the base of the tree.

In the classic desert mirage a lost and thirsty wanderer stumbles on a mirage consisting of an oasis of palm trees and a shimmering water surface on which he can see a reflection of the palms. The palm trees are real, but the water and the reflected palms are part of the mirage. Light traveling to the observer through the cooler air above produces the image of the actual tree. The reflected image of the palms is produced, as stated earlier, from the light that traveled downward from the trees and was gradually bent upward as it traveled through the hot (less dense) air near the ground. The image of water is produced in the same manner as the reflection of the palms. Light that traveled downward from the sky is bent upward to generate the mirage of water. Such desert mirages are called **inferior mirages** because the images appear below the true location of the observed object.

Inferior Mirage

You have undoubtedly experienced a form of desert mirage while traveling down a highway on a hot afternoon. Occasionally "wet areas" will appear on the

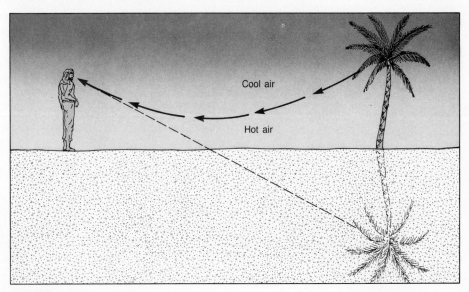

FIGURE 12–6
Light travels more rapidly in the hot air near the surface. Thus as downward-directed rays enter this warm zone they are bent upward so that they reach the observer from below eye level.

pavement ahead only to disappear as you approach. The "wet areas" are the result of bent light that originated from the sky some distance ahead of you and, like the desert mirage, will disappear as you approach. It is important to remember that such mirages are real images and can be photographed. They are not, as is often contended, "tricks played on the mind."

In addition to the "desert mirage," another common type of mirage occurs when the air near the ground is much cooler that the air aloft. Consequently, this effect is observed most frequently in polar regions or over cool ocean surfaces. When the air near the ground is substantially colder than the air aloft, the light rays bend with a curvature having the same direction as the earth's curvature. As shown in Figure 12–7, this effect allows ships to be seen where ordinarily the earth's curvature would block them from view. This phenomenon is often referred to as **looming** because sometimes the refraction of light is so great that the object appears suspended above the horizon. In contrast to a desert mirage, looming is considered a **superior mirage** because the image is seen above its true position.

Looming

Superior Mirage

In addition to the rather easily explained inferior and superior mirages, a number of much more complex variations have been observed. They occur when the atmosphere develops a temperature profile in which rapid temperature changes are observed with height. Under these conditions each thermal layer acts like a glass lens. Because each layer will bend the light rays somewhat differently, the size and shape of the objects observed through these thermal layers are greatly distorted.

You may have observed an analogous sight if you have ever entered a "House of Mirrors" at the county fair. Here one of the mirrors makes you look taller, whereas others stretch the image of part of your body and compress other portions. Mirages are capable of similarly distorting objects and occasionally will even form a mountainlike image over a barren ice cap or over an open ocean.

Towering

One mirage that changes the apparent size of an object is called **towering.** As the name implies, towering results in a much larger object. An interesting type of towering is called the **Fata Morgana.** It is named for the legendary sister of King Arthur, who was credited with the magical power of being able to create

Fata Morgana

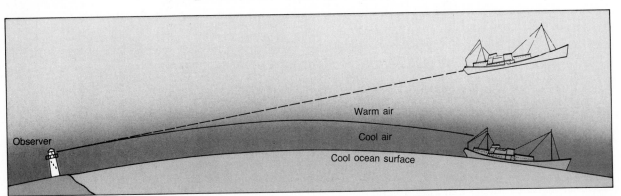

FIGURE 12–7
As light enters a cool layer of air, it will slow and bend downward. This results in objects which appear to loom above their true position.

towering castles out of thin air. This optical phenomenon is most frequently observed in coastal areas, where sharp temperature contrasts are common. In addition to generating magical castles, the Fata Morgana probably explains the towering mountains observed by early explorers of the north polar region that never materialized.

RAINBOWS

Rainbow

Probably the most spectacular and best known of all optical phenomena that occur in our atmosphere is the **rainbow** (Figure 12–8). An observer on the ground sees the rainbow as an arch-shaped array of colors that trail across a large segment of the sky. Although the clarity of the colors varies with each rainbow, the observer can usually discern six rather distinct bands of color. The outermost band of the bow is always red and blends gradually to orange, yellow, green, blue, and eventually ends with an innermost band of violet. Typically these spectacular splashes of color are seen when the observer is situated with the sun on one side and a rain shower occurring in the opposite part of the sky. However, a fine mist of water droplets generated by a waterfall or lawn sprinkler can also generate a miniature rainbow.

Like other optical phenomena, rainbows have been used by people as a means of predicting the weather. A well-known weather proverb illustrates this point:

Rainbow in the morning, sailors take warning.
Rainbow at night, sailors delight.

FIGURE 12–8
Double rainbow. (Photo by James E. Patterson)

This bit of weather lore relies on the fact that weather systems in the midlatitudes usually move from west to east. Remember that an observer must be positioned with his back to the sun and facing the rain in order to see the rainbow. When a rainbow is seen in the morning, the sun is located to the east of the observer and the raindrops that are responsible for its formation must therefore be located to the west. In the early evening the opposite situation exists—the rain clouds are located to the east of the observer. Thus we predict the advance of foul weather when the rainbow is seen in the morning because the rain is located to the west of the observer and is traveling toward him. On the other hand, when the rainbow is seen late in the day, the rain has already passed. Although this famous proverb does have a scientific basis, a small break in the clouds, which lets the sunshine through, can generate a late-afternoon rainbow. In this situation, a rainbow may certainly be followed shortly by more rainfall.

On those occasions when a rather spectacular rainbow is visible, an observer will occasionally be treated to a view of a dimmer secondary rainbow. The secondary bow will be visible about 8 degrees above the primary bow and will suspend a larger arc across the sky (Figure 12–8).[2] The secondary bow also has a slightly narrower band of colors than the primary rainbow, and the colors are in reverse order. Red makes up the innermost band of the secondary rainbow, and violet the outermost.

Although primary and secondary rainbows are produced in an almost identical manner, for clarity we consider first the formation of the primary bow. It should be apparent that sunlight and water droplets are needed for the generation of a rainbow. Also, let us not forget the observer, who must be located between the sun and the rain.

To understand how raindrops disperse sunlight to generate the primary rainbow, recall our discussion of refraction. Remember that as light passes obliquely from the atmosphere to water, its speed is slowed, which causes it to be refracted (bent). In addition, each color of light travels at a different velocity in water; consequently, each color will be bent at a slightly different angle. Violet-colored light, which interacts most with the intervening material, travels at the slowest rate and is therefore refracted the most, whereas red light travels most rapidly and is therefore bent the least. Thus when sunlight, which consists of all colors, enters water, the effect of refraction is to separate it into colors according to their velocity. Sir Isaac Newton is credited with demonstrating the concept of color separation, using a prism.

Light that is transmitted through a prism is refracted twice, once as it passes from the air into the glass and again as it leaves the prism and reenters the air. Newton noted that when light is refracted twice, as by a prism, the separation of sunlight into its component colors is quite noticeable (Figure 12–9). We refer to this separation of colors by refraction as **dispersion.**

Dispersion

When a rainbow forms, water drops act as a prism, dispersing sunlight into the spectrum of colors we see (Figure 12–10). On impacting with the droplet, the sunlight is refracted, with violet light bent the most and red the least. On

---

[2] For reference, the diameter of the sun is equal to about ½ degree of arc.

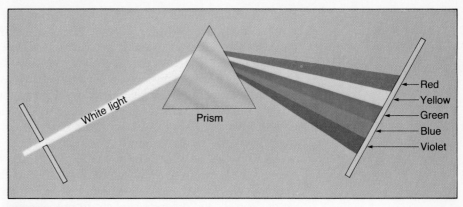

FIGURE 12–9
The spectrum of colors is produced when sunlight is passed through a prism and each wavelength of light is bent differently.

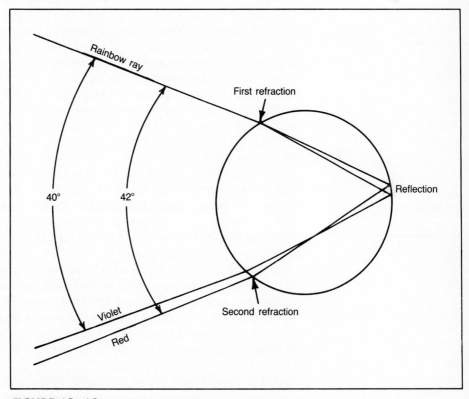

FIGURE 12–10
Color separation results because sunlight is refracted and reflected by a raindrop to produce rainbow rays.

reaching the opposite side of the droplet, the rays are reflected and exit the droplet on the same side they entered. After leaving the droplet, further refraction increases the dispersion already produced and accounts for the complete color separation.

The angle between the incident sunlight and the dispersed colors that constitute the rainbow is 42 degrees for red and 40 degrees for violet. The other colors—orange, yellow, green, and blue—are dispersed at intermediate angles. Although each droplet disperses the full spectrum of colors, an observer will see only one color from any single raindrop. For example, if green light from a particular droplet reaches an observer's eye, the violet light from the droplet will pass over his head while the red light will fall toward the ground in front of him. Consequently, each observer sees his or her "own" rainbow generated by a different set of droplets and different sunlight from that which produces another person's rainbow.

The curved shape of the rainbow results because the rainbow rays always travel toward the observer at an angle between 40 and 42 degrees from the path of the sunlight. Consequently, when an observer looks upward at 42 degrees from the path of the sunlight, he or she will see the color red. When the observer looks to either side at an angle of 42 degrees, the color red will also be visible. In any direction at an angle of 42 degrees from the path of the sun's rays, droplets will be directing red light toward the observer. Thus we experience a 42-degree semicircle of color across the sky that we identify as the arch shape of the rainbow. Because an observer in an airplane can also look downward at an angle of 42 degrees, under ideal conditions he or she can see the rainbow as a full circle. On the other hand, if the sun is higher than 42 degrees above the horizon, an earthbound observer will not see a rainbow. The raindrops needed to form a rainbow when the sun is above 42 degrees have already hit the ground. So if you live in the midlatitudes, do not look for a rainbow during a summer rainshower occurring at midday.

As stated earlier, the secondary rainbow is generated in much the same way as the primary rainbow. The main difference is that the dispersed light that constitutes the secondary bow is reflected twice within a raindrop before it exits, as shown in Figure 12–11. The extra reflection results in a 50-degree angle for the dispersion of the color red (about 8 degrees higher than the primary rainbow) and a reverse order of the colors.

In addition, the extra reflection accounts for a dimmer and therefore less frequently observed secondary bow. Each time light strikes the inner surface of the droplet, some of the light is reflected while the remainder is transmitted through the reflecting surface. The light that is transmitted through the back surface of the droplet does not contribute to the rainbow. Because the rays that form the secondary rainbow experience an additional reflection, they are not as bright as those that form the primary rainbow. Ideally the secondary rainbow always forms; it is just not often discernible by the observer. In addition, other rainbows result because of three or even more internal reflections. These higher-order rainbows are too dim to be seen.

Early workers who pondered the nature of rainbows had one more question

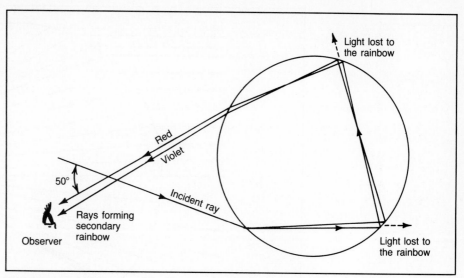

FIGURE 12–11
Idealized drawing showing the geometry of the rays which contribute to the secondary
rainbow. By comparing this sketch with Figure 12–10 you will see that the positions of
the red and violet rays are reversed, which accounts for the order of the colors observed.

to answer. They were aware of the fact that, in addition to the rays that are
scattered at 40 to 42 degrees to form the primary rainbow, the spheroidal shape
of raindrops causes light to be scattered in virtually all other directions as well.
Why, then, are the rays that constitute the rainbow more intense than the light
scattered at other angles? This question was answered by meticulously plotting
the paths of rays that strike a raindrop at various points along its surface. Rays
that penetrate the center of the droplet are reflected off the back surface of the
droplet and return by the same path; that is, they are backscattered directly toward
the sun. Examination of rays that penetrate the droplet at increasing distances
from the center reveal that they are bent more intensely. Consequently, these
rays are backscattered at ever-increasing angles from the incoming rays, as is
shown in Figure 12–12. However, this trend does not continue to the outermost
edge of the droplet. About three-fourths of the way from the center of the raindrop,
the angle between the impacting sunlight and the outgoing scattered light reaches
a maximum, which happens to be 42 degrees.[3] Rays that strike the droplet above
this point are once again scattered at lesser angles.

These early workers also discovered that around the point of maximum scattering there is a narrow zone where a change in the angle of incidence does not
change the amount of dispersion. Thus all light that strikes within this zone leaves
the droplet traveling in roughly the same direction. Consequently, we would
expect the light to be brightest at the angle (42 degrees) where the greatest
number of rays are concentrated.

---

[3] It is common practice to measure the angle of scattering from an extension of the reflected ray.
Following this practice, 42 degrees would equal 138 degrees.

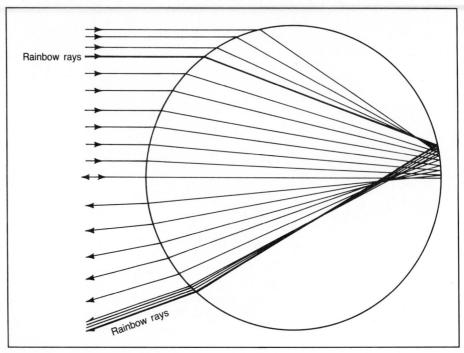

FIGURE 12–12
Schematic diagram illustrating a number of the possible paths which light rays may take through a raindrop. Near the area of the rainbow rays, the amount of dispersion experienced by the incident rays changes; thus in this region we find a concentration of light.

HALOS, SUN DOGS, AND SOLAR PILLARS

Halo

Although a fairly common occurrence, halos are rarely seen by the casual observer. When noticed, the **halo** appears as a rather narrow whitish ring having a quite large diameter centered on the sun (Figure 12–13). Look for halos on days when the sky is covered with a thin layer of cirrus clouds. In addition, this optical phenomenon is generally more often viewed in the morning or late afternoon, when the sun is near the horizon. Occupants of polar regions, where a low sun and cirrus clouds are common, are frequently treated to views of halos and associated phenomena. Occasionally halos may also be seen around the moon.

The most common halo is the *22-degree halo*, so named because its radius subtends an angle of 22 degrees from the observer. Less frequently observed is the larger 46-degree halo. Like the rainbow, the halo is produced by dispersion of sunlight. In the case of the halo, however, it is ice crystals, rather than raindrops, that refract light. Thus as we stated earlier, the clouds most often associated with halo formation are cirrus clouds. Because cirrus clouds often form as a result of frontal lifting, which, in turn, is associated with cyclonic storms, halos have been

FIGURE 12–13
A 22-degree halo produced by the dispersion of sunlight by cirrostratus clouds. (Courtesy of Ward's Natural Science Establishment, Inc., Rochester, N.Y.)

accurately described as harbingers of foul weather, as the following weather proverb attests:

The moon with a circle brings water in her beak.

Four basic types of ice crystals are believed to contribute to the formation of halos: plates, columns, bullets, and capped columns (Figure 12–14). All these crystals are hexagonal (six sided), as we saw is also the case for snowflakes, and all except the bullets have a flat top and a basal surface.

Halos form when the ice crystals that compose cirrus clouds have random orientation. Because sunlight will strike the faces of these crystals at every possible angle, we might expect the scattered light to be dispersed equally in all directions. Yet just as we noted in the formation of the rainbow, as the angle at which the rays strike the surface changes so does the amount of dispersion, but only up to a point. After this point a change in the angle of incidence does not appreciably change the direction of scattering. Consequently, a larger portion of the light will be scattered in one direction than in any other. For a six-sided ice crystal, this angle of maximum scattering is 22 degrees; thus we have the 22-degree halo.

The primary difference between the 22-degree and 46-degree halos is the path that the light takes through the ice crystals. The scattering sunlight that is responsible for the 22-degree halo strikes one of the sides of the ice crystal and exits from an alternating side, as shown in Figure 12–15(a). The angle of separation between the alternating faces of an ice crystal is 60 degrees, which is the same

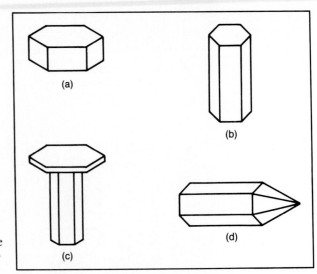

FIGURE 12–14
Common ice crystal configurations which contribute to the formation of certain optical phenomena. (a) Plate. (b) Column. (c) Capped column. (d) Bullet.

as a common glass prism. Consequently, an ice crystal disperses light in a manner similar to a prism in order to produce the 22-degree halo. The 46-degree halo, on the other and, is formed from light that passes through one side of the crystal and exits at the base or top (Figure 12–15b). The angle separating these two surfaces is 90 degrees. Light that passes through two ice faces separated by 90 degrees is concentrated at an angle of 46 degrees, which accounts for the latter halo.

Although ice crystals disperse light in the same manner as a raindrop (or prism), halos are generally whitish in color, partly because of the rather imperfect shape and size of ice crystals compared to rain droplets. The colors produced

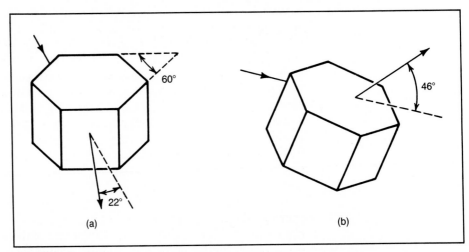

FIGURE 12–15
Illustration of the paths which are taken by light to generate the (a) 22° halo (b) 46° halo.

FIGURE 12–16
Sun dogs, or parhelia, produced by the dispersion of sunlight by the ice crystals of cirrus clouds. (Photo by E. J. Tarbuck)

by this dispersion overlap and wash each other out. Occasionally, however, halos will be colored. Most commonly, a reddish band will be seen in the inner portion of the ring. Because red is refracted the least of all colors, we would expect to find it located on the inner edge of the halo, which is nearest the sun. The other colors, which are refracted more than red, will tend to wash each other out, leaving the red surrounded by a whitish ring.

Sun Dogs

One of the most spectacular effects associated with a halo is termed **sun dogs** or **parhelia.** These two bright regions, or mock suns as they are often called, can be seen adjacent to the 22-degree halo and usually slightly below the elevation of the sun (Figure 12–16). Sun dogs form under the same conditions as, and in conjunction with, the halo, except that their existence depends on numerous ice crystals oriented vertically (see Figure 12–15). This particular orientation results when ice crystals are allowed to descend slowly. Then a large portion of the striking rays will be concentrated in two areas at a distance usually slightly greater than 22 degrees from the sun. When the sun is near the horizon, so that the impact angle is perpendicular to the vertical crystal faces, the mock suns will appear directly on the 22-degree halo, with the sun positioned between them.

Sun Pillar

Another optical phenomenon that is related to the halo is the **sun pillar.** These vertical shafts of light are most often viewed around sunset or sunrise, when they appear to extend upward from the sun. These bright pillars of light are created when sunlight is reflected from the lower sides of descending plates and capped columns, which are oriented like slowly falling leaves. Because direct sunlight is often reddish at that time of day, pillars will appear similarly colored as well. Occasionally pillars that extend below the sun can be viewed.

THE GLORY

Glory

To the earthbound observer, the **glory** is a spectacle that is rarely witnessed. The next time you are in an airplane and fortunate enough to have a window seat, however, look for the shadow of the aircraft projected on the clouds below. The airplane shadow will often be surrounded by one or more colored rings that constitute the glory. Each ring will be colored in a manner similar to the rainbow, with red being the outermost band and violet the innermost. Generally, however, the colors are not as discernible as those of the primary rainbow. When two or more sets of rings are seen, the inner one will be the brightest and thinnest.

Although the glory is most commonly seen by pilots, its name comes from its appearance when viewed by an observer located on the ground. The glory can be seen if an observer is located so that he or she is above a layer of fog with the sun at his or her back. Should the observer's shadow be cast on the fog bank, the glory will enshroud his or her head. When two persons simultaneously witness such a sight, only the observer's head will appear within the glory. Consequently, this type of "halo" has been represented in many ancient artistic works to glorify the wearer.

The glory is formed in a manner not unlike that of the rainbow. But the cloud droplets that are responsible for the glory are much smaller and more uniform in size than the raindrops that scatter the rainbow rays. The light that becomes the glory strikes the very edge of the droplets. These rays then travel to the opposite side of the droplet, partly by one internal reflection, and the remaining distance along the surface of the droplet. This path causes the rays to be backscattered directly toward the sun. Because the glory always forms opposite the sun's position, the observer's shadow will always be found within the glory.

THE CORONA

Corona

The only optical phenomenon more commonly witnessed in association with the moon than the sun is the corona. Typically the **corona** appears as a bright whitish disk centered on the moon or sun. Whenever colors are discernible, the corona appears as several concentric rings, each with a red outer band and a bluish inner region (Figure 12–17).

The corona is produced when a thin layer of water-laden clouds, usually altostratus, veil the illuminating body. Although water droplets are responsible for scattering the light that produces the corona, the colors are not the result of reflection and refraction, as were the colors of the rainbow. Instead the corona forms because of a slight bending of light that occurs as light passes near the edge of cloud droplets. Although we described light as traveling in a straight line, which it generally does, light will bend very slightly around sharp edges, a **Diffraction** process referred to as **diffraction.** It is because of diffraction that even the sharpest shadow appears blurred at the edges when examined carefully.

FIGURE 12–17
The corona. Typically the corona appears as a bright whitish disk centered on the moon or sun. Those like the one shown above that display the colors of the rainbow are rare. (Photo by Henry Lansford)

Interference

Because of their small size, cloud droplets are particularly effective at bending light. From all sides of a cloud droplet diffracted light will be directed into the "shadow" of the droplet. Here light rays will meet and interfere with each other. It is the **interference** of the various components of white light that generates the colors that make up the corona.

To help understand how interference produces color, we will need to recall our discussion of radiation in Chapter 2. Remember that light energy exhibits wave motion, much like ocean swells. White light consists of an array of colors, each with a different wavelength (distance from one crest to the next). That portion of visible light with the shortest wavelength appears violet, whereas the portion with the longest wavelength appears red.

When a light wave of any type becomes superimposed on another, interference results in a new wave having characteristics different from either of the interacting waves. In the generation of the corona, interference occurs as light waves, which are diffracted around the edge of cloud droplets, converge out of phase. Two waves are out of phase when the crest of one is aligned with the trough of the other. When two similar but out-of-phase waves become superimposed, the result is the cancellation of both waves. When white light is diffracted, such that one color is out of phase while the other colors are in phase, the out-of-phase color will be subtracted from the light. When yellow light, for example, is subtracted from white light, the remaining light will appear blue. Because each color is bent a different amount, each color will experience destructive interference at a different angle. The result of the diffraction and interference caused by cloud droplets is an array of colors, each produced by the cancellation and subtraction of a different color.

REVIEW

1. Which of the six colors of the rainbow is refracted at the greatest angle? Why?

2. State the law of reflection.

3. If you have ever been close to a large movie screen, you might have noticed that the screen is made of a number of small pieces oriented at slightly different angles rather than one very smooth surface. Why do you think this is so?

4. When light travels from warm air into a region of colder air, its path will curve. Will it curve away from the cold and toward the warm, or vice versa?

5. Why does a mirage always disappear when the observer gets near?

6. What is meant by an inferior mirage? a superior mirage?

7. If you were looking for a rainbow in the morning, which direction would you look? Why that direction?

8. Explain why the secondary rainbow is dimmer than the primary rainbow.

9. How are halos and rainbows similar? How are they different?

10. What is the orientation of the ice crystals that produce a halo? sun dogs?

11. What gives the glory its name?

12. How are the colors of a corona produced?

13. Describe the relative positions of the observer, the optical phenomenon, and the illuminating body for each of the following:
    a. rainbow
    b. halo
    c. glory
    d. corona
    e. sun dogs

14. At what time of day, if any, can each of the optical phenomena listed in question 13 be best observed?

15. What types of particles (water droplets or ice crystals) are found in the clouds (or fog) that generates each of the optical phenomena listed in question 13?

VOCABULARY REVIEW

law of reflection
internal reflection
refraction
mirage
inferior mirage
looming
superior mirage
towering
Fata Morgana

rainbow
dispersion
halo
parhelia or sun dogs
sun pillar
glory
corona
diffraction
interference

13

The Changing Climate

Not many years ago the subject of climatic change was believed to be of only academic importance. Research in this area focused primarily on the remote past, and changes were believed to occur only very gradually over vast time spans. Today we know that climate is inherently variable on virtually all time scales. Moreover, it has been recognized that there are many factors responsible for such shifts, including human activities. As we shall see, people not only alter local climates, but they are probably responsible for modifying global climate as well. In Chapter 13 we will survey the factors, both natural and human induced, that cause or contribute to climatic change.

THE CLIMATE OF CITIES

The most apparent human impact on climate is the modification of the atmospheric environment by the building of cities. The construction of every factory, road, office building, and house destroys existing microclimates and creates new ones of great complexity. As far back as the early nineteenth century, Luke Howard, the Englishman who is most remembered for his cloud classification scheme, recognized that the weather in London was different from that of the surrounding rural countryside, at least in terms of reduced visibility and increased temperature. Indeed, with the coming of the Industrial Revolution in the 1800s, the trend toward urbanization accelerated, leading to significant changes in the climate in and near most cities. At the beginning of the last century only about 2 percent of the world's population lived in cities of more than 100,000 people. Today not only is the total world population dramatically larger, but a far greater percentage of people reside in cities. On a worldwide basis, perhaps a quarter of the population is urban, and in many regions (including the United States, Western Europe, and Japan) the percentage is more than two or three times that figure.

As Table 13–1 illustrates, the climatic changes produced by urbanization involve all major surface conditions. Some changes are obvious and relatively easy to measure. Others are more subtle and sometimes difficult to measure. The amount of change in any of these elements, at any time, depends on several variables, including the extent of the urban complex, the nature of industry, site factors, such as topography and proximity to water bodies, time of day, season of the year, and existing weather conditions.

TABLE 13–1 Average Climatic Changes Produced by Cities

Element	Comparison with Rural Environment
Particulate matter	10 times more
Temperature	
Annual mean	0.5–1.5°C higher
Winter	1–2°C higher
Heating degree-days	10% fewer
Solar radiation	15–30% less
Ultraviolet, winter	30% less
Ultraviolet, summer	5% less
Precipitation	5–15% more
Thunderstorm frequency	16% more
Winter	5% more
Summer	29% more
Relative humidity	6% lower
Winter	2% lower
Summer	8% lower
Cloudiness (frequency)	5–10% more
Fog (frequency)	60% more
Winter	100% more
Summer	30% more
Wind speed	25% lower
Calms	5–20% more

SOURCE: Helmut E. Landsburg, "City Air—Better or Worse," *Symposium: Air Over Cities*, U.S. Public Health Service, Taft Sanitary Eng. Center, Cincinnati, Ohio, Tech. Rept. A62-5; Helmut E. Landsburg, "Man-Made Climate Changes," *Science*, 170, 3964; Stanley A. Changnon, Jr., "Atmospheric Alterations from Man-Made Biospheric Changes," *Modifying the Weather*, Western Geographical Series, Vol. 9 (Toronto: University of Victoria, 1973).

The Urban Heat Island

Urban Heat Island

Certainly the most studied and well-documented urban climatic effect is the **urban heat island.** The term simply refers to the fact that temperatures within cities are generally higher than in rural areas. One easy way to view this phenomenon is by examining satellite images. Figure 13–1 reveals the presence of 17 urban heat islands in New Hampshire, Massachusetts, and Rhode Island. It shows us that small as well as large cities produce this effect. Figure 13–2, which is a close-up view of the St. Louis area, shows us that the strength of the heat island varies within an urban area. The highest temperatures (darkest shading) occur where the highest building density and industrialization exist.

The heat island is also revealed when we examine statistical data. Table 13–2 shows the heat island by using mean temperatures, and Table 13–3 by using a less common statistic, the number of days on which the temperature either equals or exceeds 32°C or is lower than or equal to 0°C. Although used less frequently,

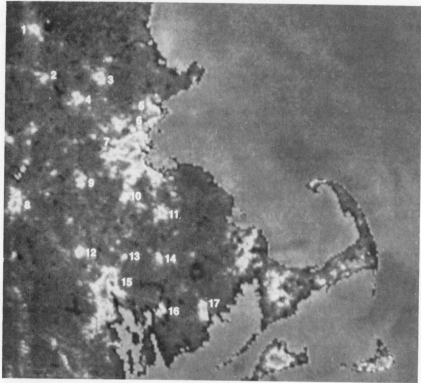

FIGURE 13–1
Urban heat islands detected by the NOAA-5 satellite on May 23, 1978. Seventeen heat islands for cities with populations ranging from 32,000 (Norwood, Mass.) to 625,000 (Boston, Mass.) are evident in the image. At the time of the satellite overpass high pressure was centered over the northeastern United States, bringing clear skies, low-speed surface winds, and moderate relative humidity to the area. Such conditions are ideal for heat island formation. (Courtesy of Michael Matson, National Environmental Satellite Service, NOAA)

the data in Table 13–3 are nonetheless revealing. They show us that in addition to the fact that mean temperatures are higher in the city, these locales also have a greater number of hot days and fewer cold days than outlying areas. The magnitude of the temperature differences as shown by these tables is probably even greater than the figures indicate because studies have shown that temperatures observed at suburban airports are usually higher than those in truly rural environments.

As is typical, the data for Philadelphia show that the heat island is most pronounced when minimum temperatures are examined. Although mean maximum temperatures are only 0.3 to 0.4°C higher in the city, minimums are from 1.2 to 1.7°C higher. Figure 13–3 shows the distribution of average minimum temperatures in the Washington, D.C., metropolitan area for the 3-month winter period over a 5-year span, and it also illustrates a well-developed heat island. The warmest winter temperature occurred in the heart of the city, whereas the suburbs and surrounding countryside experienced average minimum temperatures that were

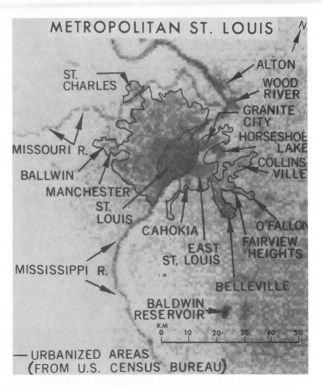

FIGURE 13–2
A computer-enhanced thermal infrared satellite image of
the St. Louis urban heat island on July 28, 1977. Temperature
variations are depicted in shades of gray ranging from black
(warmest) to white (coldest). The highest temperatures occur
along the west bank of the Mississippi River where the highest
building density and industrialization are found. (Courtesy
of Michael Matson, National Environmental Satellite Service,
NOAA)

TABLE 13–2 Average Temperatures (in °C) for
Philadelphia Airport and Downtown Philadelphia
(10-Year Averages)

	Downtown	Airport
Annual mean	13.6°	12.8°
Mean June maximum	28.2°	27.8°
Mean December maximum	6.7°	6.4°
Mean June minimum	17.7°	16.5°
Mean December minimum	−0.4°	−2.1°

SOURCE: Hans Neuberger and John Cahir, *Principles of Climatology*
(New York: Holt, Rinehart and Winston, Inc., 1969), p. 128.

as much as 3.3°C lower. It should be remembered that these temperatures are
averages, and that on many clear, calm nights the temperature difference between
the city center and the countryside was considerably greater, often 11°C or more.
Conversely, on many overcast or windy nights the temperature differential approached 0°C. The same study of metropolitan Washington, D.C., also revealed
another interesting aspect of the heat island effect. It found that over the 5-year
study period the last winter freeze occurred 18 days earlier in the central part of
the city than in outlying areas.

TABLE 13–3 Average Annual Number of Hot Days (≥32°C) and Cold Days (≤0°C) at Downtown (D) and Airport (A) Stations for Five American Cities

City	≤0°C	≥32°C
Philadelphia, PA		
D	73	32
A	89	25
Washington, DC		
D	68	39
A	72	33
Indianapolis, IN		
D	106	36
A	124	23
Baltimore, MD		
D	62	35
A	96	33
Pittsburgh, PA		
D	96	19
A	124	9

SOURCE: Hans Neuberger and John Cahir, *Principles of Climatology* (New York: Holt, Rinehart, and Winston, Inc., 1969), p. 128.

The radical change in the surface that results when rural areas are transformed into cities is a significant cause of the urban heat island. First, the tall buildings and the concrete and asphalt of the city absorb and store greater quantities of solar radiation than do the vegetation and soil typical of rural areas. In addition, because the city surface is impermeable, the runoff of water following a rain is rapid, resulting in a severe reduction in the evaporation rate. So heat that once would have been used to convert liquid water to a gas now goes to increase the surface temperature further. At night, although both city and countryside cool by radiation losses, the stonelike surface of the city gradually releases the additional heat accumulated during the day, keeping the urban air warmer than that of the outlying areas.

Part of the urban temperature rise must also be attributed to waste heat from such sources as home heating and air conditioning, power generation, industry, and transportation. Many studies have shown that the magnitude of human-made energy in metropolitan areas can be a significant fraction of the energy received from the sun at the surface. Investigations in Sheffield, England, and Berlin showed that the annual heat production in these cities was equal to approximately one-third of that received from solar radiation. Another study of densely built-up Manhattan revealed that during the winter the quantity of heat produced from combustion alone was 2½ times greater than the amount of solar energy reaching the ground. In summer the figure dropped to one-sixth. It is interesting to note that during the summer there is a mutual reinforcement between the

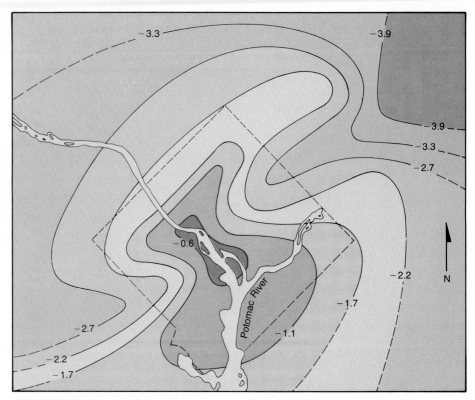

FIGURE 13–3
The heat island of Washington, D.C. as shown by average minimum temperatures (°C) during the winter season (December through February). The city center had an average minimum that was nearly 4°C higher than some outlying areas. (After Clarence A. Woolum, "Notes from a Study of the Microclimatology of the Washington, D.C. Area for the Winter and Spring Seasons," *Weatherwise*, 17, no. 6 [1964], 264, 267)

higher nighttime temperatures of the city and the human-made heat that helped create them. That is, the higher temperatures result in the increased use of air conditioners, which, in turn, use energy and further increase the amount of urban heat. During the winter the nighttime warmth of urban areas, produced in large part by heavy energy consumption, is beneficial because less energy is needed to heat buildings.

In addition to the primary reasons for the heat island just described, other factors are also influential and worthy of mention. For example, the "blanket" of pollutants over a city, including particulate matter, water vapor and carbon dioxide, contributes to the heat island by absorbing a portion of the upward-directed long-wave radiation emitted at the surface and reemitting some of it back to the ground. A somewhat similar effect results from the complex three-dimensional structure of the city. The vertical walls of office buildings, stores, and apartments do not allow radiation to escape as readily as in outlying rural areas where surfaces are relatively flat. As the sides of these structures emit their stored heat, a portion is reradiated between buildings instead of upward and is therefore slowly dissipated.

In addition to retarding the loss of heat from the city, tall buildings also alter the flow of air. Because of the greater surface roughness, wind speeds within an urban area are reduced. Estimates from available records suggest a decrease on the order of about 25 percent from rural values. The lower wind speeds decrease the city's ventilation by inhibiting the movement of cooler outside air that, if allowed to penetrate, would reduce the higher temperatures of the city center. Indeed, when regional winds are strong enough, the dampening effect of the city's vertical structures is not sufficient to stop the ventilating effect. Studies have shown that when the regional wind exceeds a critical value, the heat island can no longer be detected. This critical value appears to be closely related to the size of the urban area; the larger the city, the greater the wind speed necessary to eliminate the heat island.

In summary, the existence of the urban heat island can be linked to the following causative factors:

1. The rocklike materials from which the city is made, which have large thermal capacities (ability to store heat) and create impervious surfaces that lead to the rapid removal of precipitation
2. Heat generated by artificial sources, such as industry, motor vehicles, and domestic heating
3. Increased atmospheric pollution, which inhibits the loss of upward-directed radiation from the surface
4. The tall buildings of the cities, which create a three-dimensional structure that alters the flow of air and creates a complex geometry for heat exchange

Urban-Induced Precipitation

Most climatologists agree that cities influence the occurrence and amount of precipitation in their vicinities. Several reasons are given to explain why an urban complex might be expected to increase precipitation.

1. The urban heat island creates thermally induced upward motions that act to increase the atmosphere's instability.
2. Clouds may be modified by the addition of condensation nuclei and freezing nuclei from industrial discharges.
3. The rougher city surface leads to low-level convergence and increased upward air motions. In addition, the rougher urban landscape creates an obstacle effect that impedes the progress of weather systems. Therefore when rain-producing processes are taking place, they may linger over the urban area and increase the city's rainfall.

Several studies comparing urban and rural precipitation have concluded that the amount of precipitation over a city is about 10 percent greater than over the nearby countryside. Later investigations showed that although cities may indeed increase their own rainfall totals, the greatest effects may occur downwind of the city center.

A striking and controversial example of such a downwind effect was examined by Stanley Changnon in the late 1960s.[1] Records indicated that since 1925 La Porte, Indiana, located 48 kilometers downwind of the large complex of industries at Chicago, experienced a notable increase in total precipitation, number of rainy days, number of thunderstorm days, and number of days with hail. The magnitude of the changes and the absence of such change in the surrounding area led to widespread public attention. Many questioned whether the anomaly was real or simply the result of such factors as observer error and changes in the exposure of instruments. But after completing his study, Changnon concluded that the observed differences were real and were probably produced by the large industrial complex west of La Porte. Among the reasons for this conclusion was the fact that the number of days with smoke and haze (a measure of atmospheric pollution) at Chicago after 1930 corresponded very well with the La Porte precipitation curve. An examination of Figure 13–4, for example, shows a marked increase in smoke and haze days after 1940 when the La Porte curve began its sharp rise. Also, the decrease in smoke and haze days after a peak in 1947 also generally matches a drop in the La Porte curve. A second urban-related factor, steel production, was also found to correlate with the La Porte precipitation curve. Records showed that seven peaks in steel production between 1923 and 1962 were all associated with highs in the La Porte curve.

The conclusions were accepted cautiously by some and were criticized by others who believed that because of the high natural rainfall variability of the area, there was neither sufficient nor accurate enough data to make a sound determination. Nevertheless, this study was a pioneering effort that illustrated the possible effect of human activity on local climates.

Since the La Porte investigation subsequent studies of the complex problem of urban effects on precipitation have generally confirmed the view that cities increase rainfall in downwind areas. The results of one study involving eight American cities revealed increases ranging from 9 to 27 percent over outlying rural areas at six of the cities. In addition, the incidence of thunderstorms and hailstorms also showed an increase.

A second, more comprehensive urban rain investigation was centered in St. Louis, Missouri. Known as METROMEX (for *Metro*politan *Met*eorological *Ex*periment), it was a major program, involving a wide variety of equipment and several research groups. The goal of METROMEX was to determine how, when, and where an urban complex influences atmospheric behavior, especially precipitation. The results of this important multiyear study indicated that the St. Louis urban-industrial complex definitely affects the distribution of precipitation in and downwind of the city's center. Data showed that there was increased average precipitation of about 10 percent in the downwind area, with values for individual stations ranging from 6 to 15 percent. Although such factors as topography were examined as possible causes for the significant downwind increase in rainfall, the evidence

---

[1] Stanley A. Changnon, Jr., "The La Porte Weather Anomaly—Fact of Fiction?" *Bulletin of the American Meteorological Society*, 49 no. 1 (1968), 4–11. (*Note*: The term anomaly refers to something that deviates in excess of normal variation.)

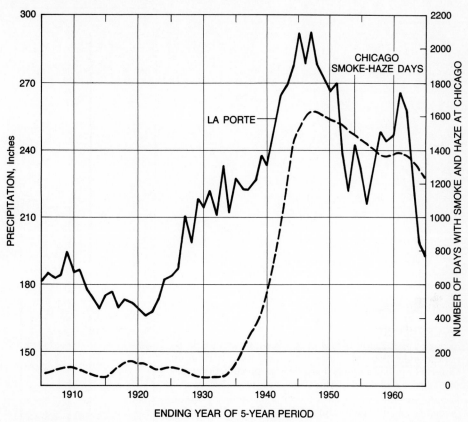

FIGURE 13–4

Precipitation values at La Porte, Indiana, and smoke-haze days at Chicago, both plotted as 5-year moving totals. (From Stanley A. Changnon, Jr., "The La Porte Weather Anomaly—Fact or Fiction?" *Bulletin of the American Meteorological Society*, 49, no. 1 [1968], 5)

seems to point to the city as the primary cause. Discussing the higher summer rainfall at a downwind station relative to the urban area, Huff and Changnon state, "The trend was most pronounced in the last 15 years of the sampling period, 1954–1968. If urban-induced, this observed trend would be expected with gradual expansion of the urban-industrial complex; if topographically related, the trend should not occur."[2]

In addition to the fact that the magnitude of the downwind precipitation increase climbed as industrial development expanded, a second line of evidence also supports the hypothesis that cities are responsible for these increases. If cities are really modifying precipitation, the effects should be greater on weekdays, when urban activities are most intense, than on weekends, when much of the activity ceases. Such is indeed the case. In the St. Louis study analysis of precipitation on weekdays versus weekends revealed a significantly greater frequency of rain per weekday than per weekend day in the affected downwind area. The graph in

[2] F. A. Huff and S. A. Changnon, Jr., "Climatological Assessment of Urban Effects on Precipitation at St. Louis," *Journal of Applied Meteorology*, no. 3 (1972), 829.

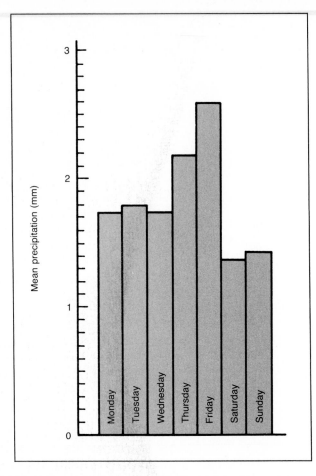

FIGURE 13–5
Average precipitation by day of the week in Paris for an 8-year period (1960–1967). Notice the gradual increase from Monday to Friday and the sharp drop in precipitation totals on Saturday and Sunday. The average for weekdays was 1.93 millimeters compared with only 1.47 on weekends, a difference of 24 percent. (After I. Dettwiller, in Helmut E. Landsberg, "Inadvertent Atmospheric Modification," in *Weather and Climate Modification*, edited by Wilmot N. Hess, John Wiley & Sons, Inc., 1974, p. 755)

Figure 13–5 shows similar findings that have been reported for Paris. Here a rise in precipitation can be noted from Monday through Friday, whereas the amounts for Saturday and Sunday are considerably smaller. Referring to this relationship and to city-induced precipitation in general, Changnon has stated: "The reality of these urban rain changes has been questioned as has the importance of the various mechanisms involved, but urban differences in rain on weekends and weekdays have been important in proving the reality of such urban effects.[3]

Other Urban Effects

Earlier we saw that the great air pollution in cities contributes to the heat island by inhibiting the loss of long-wave radiation at night and may have a "cloud-seeding" effect that increases precipitation in and downwind of cities. These influences, however, are not the only ways in which pollutants influence urban climates. The blanket of particulates over most large cities significantly reduces the amount of solar radiation reaching the surface. As Table 13–1 indicates, the overall reduction

[3] Stanley A. Changnon, Jr., *"Atmospheric Alterations from Man-Made Biospheric Changes,"* *Modifying the Weather*, Western Geographical Series, 9 (Toronto: University of Victoria, 1973), 143.

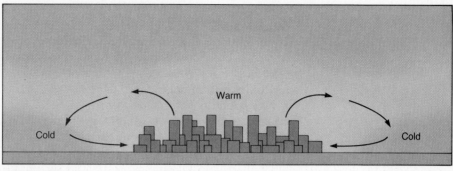

FIGURE 13–6
Idealized scheme of nighttime circulation above a city on a clear, calm night. The arrows represent the wind (country breeze). (After Helmut E. Landsberg, "Man-Made Climatic Changes," *Science*, 170 [1970], 1271. © 1970 by the American Association for the Advancement of Science)

in the receipt of solar energy is 15 percent, whereas short-wavelength ultraviolet is decreased by up to 30 percent. This weakening of solar radiation is, of course, variable. During air pollution episodes the decrease will be much greater than for periods when ventilation is good. Furthermore, particles are most effective in reducing solar radiation near the surface when the sun angle is low, because the length of the path through the dust increases as the sun angle drops. So for a given quantity of particulate matter, solar energy will be reduced by the largest percentage at higher-latitude cities and during the winter.

Relative humidities are generally lower in cities, because temperatures are higher and evaporation is reduced (see Table 13–1). Nevertheless, the frequency of fogs and the amount of cloudiness are greater. It is likely that the large quantities of condensation nuclei produced by human activities in urban areas lead to this increase in cloudiness and fog. When hygroscopic (water-seeking) nuclei are plentiful, water vapor readily condenses on them, even when the air is not yet saturated.

Country Breeze

Finally, mention should be made of another urban-induced phenomenon termed the **country breeze.** As the name implies, this circulation pattern is characterized by a light wind blowing into the city from the surrounding countryside. It is best developed on relatively clear, calm nights—that is, on nights when the heat island is most pronounced. Heating in the city creates upward air motion, which, in turn, initiates the country-to-city flow (Figure 13–6). One investigation in Toronto showed that the heat island created a rural–city pressure difference that was sufficient to cause an inward and counterclockwise circulation centered on the downtown area. When this circulation pattern exists, pollutants emitted near the urban perimeter tend to drift in and concentrate near the city's center.

Is Our Climate Changing?

Not too many years ago the concept of *climatic change* was perceived as a subject that had little but academic importance, for the problems most often investigated related primarily to the remote past. "What caused the Ice Age?" seemed to be the most debated question.[4] Indeed, in the nineteenth and early twentieth centuries it was assumed that climatic changes were a thing of the past or at least that such changes occurred only over vast periods of geologic time. All the observed departures from mean values were thought to be nothing more than what we now call statistical noise. Since then, however, and especially in the past decade or two, scientists have come to recognize that climate is inherently variable on virtually all time scales. It is no longer described as static but rather as dynamic. Furthermore, it is not just the scientific community that has shown increasing interest in climatic change. In recent years governments, as well as the general public, have also become aware of, and shown an interest and concern in, the possible variability of our planet's climate.

As to why climatic change has become a much discussed topic and a matter of considerable concern for the future, we can point to the following reasons.

1. Detailed reconstructions of past climates show that the climate has varied on all time scales, which suggests that climate in the future will more likely differ from the present than stay the same.
2. Increased interest and research into human activities and their effect on the environment have created suspicions that people have or will inadvertently change the climate.
3. There is observational evidence that, at least in some repects, world climate has become more variable.

At this point it should be noted that the fact that a given winter was warmer than the previous one or that a given summer was among the driest on record does not prove that climate is changing. The list of weather peculiarities over the past several years is lengthy; yet it provides no clear-cut indication of our planet's immediate or long-term climatic future. Many years of data are required to indicate a trend in the climate, but even then there is often considerable disagreement among atmospheric scientists about its meaning and cause.

Because instrumental records of climatic elements go back only a couple of hundred years (at best), how do scientists find out about climates and climatic changes prior to that time? The obvious answer is that they must reconstruct past climates from indirect evidence; that is, they must examine and analyze phenomena that respond to and reflect changing atmospheric conditions. In the following discussion we will briefly look at some of these techniques. Keep in mind that these reconstructions may capture no more than the most general features of climate.

Among the most interesting and important techniques for analyzing the earth's climatic history on a scale of hundreds to thousands of years are the study of ocean floor sediments and oxygen isotope analysis. Both methods are relatively recent developments used to reconstruct past temperatures, and each, in part, is related to the other.

---

[4] It is still a much discussed problem and an important focus of study.

Although seafloor sediments are of many types, most contain the remains of organisms that once lived near the sea surface (the ocean–atmosphere interface). When such near-surface organisms die, their shells slowly settle to the floor of the ocean, where they become part of the sedimentary record. One reason why seafloor sediments are useful recorders of worldwide climatic change is that the numbers and types of organisms living near the sea surface change as the climate changes. This principle is explained by Richard Foster Flint as follows:

> We would expect that in any area of the ocean/atmosphere interface the average annual temperature of the surface water of the ocean would approximate that of the contiguous atmosphere. The temperature equilibrium established between surface seawater and the air above it should mean that . . . changes in climate should be reflected in changes in organisms living near the surface of the deep sea. . . . When we recall that the sea-floor sediments in vast areas of the ocean consist mainly of shells of pelagic foraminifers, and that these animals are sensitive to variations in water temperature, the connection between such sediments and climatic change becomes obvious.[5]

Thus in seeking to understand climatic change, as well as other environmental transformations, scientists have become increasingly interested in the huge reservoir of data in seafloor sediments. Since the late 1960s the National Science Foundation has been involved in a major international ocean-drilling program. The Deep Sea Drilling Project and its successor, the Ocean Drilling Program, have used specially designed research vessels capable of drilling into the ocean floor and collecting cores of deep-sea sediments (Figure 13–7). The sediment cores have proven to be excellent sources of useful data that have greatly expanded our understanding of past climates.

Oxygen-Isotope Analysis

The second technique, **oxygen isotope analysis,** is based on precise measurement of the ratio between two isotopes of oxygen: ^{16}O, which is the most common, and the heavier ^{18}O. Because the lighter isotope, ^{16}O, evaporates more readily from the oceans, precipitation (and hence the glacial ice that it may form) is enriched in ^{16}O. Of course, this leaves a greater concentration of the heavier isotope, ^{18}O, in the ocean water. Thus during periods when glaciers are extensive, the concentration of ^{18}O in seawater increases; conversely, during warmer interglacial periods when the amount of glacial ice drops dramatically, the amount of ^{18}O relative to ^{16}O in ocean water also drops. As certain microorganisms secrete their shells of calcium carbonate ($CaCO_3$), the prevailing $^{18}O/^{16}O$ ratio is reflected in the composition of these hard parts. Consequently, periods of glacial activity can be determined from variations in the oxygen isotope ratio found in shells of certain microorganisms buried in deep-sea sediments.

A second use of the $^{18}O/^{16}O$ ratio technique is applied to the study of cores taken from ice sheets, such as the one that covers Greenland. Here another cause for variation in the oxygen isotope ratio is used—namely, that the ratio is influenced by temperature. More ^{18}O is evaporated from the oceans when temperatures are

---

[5] *Glacial and Quaternary Geology* (New York: John Wiley & Sons, Inc., 1971), p. 718.

(a)

(b)

FIGURE 13–7
(a) The *JOIDES Resolution*, the drilling ship of the Ocean Drilling Program. (b) The cores of deep-sea sediments collected by this ship are very important sources of data on past climates. (Photos courtesy of the Ocean Drilling Program)

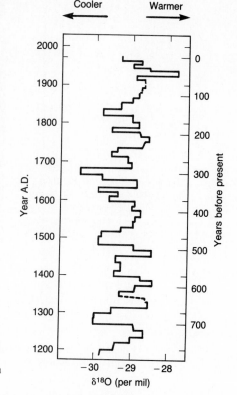

FIGURE 13–8
Temperature variations as revealed by differences in the $^{18}O/^{16}O$ ratio in Greenland ice cores. A decrease of 1 per mil in the ^{18}O content corresponds to a 1.5°C drop in air temperature. (After Wallace S. Broecker, "Climate Change: Are We on the Brink of a Pronounced Global Warming?" *Science*, 189 no. 3859 [1975], 462. © 1975 by the American Association for the Advancement of Science)

high and less is evaporated when temperatures are low. Thus the heavy isotope is more abundant in the precipitation of warm eras and less abundant during colder periods. Using this principle, scientists studying the layers of ice and snow in Greenland have been able to produce a record of past temperature changes. A portion of this record is shown in Figure 13–8.

Many other methods have been used to gain insight into past climates. Because climate has a major effect on soil development and the growth and nature of vegetation, the study of buried soils (*paleosols*) and the analysis of the yearly growth rings of trees have been used to examine climate history (Figure 13–9).

Still another method is the use of historical documents. Although it might seem that such records should readily lend themselves to climatic analysis, such is not the case. Most manuscripts were written for purposes other than climatic description. Furthermore, writers often had a tendency to neglect periods of relatively stable atmospheric conditions and to mention only such events as droughts, severe storms, and other extremes. Nevertheless, records of such items as crops, floods, and the migration of people have furnished useful evidence of the possible influence of a changing climate.

Before ending this discussion about how past changes in climatic elements are deciphered, mention should be made of the time period represented by records of instrumental observations. Although such records might be expected

FIGURE 13–9
Each year a growing tree produces a layer of new cells beneath the bark. When a tree is cut down and the trunk examined, each year's growth appears as a ring. Since the amount of growth (thickness of a ring) depends upon precipitation and temperature, tree rings are useful sources of information on past climates. (Photo courtesy of Laboratory of Tree Ring Research, University of Arizona)

to cause no speculation about the nature of recent climatic trends and fluctuations, this is not true. The reasons for this unexpected problem are summarized by Howard J. Critchfield:

> Climatic records are not readily subjected to objective study. In the first place, weather observers are subject to human failings, and even small errors affect calculations that may involve equally small trends. The exposure and height above ground of instruments also materially affect results. Removal of a weather station to a new location practically destroys the value of its records for purposes of studying climatic change. But even if a station remains in the same location for a century, the changes in vegetation, drainage, surrounding buildings, and atmospheric pollution are likely to produce a greater effect on climatic record than any true climatic changes. Thus, very careful checking and comparison of climatic records are necessary to detect climatic fluctuations.[6]

NATURAL CAUSES OF CLIMATIC CHANGE

The hypotheses that have been proposed to explain climatic change are many and varied, to say the least. Several have gained relatively wide support, only to subsequently lose it and then, in some cases, regain it again. It is safe to say that most, if not all, explanations are controversial. This situation is to be expected if we consider that there is presently no deterministic predictive model of the earth's climate; therefore each proposal is as valid as the logic behind it. In this section we examine several current hypotheses that to varying degrees have gained support from a portion of the scientific community. Four of the discussions deal with

---

[6] *General Climatology*, 3rd ed. (Englewood Cliffs, N.J.: Prentice-Hall, Inc., 1974), p. 376.

"natural" mechanisms of climatic change—that is, causes unrelated to human activities. These include changes brought about by continental drift and volcanic activity as well as by variations caused by fluctuations in solar output and changes in the earth's orbit. The other major proposal relates to possible human-made climatic changes. Here we examine the possible effects of rising carbon dioxide levels caused primarily by combustion processes, as well as the impact of increasing concentrations of several trace gases.

As you read this section you will find that there may be more than one logical way to explain the same climatic change. Also, it should be pointed out that no single hypothesis or theory explains climatic change on all time scales. A proposal that explains variations over millions of years, for example, is generally not satisfactory when dealing with fluctuations over a span of just hundreds of years. When (or perhaps if) the time comes that our atmosphere and its changes through time are fully understood, we will probably see that many of the ideas discussed here, plus others, will be found to influence climatic change.

Plate Tectonics and Climatic Change

Plate Tectonics Theory

Over the past 25 years a revolutionary theory has emerged from the science of geology: the **plate tectonics theory.** This theory, which has gained wide acceptance among scientists, states that the outer portion of the earth is made up of several individual pieces, called plates, which move in relation to one another on a partially molten zone below. With the exception of the plate that encompasses the Pacific Ocean Basin, the other large plates are made up of both continental and oceanic crust. Thus as plates move, the continents also change positions. Not only does this theory provide the geologist with explanations for many previously misunderstood processes and features, it also provides the climatologist with a probable explanation for some hitherto unexplainable climatic changes. Glacial features in present-day Africa, Australia, South America, and India, for example, indicate that these regions experienced an ice age near the end of the Paleozoic Era, about 230 million years ago. This finding puzzled scientists for many years. Was the climate in these relatively tropical latitudes once like it is today in Greenland and Antarctica? Until the plate tectonic theory was formulated and proven, no reasonable explanation existed. Today scientists realize that the areas containing these ancient glacial features were joined together as a single "supercontinent" that was located at high latitudes far to the south of their present positions. Later this landmass broke apart, and its pieces, each moving on a different plate, slowly migrated toward their present locations. Thus large fragments of glaciated terrain have ended up in widely scattered subtropical locations (Figure 13–10).

It is now believed that during the geologic past continental drift accounted for many other equally dramatic climate changes as landmasses shifted in relation to one another and moved to different latitudinal positions. Changes in oceanic circulation must also have occurred, altering the transport of heat and moisture, and hence the climate as well. Because the rate of plate movement is very slow, on the order of a few centimeters per year, appreciable changes in the positions

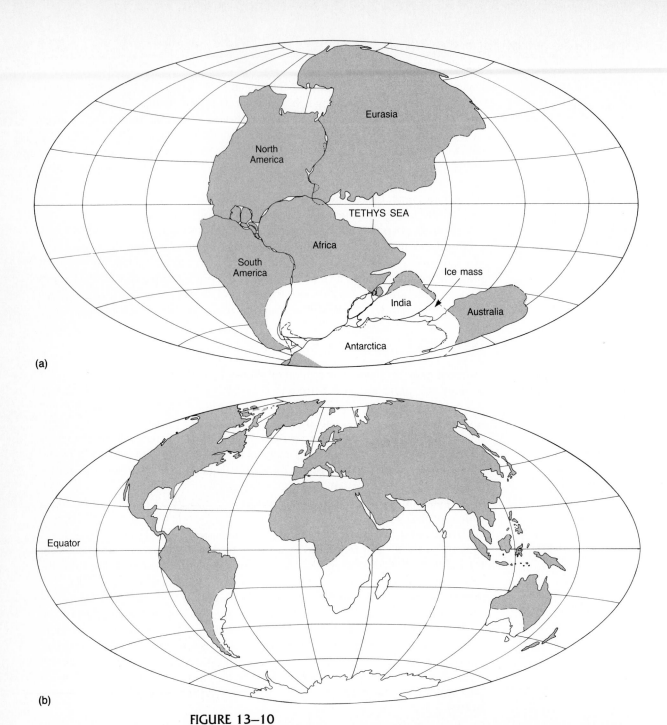

(a)

(b)

FIGURE 13–10
(a) The supercontinent Pangaea showing the area covered by glacial ice 300 million years ago. (b) The continents as they are today. The shading outlines areas where evidence of the old ice sheets exists. (After R. F. Flint and B. J. Skinner, *Physical Geology*, 2nd ed. © 1977 by John Wiley & Sons, Inc.; reprinted by permission of the publisher)

of the continents occur only over great spans of geologic time. Thus climatic changes brought about by continental drift are extremely gradual and happen on a scale of millions of years. As a result, the theory of plate tectonics is not useful for explaining climatic variations that occur on shorter time scales, such as tens, hundreds, or thousands of years. Other explanations must be sought to explain these changes.

Volcanic Activity and Climate

The idea that explosive volcanic eruptions may cause changes in the earth's climate was first proposed many years ago and is still regarded as a plausible explanation for some aspects of climatic variability. Explosive eruptions emit huge quantities of gases and fine-grained debris into the atmosphere. The greatest eruptions are sufficiently powerful to inject material high into the stratosphere, where it spreads around the globe and remains for many months or even years. The basic premise is that this suspended volcanic material will filter out a portion of the incoming solar radiation, which, in turn, will lower air temperatures.

Perhaps the most notable cool period linked to a volcanic event is the "year without a summer" that followed the 1815 eruption of Mount Tambora in Indonesia. In many northern hemisphere locations, including New England, the abnormally cold spring and summer of 1816 were believed to be caused by the cloud of volcanic debris ejected from Tambora. Similar, although apparently less dramatic effects were associated with other explosive volcanoes, including Krakatoa in 1883 and Mount Agung in 1963. The later eruptions of Mount St. Helens in 1980 and the Mexican volcano El Chichón in 1982 provided scientists with an opportunity to study the atmospheric effects of volcanic eruptions with the aid of more sophisticated technology than was available in the past. Satellite images and remote sensing instruments allowed scientists to monitor closely the effects of the clouds of gases and ash that these volcanoes emitted (Figure 13–11).

When Mount St. Helens erupted, there was almost immediate speculation about the possible effects of this event on our climate. Can such eruptions cause our climate to change? There is no doubt that the large quantity of volcanic ash emitted by the explosive eruption had significant local and regional effects for a short period. Still, studies indicated that any longer-term lowering of hemispheric temperatures was negligible. That is, the cooling was so slight, probably less than 0.1°C, that it could not be distinguished from natural temperature fluctuations.

In contrast, two years of monitoring and studies following the El Chichón eruption indicated that its cooling effect on global mean temperatures would be on the order of 0.3° to 0.5°C. Because the eruption of El Chichón was less explosive than the Mount St. Helens blast, why was it expected to have a greater impact on global temperatures? The reason seems to be that the material emitted by Mount St. Helens was largely fine ash that settled out in a relatively short time. El Chichón, on the other hand, emitted far greater quantities of sulfur-rich gases (an estimated 40 times more) than Mount St. Helens. These gases combined with water vapor in the stratosphere to produce a dense cloud of tiny sulfuric acid droplets. Such clouds take several years to settle out completely and are capable

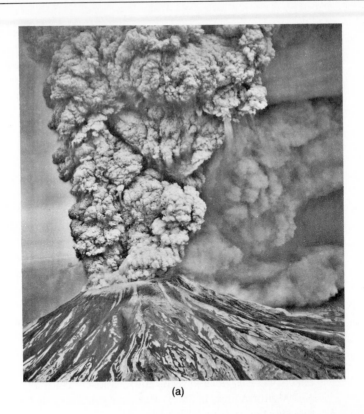

(a)

FIGURE 13–11
(a) When Mount St. Helens erupted on May 18, 1980,
huge quantities of volcanic ash were blown into the
atmosphere. (b) This satellite image was taken less
than 8 hours after the eruption. The ash cloud had
already spread as far as western Montana. (Courtesy
of the National Environmental Satellite Service)   **(b)**

of decreasing the mean global temperature because the droplets both absorb solar radiation and scatter it back to space. It now appears that long-lived volcanic clouds are composed largely of sulfuric acid droplets and not of dust as was once thought. Thus explosiveness alone is a poor criterion for predicting the global atmospheric effects of an eruption. Although Mount St. Helens had the power to lift great quantities of ash into the stratosphere, it lacked the gases needed for a long-lasting effect.

It may be true that the impact on global temperatures of an eruption like that of El Chichón is relatively minor, but many scientists agree that the cooling produced could alter the general pattern of atmospheric circulation for a limited period. Such a change, in turn, could have an effect on the weather in some regions. Predicting or even identifying specific regional effects still presents a considerable challenge to atmospheric scientists, however.

The preceding examples illustrate that the impact on climate of a single volcanic eruption, no matter how great, is relatively small and short lived. Therefore if volcanism is to have a pronounced impact over an extended period, many great eruptions, closely spaced in time, would need to occur. If they did, the stratosphere would be loaded with enough gases and volcanic dust to seriously diminish the amount of solar radiation reaching the surface. Because no such period of explosive volcanism is known to have occurred in historic times, it is most often mentioned as a possible contributor to such prehistoric climatic shifts as the Ice Age. However, as we shall see in the following section, there is convincing evidence that other mechanisms can better explain Ice Age climates. At present, there is little support in the scientific community for the view that explosive volcanism can significantly contribute to an ice age.

Astronomical Theory

Astronomical Theory

The **astronomical theory** is based on the idea that "variations in the earth's orbit influence climate by changing the seasonal and latitudinal distribution of incoming solar radiation."[7] Although proposals linking orbital variations and climatic changes have been around since early in the nineteenth century, credit for developing the modern astronomical theory is given to the Yugoslavian astronomer Milutin Milankovitch. He formulated a comprehensive mathematical model based on the following elements:

Eccentricity

1. Variations in the shape (**eccentricity**) of the earth's orbit about the sun

Obliquity

2. Changes in **obliquity**—that is, changes in the angle that the axis makes with the plane of the ecliptic (plane of the earth's orbit)

Precession

4. Precession—that is, the wobbling of the earth's axis

Although variations in the distance between the earth and sun are of minor significance in understanding current seasonal temperature fluctuations, they may

---

[7]John Imbrie and John Z. Imbrie, "Modeling the Climatic Response to Orbital Variations," *Science*, 207, no. 4434 (1980), 943.

play an important role in producing global climatic changes on a time scale of tens of thousands of years. A difference of only 3 percent exists between aphelion, which occurs on about July 4 in the middle of the northern hemisphere summer, and perihelion, which takes place in the midst of the northern hemisphere winter on about January 3. This small difference in distance means that the earth receives about 6 percent more solar energy in January than in July. Such is not always the case, however. The shape of the earth's orbit changes during a cycle that astronomers say takes between 90,000 and 100,000 years: It stretches into a longer ellipse and then returns to a more circular shape. When the orbit is very eccentric, the amount of radiation received at closest approach (perihelion) would be on the order of 20 to 30 percent greater than at aphelion. This would most certainly result in a substantially different climate from what we now have.

In Chapter 2 the inclination of the earth's axis to the plane of the ecliptic was shown to be the most significant cause for seasonal temperature changes. At present, the angle that the earth's axis makes with the plane of its orbit is about 23.5 degrees. But this angle changes. During a cycle that averages about 41,000 years, the tilt of the axis varies between 22.1 and 24.5 degrees. Because this angle varies, the severity of the seasons must also change. The smaller our tilt, the smaller is the temperature difference between winter and summer. It is believed that such a reduced seasonal contrast could promote the growth of ice sheets. Because winters could be warmer, more snow would fall because the capacity of air to hold moisture increases with temperature. Conversely, summer temperatures would be cooler, meaning that less snow would melt. The result could be the growth of ice sheets.

Like a partly run-down top, the earth is wobbling as it spins on its axis. At present, the axis points toward the star Polaris (often called the North Star). However, about the year A.D. 14,000 the axis will point toward the bright star Vega, which will then be the North Star (Figure 13–12). Because the period of precession is about 26,000 years, Polaris will once again be the North Star by the year 27,000. As a result of this cyclical wobble of the axis, a climatically significant change must take place. When the axis is tilted toward Vega in about 12,000 years, the orbital positions at which the winter and summer solstices occur will be reversed. Consequently, the northern hemisphere will experience winter near aphelion (when the earth is farthest from the sun) and summer will occur near perihelion (when our planet is closest to the sun). Thus seasonal contrasts will be greater because winters will be colder and summers will be warmer than at present.

Using these factors, Milankovitch calculated variations in insolation and the corresponding surface temperature of the earth back into time in an attempt to correlate these changes with the climatic fluctuations of the Ice Age. In explaining climatic changes that result from these three variables, it should be pointed out that they cause little or no variation in the total annual amount of solar energy reaching the ground. Instead their impact is felt because they change the degree of contrast between the seasons.

Since Milankovitch's pioneering work the time scales for orbital and insolational changes have been recalculated several times. Past errors have been corrected and measurements have been made with greater precision. Over the years the

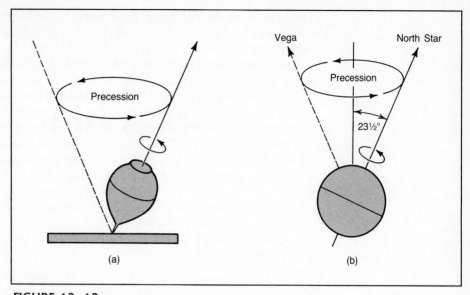

FIGURE 13–12
(a) Precession as illustrated by a spinning top. (b) Precession of the earth will eventually result in the northern hemisphere experiencing winter near aphelion.

astronomical theory has been widely accepted, then largely rejected, and now, in light of recent investigations, is popular again.

Among recent studies that added credibility and support to the astronomical theory is one in which deep-sea sediments were examined.[8] Through oxygen isotope analysis and other statistical analyses of certain climatically sensitive microorganisms, a chronology of temperature changes going back 450,000 years was established. This time scale of climatic change was then compared to astronomical calculations of eccentricity, obliquity, and precession in order to determine if a correlation did indeed exist. It should be noted that the study was not aimed at identifying or evaluating the mechanisms by which the climate is modified by the three orbital variables. The goal was to see if past climatic and astronomical changes corresponded.

Although the study was involved and mathematically complex, its conclusions were straightforward. The authors found that major variations in climate over the past several hundred thousand years were closely associated with changes in the geometry of the earth's orbit; that is, cycles of climatic change were shown to correspond closely with the periods of obliquity, precession, and orbital eccentricity. More specifically, they stated: "It is concluded that changes in the earth's orbital geometry are the fundamental cause of the succession of Quaternary ice ages."[9] Also, the study went on to predict the future trend of climate, but with

---

[8] J. D. Hays, John Imbrie, and N. J. Shackelton, "Variations in the Earth's Orbit: Pacemaker of the Ice Ages," *Science*, 194, no. 4270 (1976), 1121–1132.
[9] Ibid., p. 1131. The term "Quaternary" refers to the period on the geologic time scale that encompasses the last two million years.

two qualifications: (1) that the prediction apply only to the natural component of climatic change and ignore any human influence and (2) that it be a forecast of long-term trends because it must be linked to factors that have periods of 20,000 years and longer. Thus even if the prediction is correct, it contributes little to our understanding of climatic changes over periods of tens to hundreds of years because the cycles in the astronomical theory are too long for this purpose. With these qualifications in mind, the study predicts that the long-term trend (over the next 20,000 years) will be toward a cooler climate and extensive glaciation in the northern hemisphere. Since the time of this study subsequent research has supported its basic conclusions, and the astronomical theory has gained ever-increasing acceptance. A 1982 report states:

> . . . orbital variations remain the most thoroughly examined mechanism of climatic change on time scales of tens of thousands of years and are by far the clearest case of a direct effect of changing insolation on the lower atmosphere of Earth.[10]

If the astronomical theory does indeed explain alternating glacial-interglacial periods, a question immediately arises as to why glaciers have been absent throughout most of earth's history. Prior to the plate tectonic theory, there was no widely accepted answer. In fact, this question was a major obstacle for the supporters of Milankovitch's hypothesis. Today, however, there is a plausible answer. Because glaciers can form only on the continents, landmasses must exist somewhere in the higher latitudes before an ice age can commence. Long-term temperature fluctuations are not great enough to create widespread glacial conditions in the tropics. Thus many now believe that ice ages have occurred only when the earth's shifting crustal plates carried the continents from tropical latitudes to more poleward positions.

Solar Variability and Climate

Among the most persistent hypotheses of climatic change have been those based on the idea that the sun is a variable star and that its output of energy varies through time. The effect of such changes would seem direct and easily understood: increases in solar output would cause the atmosphere to warm and reductions would result in cooling. This notion is appealing because it can be used to explain climatic changes of any length or intensity. Still, there is at least one major drawback. No major long-term variations in the total intensity of solar radiation have yet been measured outside the atmosphere. Such measurements were not possible until satellite technology became available. Now that it is possible, records will be necessary for many years before we begin to get a feeling of how variable (or invariable) energy from the sun really is.

Several proposals for climatic change, based on a variable sun, relate to sunspot cycles. The most conspicuous and best-known features on the surface of the sun

---

[10] National Research Council, *Solar Variability, Weather and Climate* (Washington, D.C.: National Academy Press, 1982), p. 7.

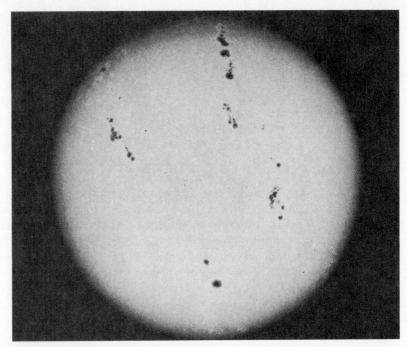

FIGURE 13–13
Sunspots. (Courtesy of Mount Wilson and Las Campanas
Observatories, Carnegie Institution of Washington)

Sunspot

are the dark blemishes called **sunspots** (Figure 13–13). Although their origin is uncertain, it has been established that sunspots are huge magnetic storms that extend from the sun's surface deep into the interior. Moreover, these spots are associated with the sun's ejection of huge masses of particles that, on reaching the upper atmosphere, interact with gases there to produce auroral displays.

Along with other solar activity, the number of sunspots increases and decreases on a regular basis, creating a cycle of about 11 years. A curve of the annual number of sunspots, beginning in the early 1700s, appears to be very regular (Figure 13–14). In fact, until recently few scientists doubted that sunspots and

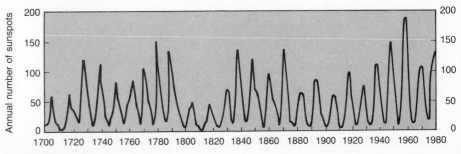

FIGURE 13–14
Mean annual sunspot numbers.

the 11-year cycle were enduring features of the sun. Today, however, we know that there have been periods when the sun was essentially free of these blemishes. In addition to the well-known 11-year cycle, there is a 22-year cycle as well. This longer cycle is based on the fact that the magnetic polarities of sunspot clusters reverse every successive 11 years.

Interest in possible sun-climate effects has been sustained over the years by an almost continuous series of efforts to find evidence of correlations on time scales ranging from days to tens of thousands of years. Two relatively recent and widely debated examples are briefly described here.

Studies in the late 1970s verified that there have been prolonged periods when sunspots were absent or nearly so. Moreover, it was found that these events corresponded closely with cold periods in Europe and North America. Conversely, periods characterized by plentiful sunspots were found to correlate well with warmer times in these regions. Referring to these excellent matches, the solar astronomer John Eddy stated: "These early results in comparing solar history with climate make it appear that changes on the sun are the dominant agent of climatic changes lasting between 50 and several hundred years."[11] But other scientists seriously questioned this conclusion. Their hesitation stems in part from subsequent investigations using different climate records from around the world that failed to find a significant correlation between variations in solar activity and climate. Even more troubling seems to be the fact that no testable physical mechanism exists to explain the purported effect.

A second possible sun-climate connection, on a time scale different from the preceding example, relates to variations in precipitation rather than in temperature. An extensive study of tree rings revealed a recurrent period of about 22 years in the pattern of droughts in the western United States. This periodicity coincides with the 22-year magnetic cycle of the sun mentioned earlier. Commenting on this possible connection, a panel of the National Research Council pointed out that "no convincing mechanism that might connect so subtle a feature of the sun to drought patterns in limited regions has yet appeared. Moreover, the cyclic pattern of droughts found in tree rings is itself a subtle feature that shifts from place to place within the broad region of the study."[12] The essay by Henry Lansford that follows this section explores the sunspot-drought hypothesis further.

It is apparent that the possible connections between solar variability and climate would be much easier to determine if researchers could identify probable physical linkages between the sun and the lower atmosphere. The National Research Council sums up the situation this way:

> Despite much research, no connection between solar variations and weather has ever been unequivocally established. Apparent correlations have almost always faltered when put to critical statistical examination or have failed when tested with different data sets. As a result the subject has been one of continual controversy and debate.[13]

---

[11] The Case of the Missing Sunspots," *Scientific American*, 236, no. 5 (1977), 88, 92.
[12] *Solar Variability, Weather, and Climate* (Washington, D.C.: National Academy Press, 1982), p. 7.
[13] *Op. cit.,* p. 4.

TREE RINGS: PREDICTORS OF DROUGHT?

Henry Lansford[*]

Can we predict drought? Ask most meteorologists this question and the answer will be an emphatic no. Without a thorough scientific understanding of the mechanisms that cause climatic fluctuations such as drought, according to the majority of atmospheric scientists, we cannot predict their occurrence in any intelligent way.

And yet, in the late 1960s and early 1970s, a good many people were warning that drought was likely to strike the high plains of Colorado and surrounding states in the mid-1970s. And they were right. Although the drought of the 1970s was not another Dust Bowl episode like the one of the 1930s, a lot of topsoil blew away and there were crop failures in many places in the high plains region.

When can we expect drought to strike the High Plains again? The meteorologists are not saying; they still do not understand droughts well enough to try to predict them scientifically. But a good many farmers and other experienced observers of the High Plains climate believe that we are not due for another big drought until around the mid-1990s.

Were the people who warned us about the drought of the 1970s just intelligent guessers who happened to hit it right? Or is there some basis for such forecasts, even if it is not scientific enough for the atmospheric scientists?

The most useful tool in future policymaking for the western United States would be accurate long-range climate forecasts specifying the years when drought will come. But atmospheric scientists do not hold out much hope that such predictions will be possible, at least within the next decade.

The second-best tool for western policymakers would be long-term records of precipitation and stream flow over the last several centuries. By indicating the severity, duration, and geographical extent of past droughts, such records could at least give us a rough idea of what to expect in the future. But instrumental measurements do not go nearly far enough into the past to define long-term climatic trends.

There is one natural record of rainfall that can be found throughout the western United States as well as in some other parts of the world—the annual growth rings in tree trunks.

Dendroclimatology—the study of climate through tree rings—was pioneered in the early years of the twentieth century by Andrew E. Douglass, a solar astronomer at the University of Arizona.

Tree rings form because the layer of plant tissue just under the bark produces large thin-walled cells at the beginning of each growing season. This causes a sharp boundary between the last wood formed in one year and the first wood of the next year.

Douglass proposed that the relative thickness of rings in trees that are growing in harsh climates would reflect variations in climate. For a ponderosa pine growing on a dry mountain slope in southern Arizona, for example, the growth ring would be narrow in a drought year but wider in a year of adequate rainfall.

Douglass spent many years collecting core samples, taken by inserting a hollow drill into living trees, from ponderosa pines and Douglas firs in the southwestern United States. By matching early growth rings in living trees with sequences of rings in timbers taken from prehistoric Indian dwellings, he was able to extend his records hundreds of years into the past.

Douglass's work has been continued and expanded by scientists at the Laboratory of Tree-Ring Research at the University of Arizona. Equipped with high-speed computers and modern techniques of statistical analysis, they have correlated tree rings with temperature as well as precipitation and have reconstructed climatic conditions for 10-year periods over

[*]Henry Lansford is a well-known science writer and photographer. His work has appeared in many leading publications, including *Smithsonian, Science*, and *National Geographic*. With Walter Orr Roberts, Mr. Lansford authored *The Climate Mandate*, published by W. H. Freeman and Company, New York, 1979. Henry Lansford is also a consulting editor to *Weatherwise*, the publication from which this essay was excerpted (32, 5 [1979], 194–199).

many hundreds of years for much of the western United States.

Charles Stockton and his colleagues at the tree-ring lab have developed estimates of historical stream flow for the Colorado and other rivers that indicate a much lower average annual flow than many hydrologists had assumed. More recently, Stockton and his co-workers have used tree-ring data from 40 sites across the western United States to reconstruct the occurrence of drought back to A.D. 1700. This reconstruction does not specify locations, but emphasizes the geographic extent of drought over the western United States. It shows a repeated pattern of widespread drought every 22 years.

An interesting aspect of Stockton's reconstruction is that it seems to confirm a correlation that other scientists have observed between drought and sunspots. The droughts coincide with the so-called double sunspot cycle—a 22-year pattern of rising and falling numbers of sunspots, which are dark areas on the face of the sun.

Although there is no convincing scientific explanation for a cause-and-effect relationship between sunspots and drought, Stockton's work confirms that the correlation, coincidence, or whatever you want to call it goes back more than 250 years. Stockton has identified 13 droughts in his reconstructed record, and they all match the 22-year sunspot cycle. One climatologist who was formerly skeptical of the sunspot-drought link, J. Murray Mitchell of the National Oceanic and Atmospheric Administration, has analyzed Stockton's data carefully and cannot find any evidence to refute the apparent connection between sunspots and droughts.

To those outside the scientific community, and particularly to water policymakers in the western states, the key question is whether or not this tree-ring research has provided a reliable and credible method for predicting future droughts. At this point, the answer is no. Stockton's work provides good evidence of a historical pattern of droughts every 20 to 22 years. But unless we understand why this pattern occurred, it is impossible to say with scientific certainty that it will repeat itself again.

However, most of us who are not scientists are likely to feel that a pattern that has recurred 13 times over nearly 300 years is very likely to repeat itself again. Even though the meteorologists and climatologists will not predict drought unless they understand the physical mechanisms that are involved, the odds seem to be good that the West will be hit by another big drought in the mid-1990s. In the absence of more scientifically precise predictions, many policies and decisions affecting the region will probably be made on the basis of what the tree rings have told us.

HUMAN IMPACT ON GLOBAL CLIMATE

So far we examined four potential causes of climatic change. Although each differed considerably from the others, all four had at least one property in common—they were unaffected by human activities. They relied on "natural" variables to produce climatic fluctuations. Yet when relatively recent and future changes in our climate are considered, we must also examine the possible impact of human activities. In this section we discuss the major way in which humans are believed to contribute to global climatic change. This impact results from the addition of carbon dioxide and other gases to the atmosphere.

It should be mentioned at this point, however, that human influence on regional and global climate probably did not begin with the onset of the modern industrial period. There is good evidence that humans were modifying the environment over extensive areas for thousands of years. The use of fire, as well as the overgrazing of marginal lands by domesticated animals, has negatively affected the abundance and distribution of vegetation. By altering ground cover, such important climatological factors as surface albedo, evaporation rates, and surface winds have been, and continue to be, modified. Commenting on this aspect of human-induced climatic modification, the authors of an article dealing with this subject state, "In contrast to the prevailing view that only modern humans are able to alter climate, we believe it is more likely that the human species has made a substantial and continuing impact on climate since the invention of fire."[14]

Carbon Dioxide, Trace Gases, and Climatic Change

In Chapter 1 we learned that although carbon dioxide (CO_2) represents only 0.03 percent of the gases that make up clean, dry air, it is nevertheless a meteorologically significant component. The importance of carbon dioxide lies in the fact that it is transparent to incoming short-wavelength solar radiation, but it is not transparent to some of the longer-wavelength outgoing terrestrial radiation. A portion of the energy leaving the ground is absorbed by carbon dioxide and subsequently reemitted, part of it toward the surface, thereby keeping the air near the ground warmer than it would be without carbon dioxide. Thus along with water vapor, carbon dioxide is largely responsible for the *greenhouse effect* of the atmosphere. Carbon dioxide is an important heat absorber, and it follows logically that any change in the air's carbon dioxide content should alter temperatures in the lower atmosphere.

Paralleling the rapid growth of industrialization begun in the nineteenth century has been the consumption of fossil fuels (coal, natural gas, and petroleum). The combustion of these fuels has added great quantities of carbon dioxide to the atmosphere. Although some of this excess is taken up by plants or is dissolved in the ocean, it is believed that 45 to 50 percent remains in the atmosphere. Consequently, from the midnineteenth century until the late 1980s there was an estimated increase of between 15 and 20 percent in the carbon dioxide content of the air. Since 1958 continuous measurements of CO_2 concentrations have been made at Mauna Loa Observatory in Hawaii (Figure 13–15). These measurements and data from other sites clearly show an upward trend from about 315 parts per million (ppm) to almost 350 ppm. This increase closely matches the growth in CO_2 emissions. The seasonal fluctuations that are shown in the graph occur because CO_2 is removed from the air by plants during the growing season and returned later when the plants decay. Naturally we might expect that global temperatures should have already increased as a result of growing carbon dioxide levels. However a steady temperature rise that can be attributed to carbon dioxide has not yet been detected. Most climatologists believe that the effects of the CO_2 increase are not yet large enough to show up clearly in the climatic record. However, the time is not far off when the effects will be evident.

If we assume that the use of fossil fuels will continue to increase at projected rates, current estimates indicate that the atmosphere's present carbon dioxide content of almost 350 ppm will approach 400 ppm by the year 2000 and will

---

[14] Carl Sagan et al., "Anthropogenic Albedo Changes and the Earth's Climate," *Science*, 206 no. 4425 (1980), 1367.

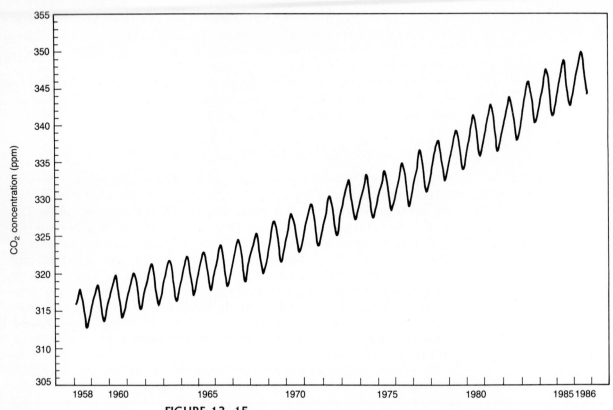

FIGURE 13–15
Monitoring at Mauna Loa Observatory in Hawaii has revealed a significant rise in the
concentration of carbon dioxide during the period shown here. The yearly oscillation is
caused by the seasonal growth and decay of vegetation. (After NOAA)

reach 600 ppm by sometime in the second half of the next century. With such
an increase in carbon dioxide, the enhancement of the greenhouse effect would
be much more dramatic and measurable than in the past. When it is assumed
that the atmosphere's carbon dioxide content will reach projected levels, the
most realistic models predict a global surface temperature increase of between
1.5 and 4.5°C.

Carbon dioxide is not the only gas contributing to a future global increase
in temperature. In recent years atmospheric scientists have come to realize that
the industrial and agricultural activities of people are causing a buildup of certain
trace gases that may also play a significant role. The substances are called *trace
gases* because their concentrations are so much smaller than that of carbon dioxide.
The trace gases that appear to be most important are methane (NH_4), nitrous
oxide (N_2O), and certain types of chlorofluorocarbons. These gases absorb wavelengths of outgoing earth radiation that would otherwise escape into space. Although
individually their impact is small, taken together the effects of these trace gases
may be as great as CO_2 in warming the earth.

Sophisticated computer models of the atmosphere show that the warming of the lower atmosphere triggered by CO_2 and trace gases will not be the same everywhere. Rather, the temperature response in polar regions could be as much as two to three times greater than the global average. Part of the reason for such a response is the fact that the polar troposphere is very stable. This stability suppresses vertical mixing and thus limits the amount of surface heat that is transferred upward. In addition, an expected reduction in sea ice would also contribute to the greater temperature increase. This topic will be explored more fully later in this section.

Climatic-Feedback Mechanisms

Because the atmosphere is a very complex interactive physical system, scientists must consider many possible outcomes when one of the system's elements is altered. These various possibilities, termed **climatic-feedback mechanisms,** not only complicate climatic modeling efforts but also add greater uncertainty to climatic predictions.

The most important and obvious of the feedback effects related to increases in carbon dioxide arises from the fact that the higher surface temperatures produce greater evaporation rates. This factor, in turn, increases the amount of water vapor in the atmosphere. Remember that water vapor is an even more powerful absorber of terrestrial radiation than carbon dioxide. Therefore with more water vapor in the air, the temperature increase caused by carbon dioxide alone is reinforced.

Recall that the temperature increase at high latitudes is expected to be as much as two to three times greater than the global average. This assumption is based in part on the likelihood that the area covered by floating pack ice will decrease as surface temperatures rise. Because ice reflects a much larger percentage of incoming solar radiation than does open water, the melting of the pack ice would replace a highly reflecting surface with a relatively dark surface. The result would be a substantial increase in the amount of solar energy absorbed at the surface. This, in turn, would feed back to the atmosphere and magnify the initial temperature increase created by higher carbon dioxide levels.

Positive-Feedback Mechanism

Negative-Feedback Mechanism

So far the climatic feedback mechanisms discussed have magnified the temperature rise caused by the buildup of carbon dioxide. Because these effects reinforce the initial change, they are termed **positive-feedback mechanisms.** As we shall see, however, other effects must be classified as **negative-feedback mechanisms** because they produce results that are just the opposite of the initial change and tend to offset it.

One possible and perhaps even probable result of a global rise in temperature would be an accompanying increase in cloud cover due to the higher moisture content of the atmosphere. Most clouds are good reflectors of solar radiation. At the same time, however, they are also good absorbers and emitters of terrestrial radiation. Consequently, clouds produce two opposite effects. They act as a negative-feedback mechanism because they increase albedo and thus diminish the amount of solar energy available to heat the atmosphere. On the other hand, clouds act as a positive-feedback mechanism by absorbing and emitting terrestrial radiation that would otherwise be lost from the troposphere.

Which effect, if either, is strongest? Atmospheric modeling shows that the negative effect of a higher albedo is more dominant. Therefore the net result of an increase in cloudiness should be a decrease in air temperature. The magnitude of this negative feedback, however, is not believed to be as great as the positive feedback caused by added moisture and decreased pack ice. Thus although increases in cloud cover may partly offset a global temperature increase, climatic models show that the ultimate effect of the projected increase in CO_2 and trace gases will still be a temperature increase.

The problem of global warming caused by human-induced changes in atmospheric composition has been and continues to be one of the most studied and discussed aspects of climatic change. Although no models yet incorporate the full range of potential influences and feedbacks, it appears that the consensus in the scientific community is that the increasing levels of atmospheric carbon dioxide and trace gases will eventually lead to a warmer planet with a different distribution of climatic regimes.

Some Possible Consequences of a Greenhouse Warming

What consequences can be expected if the carbon dioxide content of the atmosphere reaches a level that is twice what it was early in the twentieth century? Although climatic models contain many uncertainties, they are nevertheless providing us with some possible answers to this very basic question.

As noted, the magnitude of the temperature increase will not be the same everywhere. The temperature rise will probably be smallest in the tropics and increase toward the poles. Furthermore, the models indicate that some regions will experience a significant increase in precipitation and runoff, whereas others will experience a decrease in runoff either because of reduced precipitation or because of increased evaporation rates brought about by higher temperatures. Such changes could have a profound impact on the distribution of the world's water resources and hence affect the productivity of agricultural regions that depend on rivers for irrigation water. A proposed 2°C warming and 10-percent precipitation decrease in the region drained by the Colorado River, for example, could diminish the river's flow by 50 percent or more. Because the present flow of the river is barely enough to meet current demands for irrigated agriculture, the negative effect would be serious. Many other rivers form the basis for extensive systems of irrigated agriculture, and the projected reduction of their flow could have equally grave consequences. In contrast, large precipitation increases in other areas would increase the flow of some rivers and bring more frequent destructive floods to many productive agricultural regions.

The effects of increased atmospheric carbon dioxide on crops that depend on rain for moisture is complex and difficult to estimate. Some places will no doubt experience productivity losses due to decreases in rainfall or increases in evaporation rates. Still, these losses may be offset by gains elsewhere. Increased temperatures in the high latitudes could lengthen the growing season, for instance. This factor, in turn, could allow the expansion of agriculture into areas that are presently not suited to crop production. Additional gains may result because CO_2 is a basic plant nutrient. Experiments indicate that a higher concentration of CO_2 promotes photosynthesis and gives rise to faster growth, a process that could lead to the increased production of some crops. Furthermore, when CO_2 levels are high, many plants tend to partly close the leaf pores that release water vapor to the atmosphere. In areas with marginal rainfall this reduction of water loss by plants would mean that the plants would be affected less by water stress as the atmosphere's CO_2 level increased.

Another impact of a human-induced global warming may be a rise in sea level. How is a warmer atmosphere related to a global rise in sea level? The most significant connection appears to be that higher air temperatures raise the temperature of the upper layers of the ocean. This, in turn, causes the water to expand and sea level to rise. Recent research indicates that sea level has been gradually rising for nearly 100 years and that the trend will continue at an accelerated rate in the years to come. It is believed that the rise could exceed 30 centimeters by the middle of the next century. Although such a vertical rise may seem modest, many scientists believe that any rise in sea level along a gently sloping shoreline, such as the Atlantic and Gulf coasts of the United States, will lead to significant additional erosion and shoreline retreat.

Since rising sea level is a gradual phenomenon, it may be overlooked by coastal residents as an important contributor to shoreline erosion problems. Rather, the blame may be assigned to other forces, especially storm activity. Although a given storm may be the immediate cause, the magnitude of its destruction may result from the relatively small sea-level rise that allowed the storm's power to cross a much greater land area.

Finally, concern is sometimes expressed that a warmer climate will cause glaciers to melt. Of course, if this were to occur on a large scale, it could lead to a much greater rise in sea level and a major encroachment by the sea in coastal zones. It should be emphasized, however, that a significant melting of ice sheets, although possible at some future date, is not expected during the next century.

It should be emphasized that the impact on global climate of an increase in the atmoshpere's CO_2 content is obscured by many unknowns and uncertainties. A report by the National Research Council suggests that no drastic effects of a CO_2-induced global warming will occur in the next few decades.[15] The changes that do occur will take the form of gradual environmental shifts that will be imperceptible to most people from year to year. What many scientists ar urging is summarized in this quotation from an article by Roger Revelle:

> It would be prudent to begin thinking now about what the changes might be and how humankind might best avoid or ameliorate the unfavorable effects and gain the most benefit from the favorable ones.[16]

REVIEW

1. Which temperature statistic (maximums, minimums, or means) reveals the greatest rural-urban temperature difference?

2. Heat islands are associated only with large urban areas with populations in excess of 150,000. True or false? (See Figure 13–1.)

[15] Carbon Dioxide Assessment Committee, *Changing Climate* (Washington, D.C.: National Academy Press, 1983), p. 2.
[16] "Carbon Dioxide and World Climate," *Scientific American*, 247, no. 2 (August 1982), 43.

3. Compare a "typical" rural surface with a "typical" city surface. List two ways in which this difference adds to the urban heat island.

4. How do each of the following factors contribute to the urban heat island: heat production, the "blanket" of pollutants, three-dimensional city structure?

5. During what season is heat production most important as a factor affecting the heat island?

6. List three factors that are the probable causes of greater precipitation in and downwind of cities.

7. **a.** What was the La Porte anomaly?
   b. What evidence led Changnon to conclude that the La Porte situation was related to the Chicago urban-industrial complex?

8. A study of precipitation data in the St. Louis region led researchers to conclude that the urban-industrial complex caused precipitation totals to increase up to 15 percent. List two lines of evidence that supported this conclusion.

9. Describe the reasons for the following urban climatic characteristics:
   a. reduced solar radiation
   b. reduced relative humidity
   c. increased fog and cloud frequency
   d. reduced wind speeds and more days when calms prevail
   e. fewer heating degree days

10. What is a "country-breeze"? Describe its cause.

11. List and describe some methods that scientists use to gain insight into the climates of the past.

12. How does the plate tectonics theory (continental drift) help explain the previously "unexplainable" glacial features in present-day Africa, South America, and Australia?

13. Can plate tectonics explain short-term climatic changes? Explain.

14. Although the 1980 Mount St. Helens eruption was more explosive than the 1982 eruption of El Chichón, the El Chichón event had a greater impact on global temperatures. Explain.

15. List and describe each of the three variables in the astronomical theory of climatic change.

16. Do recent studies of sea floor sediments tend to confirm or refute the astronomical theory? What do these studies predict for the future?

17. List two examples of possible climatic changes linked to solar variability. Are these sun-climate connections widely accepted?

18. Why has the carbon dioxide level of the atmosphere been rising for more than 120 years?

19. How are temperatures in the lower atmosphere likely to change as carbon dioxide levels continue to increase?

20. Is carbon dioxide the only gas contributing to a future global temperature change?

21. What are climatic-feedback mechanisms? Give some examples.

22. Briefly discuss the possible effects of increased carbon dioxide on crops.

VOCABULARY REVIEW

urban heat island

country breeze

oxygen isotope analysis

plate tectonics theory

astronomical theory

eccentricity

obliquity

precession

sunspot

climatic-feedback mechanisms

positive-feedback mechanisms

negative-feedback mechanisms

14

World Climates

Although previous chapters examined the spatial and seasonal variations of the major elements of weather and climate, we have not yet investigated the combined effects of these variations in different parts of the world. The varied nature of the earth's surface and the many interactions that occur between atmospheric processes give every location on our planet a distinctive, even unique, climate. Our intention, however, is not to describe the unique climatic character of countless different locales; such a task would require many volumes. Instead the purpose of this chapter is to introduce the reader to the major climatic regions of the world. The discussion examines large areas and uses particular places only to illustrate the characteristics of these major climatic regions. In addition, for those regions that are most likely to be unfamiliar to the majority of students (in particular, the tropical, desert, and polar realms), a brief description of the natural landscape is also provided. Even so, the chapter should be considered a summary that highlights many of the most striking features in a generalized way.

In Chapter 1 it was pointed out that a common misconception is to think of climate as only "the average state of the atmosphere." Although averages are certainly important to climatic descriptions, variations and extremes must also be included in order to portray the character of an area accurately.

Temperature and precipitation are the most important elements in a climatic description because they have the greatest influence on people and their activities and also have an important impact on the broad-scale distribution of such items as vegetation and soils. Nevertheless, other factors are also important for a complete climatic description. When possible, some of these factors are introduced into our discussion of world climates.

CLIMATIC CLASSIFICATION

The distribution of the major atmospheric elements is, to say the least, complex. Because of the many differences from place to place as well as from time to time at a particular locale, it is unlikely that any two sites on the earth's surface experience exactly the same weather conditions. The fact that the number of places on earth is virtually infinite makes it readily apparent that the number of different climates must be extremely large. Of course, having a great diversity of information to investigate is not unique to the study of the atmosphere; it is a problem that is basic to all science. To cope with such variety, it is not only desirable but also essential to devise some means of classifying the vast array of data to be studied. By establishing groups consisting of items that have certain important characteristics in common, order and simplicity are introduced. Bringing order to large quantities of information not only aids comprehension and understanding, but it also facilitates analysis and explanation.

Probably the first attempt at climatic classification was made by the ancient Greeks, who divided each hemisphere into three zones: *torrid*, *temperate*, and *frigid*. The basis of this simple scheme was earth-sun relationships, and the boundaries were the four astronomically important parallels, the Tropic of Cancer and the Tropic of Capricorn and the Arctic and Antarctic circles. Thus the globe was divided into winterless climates and summerless climates and an intermediate type that had features of the two extremes.

Few other attempts were made until the beginning of the twentieth century. Since then, however, many climate classification schemes have been devised. It should be remembered that the classification of climates (or of anything else) is not a natural phenomenon but the product of human ingenuity. The value of any particular classification is determined largely by its intended use. A system designed for one purpose is not necessarily applicable to another.

Koeppen Classification

In this chapter we use a classification devised by the Russian-born German climatologist Wladimir Koeppen (1846–1940). As a tool for presenting the general world pattern of climates, the **Koeppen classification** has been the best-known and most used system for more than 50 years. It is widely accepted for many reasons. For one, it uses only easily obtained data: mean monthly and annual values of temperature and precipitation. Furthermore, the criteria are unambiguous, relatively simple to apply, and divide the world into climatic regions in a realistic way.

Koeppen believed that the distribution of natural vegetation was the best expression of the totality of climate. Consequently, the boundaries he chose were largely based on the limits of certain plant associations. Five principal groups were recognized; each group was designated by a capital letter as follows:

A Humid tropical. Winterless climates; all months having a mean temperature above 18°C.

B Dry. Climates where evaporation exceeds precipitation; there is a constant water deficiency.

C Humid middle-latitude. Mild winters; the average temperature of the coldest month is below 18°C but above −3°C.

D Humid middle-latitude. Severe winters; the average temperature of the coldest month is below −3°C and the warmest monthly mean exceeds 10°C.

E Polar. Summerless climates; the average temperature of the warmest month is below 10°C.

Notice that four of the major groups (A, C, D, E) are defined on the basis of temperature characteristics and the fifth, the B group, has precipitation as its primary criterion. Each of the five groups is further subdivided by using the

TABLE 14–1 Koeppen System of Climatic Classification

Letter Symbol			
1st	**2nd**	**3rd**	
A			Average temperature of the coldest month is 18°C or higher
	f		Every month has 6 cm of precipitation or more
	m		Short dry season; precipitation in driest month less than 6 cm but equal to or greater than $10 - R/25$ (R is annual rainfall in cm)
	w		Well-defined winter dry season; precipitation in driest month less than $10 - R/25$
	s		Well-defined summer dry season (rare)
B			Potential evaporation exceeds precipitation. The dry–humid boundary is defined by the following formulas: (NOTE: R is average annual precipitation in cm and T is average annual temperature in °C) $R < 2T + 28$ when 70% or more of rain falls in warmer 6 months; $R < 2T$ when 70% or more of rain falls in cooler 6 months; $R < 2T + 14$ when neither half year has 70% or more of rain
	S		Steppe ⎤ The BS–BW boundary is 1/2 the dry–humid boundary
	W		Desert ⎦
		h	Average annual temperature is 18°C or greater
		k	Average annual temperature is less than 18°C
C			Average temperature of the coldest month is under 18°C and above −3°C
	w		At least ten times as much precipitation in a summer month as in the driest winter month
	s		At least three times as much precipitation in a winter month as in the driest summer month; precipitation in driest summer month less than 4 cm
	f		Criteria for w and s cannot be met
		a	Warmest month is over 22°C; at least 4 months over 10°C
		b	No month above 22°C; at least 4 months over 10°C
		c	One to 3 months above 10°C
D			Average temperature of coldest month is −3°C or below; average temperature of warmest month is greater than 10°C
	s		Same as under C
	w		Same as under C
	f		Same as under C
		a	Same as under C
		b	Same as under C
		c	Same as under C
		d	Average temperature of the coldest month is −38°C or below
E			Average temperature of the warmest month is below 10°C
	T		Average temperature of the warmest month is greater than 0°C and less than 10°C
	F		Average temperature of the warmest month is 0°C or below

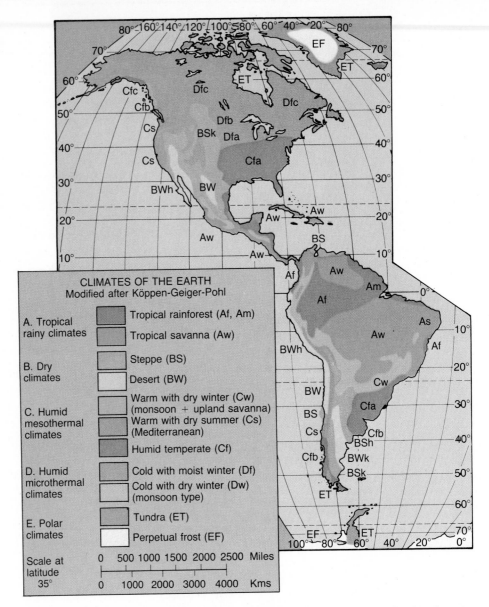

criteria and symbols presented in Table 14–1.[1] In addition, the world distribution of climates according to the Koeppen classification is shown in Figure 14–1. You will be referred to this figure several times as the climates of the earth are discussed in the following pages.

Although a strength of the Koeppen system is the relative ease with which

[1] When classifying climatic data using Table 14–1, you should first determine whether the data meet the criteria for the E climates. If the station is not a polar climate, proceed to the criteria for B climates. If your data do not fit into either the E or B groups, check the data against the criteria for A, C, and D climates, in that order.

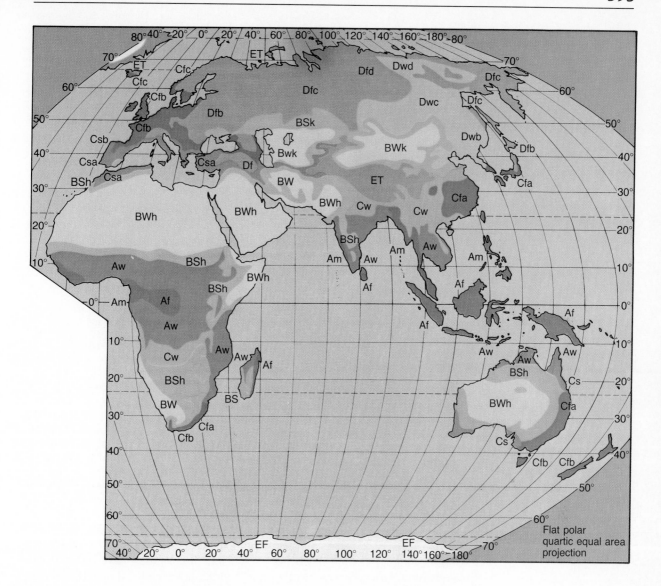

FIGURE 14–1
Climates of the earth. (From Glenn T. Trewartha, *An Introduction to Climate*, 4th ed. © 1968 by McGraw-Hill Book Company; reprinted by permission of the publisher)

boundaries are determined, these boundaries should not be viewed as fixed. On the contrary, all climatic boundaries, no matter what classification is used, shift their positions from one year to the next (Figure 14–2). The boundaries shown on maps like the world map in Figure 14–1 are simply average locations based on data collected over many years. Thus a climatic boundary should be regarded as a broad transition zone and not a sharp line.

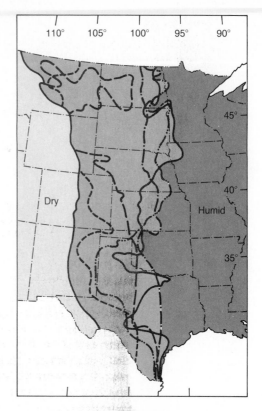

FIGURE 14–2
Yearly fluctuations in the dry–humid boundary during a 5-year period. (From C. E. Koeppe and G. C. Delong, *Weather and Climate*. © 1958 by McGraw-Hill Book Company; reprinted by permission of the publisher)

CLIMATIC CONTROLS: A SUMMARY

If the earth's surface were completely homogeneous, the map of world climates would be simple, looking much like the ancient Greeks must have pictured it—a series of latitudinal bands girdling the globe in a symmetrical pattern on each side of the equator. Such, of course, is not the case. The earth is not a homogeneous sphere and therefore many factors exist that act to disrupt the symmetry just described.

At first glance the world climate map (Figure 14–1) reveals what appears to be a scrambled or even haphazard pattern, with similar climates located in widely separated parts of the earth. A closer examination will show that although they may be far apart, similar climates generally have similar latitudinal and continental positions. This fact suggests that there is order in the distribution of climatic elements and that the pattern of climates is not a matter of chance. This is indeed the situation and reflects a regular and dependable operation of the major climatic controls. So before we begin a discussion of the earth's major climates, it will be worthwhile to review each of the major controls.

Latitude

The single greatest cause of temperature differences are fluctuations in the amount of solar radiation received at the earth's surface. Although nonperiodic variations in such factors as cloud coverage and the amount of dust in the air may be locally influential, seasonal changes in sun angle and length of daylight are the most important factors controlling the global distribution of temperature. Because all places situated along the same parallel have identical sun angles and lengths of day, variations in the receipt of solar energy are largely a function of latitude. Moreover, because the vertical rays of the sun migrate between the Tropic of Cancer and the Tropic of Capricorn, there is a latitudinal shifting of temperatures. Temperatures in the tropical realm are consistently high because the vertical rays of the sun are never far away. As one moves farther poleward, however, greater seasonal fluctuations in the receipt of solar energy are reflected in larger annual temperature ranges.

Land and Water

The distribution of land and water must be considered second in importance only to latitude as a control of temperature. Land heats more rapidly and to higher temperatures than water, and it cools more rapidly and to lower temperatures than water. Consequently, variations in air temperatures are much greater over land than over water. This differential heating of the earth's surface has led to climates being divided into two broad classes based on their position in relation to water and land. **Marine climates** are considered relatively mild for their latitude because the moderating effect of water produces summers that are warm but not hot and winters that are cool but not cold. In contrast, **continental climates** tend to be much more extreme. Although a marine station and a continental station along the same parallel in the middle latitudes may have similar annual mean temperatures, the annual temperature range will be far greater at the continental station.

Marine Climate

Continental Climate

The differential heating and cooling of land and water can also have a significant effect on pressure and wind systems and hence on seasonal precipitation distribution as well. High summer temperatures over the continents can produce low-pressure areas that allow the inflow of moisture-laden maritime air. Conversely, the high pressure that forms over the chilled continental interiors in winter causes a reverse flow of dry air toward the oceans.

Geographic Position and Prevailing Winds

To understand fully the influence of land and water on the climate of an area, the position of that area on the continent and its relationship to the prevailing winds must be considered. The moderating influence of water is much more pronounced along the windward side of a continent, for here the prevailing winds may carry the maritime air masses far inland. On the other hand, places located on the lee side of a continent, where the prevailing winds blow from the land toward the ocean, are likely to have a more continental temperature regime.

Mountains and Highlands

The location of mountains and highlands plays an important part in the distribution of climates. This impact may be illustrated by examining the situation in western North America. Here the north–south trending mountain chains serve as major barriers. They do not allow the moderating influence of maritime air masses to reach far inland. Consequently, although stations may lie within a few hundred kilometers of the Pacific, their temperature regime is essentially continental. Also, these topographic barriers trigger orographic rainfall on their windward slopes, often leaving a dry rain shadow on the leeward side. Similar effects may be seen in South America and Asia, where the towering Andes and the massive Himalayan system serve as major barriers. By way of comparison, western Europe does not have a mountain barrier to obstruct the free movement of air from the North Atlantic. As a result, moderate temperatures and sufficient precipitation mark the entire region.

Where extensive, highlands create their own climatic regions. Because of the drop in temperature with increasing altitude, areas like the Tibetan Plateau, the Altiplano of Bolivia, and the uplands of East Africa are cooler and drier than their latitudinal locations alone would indicate.

Ocean Currents

The effect of ocean currents on the temperatures of adjacent land areas can be significant. Poleward-moving currents, such as the Gulf Stream and Kuroshio currents in the northern hemisphere and the Brazilian and East Australian currents in the southern hemisphere, cause air temperatures to be warmer than would be expected. This influence is especially pronounced in the winter. Conversely, currents like the Canaries and California currents in the northern hemisphere and the Humboldt and Benguela currents south of the equator reduce the temperatures of bordering coastal zones. In addition, the chilling effect of these cold currents acts to stabilize the air masses moving across them. The result is marked aridity and often considerable advection fog.

Pressure and Wind Systems

The world distribution of precipitation shows a close relationship to the distribution of the earth's major pressure and wind systems. Although the latitudinal distribution of these systems does not generally take the form of simple "belts," it is still possible to identify a zonal arrangement of precipitation from the equator to the poles (see Figure 7–11). In the realm of the equatorial low the convergence of warm, moist, and unstable air makes this zone one of heavy rainfall. In the regions dominated by the subtropical highs general aridity prevails. Farther poleward, in the middle-latitude zone dominated by the irregular subpolar low, the influence of the many traveling cyclonic disturbances again causes precipitation totals to rise. Finally, in polar regions, where temperatures are low and the air can hold only small quantities of moisture, precipitation totals decline.

The seasonal shifting of the pressure and wind belts, which follows the movement of the sun's vertical rays, has a significant effect on areas situated in intermediate positions. Such regions are alternately influenced by two different pressure and wind systems. A station located poleward of the equatorial low and equatorward of the subtropical high, for example, will experience a summer rainy period as the low migrates poleward and a wintertime drought as the high moves equatorward. This latitudinal shifting of pressure belts is largely responsible for the seasonality of precipitation in many regions.

THE WET TROPICS

**Tropical Rain Forest
(Selva)**

The constantly high temperatures and year-round rainfall in the wet tropics combine to produce the most luxuriant vegetation found in any climatic realm: the **tropical rain forest** or **selva** (Figure 14–3). Unlike the forests that most of us living in North America are accustomed to, the tropical rain forest is made up of broadleaf trees that are green throughout the year. In addition, instead of being dominated by just a few species, the selva is characterized by many. It is not uncommon for hundreds of different species to inhabit a single square kilometer of the forest. As a consequence, the individuals of a single species are often widely spaced. Standing on the shaded floor of the forest looking upward, one would see tall, smooth-barked, vine-entangled trees, the trunks unbranching in their lower two-thirds, forming an almost continuous canopy of foliage above. A closer look would reveal a three-layered structure. Nearest the ground, perhaps 5 to 15 meters above, the narrow crowns of some rather slender trees are visible. Rising above the relatively short components of the forest, a more continuous canopy of foliage is seen in the height range 20 to 30 meters. Finally, visible through an occasional opening in the second layer, a third layer may be seen at the very top of the forest. Here the crowns of the trees tower 40 meters or more above the forest floor. Much sun streams through the highest level to the lower layers below, but

FIGURE 14–3
Unexcelled in luxuriance and characterized by hundreds of different species per square kilometer, the tropical rain forest, or selva, is a broadleaf evergreen forest that dominates the wet tropics.

little light penetrates to the ground. As a result, plant foliage is relatively sparse on the dimly lit forest floor. Where considerable light does make its way to the ground, as along river banks or in human-made clearings, an almost impenetrable growth of tangled vines, shrubs, and short trees is found. The familiar term **jungle** is used to describe such sites.

Jungle

The environment of the wet tropics just described characterizes almost 10 percent of the earth's land area. An examination of Figure 14–1 shows that Af and Am climates form a discontinuous belt astride the equator that typically extends 5 to 10 degrees into each hemisphere. The poleward margins are most often marked by diminishing rainfall, but occasionally decreasing temperatures mark the boundary. Because of the general decrease in temperature with height in the troposphere, this climatic region is restricted to elevations below 1000 meters. Consequently, the major interruptions near the equator are principally cooler highland areas. It should also be noted that the rainy tropics tend to have a great latitudinal extent along the eastern side of continents (especially South America) and along some tropical coasts. The greater width on the eastern side of a continent is due primarily to its windward position on the weak western side of the subtropical high, a zone dominated by neutral or unstable air. In other cases, as along the eastern side of Central America, the coast, backed by interior highlands, intercepts the flow of trade winds. Orographic uplift thus greatly enhances the rainfall total.

Data for some representative stations in the wet tropics are shown in Table 14–2. A brief examination of the numbers reveals the most obvious features that characterize the climate in these areas.

1. Temperatures usually average 25°C or more each month. Consequently, not only is the annual mean high, but the annual range is also very small.
2. The total precipitation for the year is high, often exceeding 200 centimeters.

TABLE 14–2 Data for Wet Tropical Stations

	J	F	M	A	M	J	J	A	S	O	N	D	Yr
Singapore 1°21′N; 10 m													
Temp. (°C)	26.1	26.7	27.2	27.6	27.8	28.0	27.4	27.3	27.3	27.2	26.7	26.3	27.1
Precip. (mm)	285	164	154	160	131	177	163	200	122	184	236	306	2282
Belém, Brazil 1°18′S; 10 m													
Temp. (°C)	25.2	25.0	25.1	25.5	25.7	25.7	25.7	25.9	25.7	26.1	26.3	25.9	25.7
Precip. (mm)	340	406	437	343	287	175	145	127	119	91	86	175	2731
Douala, Cameroon 4°N; 13 m													
Temp. (°C)	27.1	27.4	27.4	27.3	26.9	26.1	24.8	24.7	25.4	25.9	26.5	27.0	26.4
Precip. (mm)	61	88	226	240	353	472	710	726	628	399	146	60	4109

3. Although rainfall is not evenly distributed throughout the year, tropical rain forest stations are generally wet in all months. If a dry season exists, it is a very short one.

Temperature Characteristics

Because places with an Af or Am designation lie near the equator, the reason for the uniform temperature rhythm experienced in such locales is clear: the intensity of insolation is consistently high. The vertical rays of the sun are always relatively close and changes in the length of daylight throughout the year are slight; therefore seasonal temperature variations are minimal.. The small differences that exist between the warmest and coolest months often reflect changes in the amount of sky coverage rather than in the position of the sun. In the case of Belém, for example, it can be seen that the highest temperatures occur during the months when rainfall (and hence cloud cover) is least.

Daily temperature variations greatly exceed seasonal differences. Whereas annual temperature ranges in the wet tropics rarely exceed 3°C, daily temperature ranges are from two to five times greater. Thus there is a greater variation between day and night than there is seasonally. Although rapid nighttime cooling does not occur as a rule, the drop in temperature is nevertheless often sufficient to produce early morning fogs and heavy dew. Because the daytime air is often nearly saturated, only a small drop in temperature is necessary to induce condensation.

It should be remembered that the unique feature of the temperatures experienced in the rainy tropics is not their values, for monthly and daily means there are no greater than those in many U.S. cities during the summer. For example, the highest temperature recorded at Jakarta, Indonesia, over a 78-year period and at Belém, Brazil, over a 20-year period has been only 36.6°C compared with extremes of 40.5°C and 41.1°C at Chicago and New York City, respectively. Rather, the unique feature is the day-in and day-out, month-in and month-out regularity of the temperature regime. Although the thermometer may not indicate abnormal or extreme conditions, the warm temperatures combined with the high humidity and meager winds make apparent temperatures particularly high. The reputation of the wet tropics as being oppressive and monotonous is well deserved for the most part.

Precipitation Characteristics

The regions dominated by Af and Am climate normally receive from 175 to 250 centimeters of rain each year. But a glance at the data in Table 14–2 reveals more variability in rainfall than in temperature, both seasonally and from place to place. The rainy nature of the equatorial realm is partly related to the extensive heating of the region and the consequent thermal convection. In addition, this is the zone of the converging trade winds, often referred to as the **intertropical convergence zone** or simply the **ITC**. Therefore the thermally induced convection coupled with convergence leads to widespread ascent of the warm, humid, unstable

Intertropical Convergence Zone (ITC)

air. Conditions near the equator are thus ideal for the formation of precipitation.

Rain typically falls on more than 50 percent of the days each year. In fact, at some stations three-quarters of the days experience some rain. There is a marked daily regularity to the rainfall at many places. Cumulus clouds begin forming in late morning or early afternoon. The buildup continues until about 3 or 4 P.M., the time when temperatures are highest and thermal convection is at a maximum; then the cumulonimbus towers yield showers. Figure 14–4, showing the hourly distribution of rainfall at Kuala Lumpur, exemplifies this pattern. The cycle is different at many marine stations, with the rainfall maximum occurring at night. Here the lapse rate is steepest and hence instability is greatest during the dark hours instead of in the afternoon. The lapse rate steepens at night because the radiation heat loss from the air at heights of 600 to 1500 meters is greater than near the surface, where the air continues to be warmed by conduction and low-level turbulence from the water.

Portions of the rainy tropics are wet throughout the year; that is, according to the Koeppen scheme, at least 6 centimeters of rain falls each month. Yet extensive areas (those having the Am designation) are characterized by a brief dry season of 1 or 2 months. In spite of the short dry season, the annual precipitation total in Am regions closely corresponds to the total in areas that are wet year-round (Af). Because the dry period is too brief to deplete the supply of soil moisture, the rain forest is maintained. Although an explanation of the seasonal pattern of precipitation in wet tropical climates is complex and not yet fully understood, month-to-month variations are, at least in part, caused by the seasonal migration of the ITC that follows the latitudinal shift of the vertical rays of the sun.

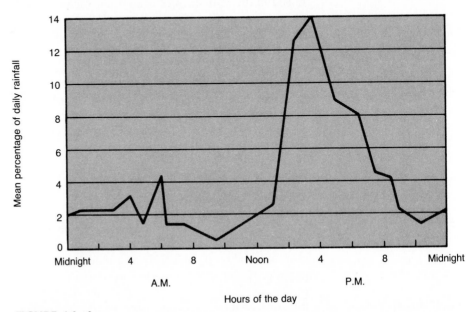

FIGURE 14–4

The distribution of rainfall by hour of the day at Kuala Lumpur, Malaya, with its midafternoon maximum, illustrates the typical pattern at many wet tropical stations. (After Ooi Jin-bee, "Rural Development in Tropical Areas," *Journal of Tropical Geography*, 12, no. 1 [1959])

FIGURE 14–5
The tropical savanna, with its stunted, drought-resistant trees scattered amid a grassland, probably resulted from seasonal burnings carried out by native human populations.

TROPICAL WET AND DRY

Tropical Wet and Dry

In the latitude zone poleward of the rainy tropics and equatorward of the tropical deserts (from about 5 or 10 degrees to 15 or 20 degrees) lies a transitional climatic region referred to as **tropical wet and dry.** Along the margins nearest the equator, the dry season is short and the boundary between Aw and the rainy tropics is difficult to define. Along the poleward side, however, the dry season is prolonged and conditions merge into those of the semiarid realm.

Savanna

Here in the tropical wet and dry climate the rain forest gives way to the **savanna,** a tropical grassland with scattered deciduous trees (Figure 14–5). In fact, Aw is often referred to as the savanna climate. This name, however, may not be appropriate, for many ecologists doubt that these grasslands are climatically induced. Instead it is believed that woodlands once dominated and that subsequently the savanna grasslands were created in response to the seasonal burnings carried out by native populations.

Temperature Characteristics

An examination of the temperature data in Table 14–3 reveals only modest differences between the wet tropics and the tropical wet and dry regions. Because of the somewhat higher latitude of most Aw stations, annual mean temperatures are slightly lower. In addition, the annual temperature range, although still small, is higher, varying from 3°C to perhaps 10°C. The daily temperature range, however,

TABLE 14–3 Data for Tropical Wet and Dry Stations

	J	F	M	A	M	J	J	A	S	O	N	D	Yr
					Calcutta 22°32′N; 6 m								
Temp. (°C)	20.2	23.0	27.9	30.1	31.1	30.4	29.1	29.1	29.9	27.9	24.0	20.6	26.94
Precip. (mm)	13	24	27	43	121	259	301	306	290	160	35	3	1582
					Cuiaba, Brazil 15°30′S; 165 m								
Temp. (°C)	27.2	27.2	27.2	26.6	25.5	23.8	24.4	25.5	27.7	27.7	27.7	27.2	26.5
Precip. (mm)	216	198	232	116	52	13	9	12	37	130	165	195	1375

still exceeds the annual variation. Because seasonal fluctuations in humidity and sky coverage are more pronounced in Aw areas, daily temperature ranges vary noticeably during the year. Generally they are small during the rainy season, when humidity and cloud cover are at a maximum, and large during periods of drought, when skies are clear and the air is dry. Furthermore, because of the more persistent summertime cloudiness, many Aw stations experience their warmest temperatures at the end of the dry season, just prior to the summer solstice. So in the northern hemisphere March, April, and May are often warmer than June and July.

Precipitation Characteristics

Because temperature regimes among the A climates are similar, the primary factor that distinguishes the Aw climate from Af and Am is precipitation. Aw stations typically receive from 100 to 150 centimeters of rainfall each year. This overall amount is often appreciably less than in the wet tropics. The most distinctive characteristic of this climate, however, is not the yearly precipitation total but the markedly seasonal character of the rainfall—wet summers followed by dry winters.

The alternating wet and dry periods are due to the intermediate position of the Aw climatic region between the intertropical convergence zone, with its sultry weather and convective thundershowers, and the stable, subsiding air of the subtropical highs. After the spring equinox, the ITC, along with the other wind and pressure belts, shifts poleward following the migration of the vertical rays of the sun (Figure 14–6). With the advance of the ITC into a region, the summer rainy season commences and features weather patterns typical of the wet tropics. Later, with the retreat of the ITC, the subtropical high advances into the region and brings with it intense drought conditions. During the dry season the landscape takes on a parched appearance and nature seems to become dormant as trees

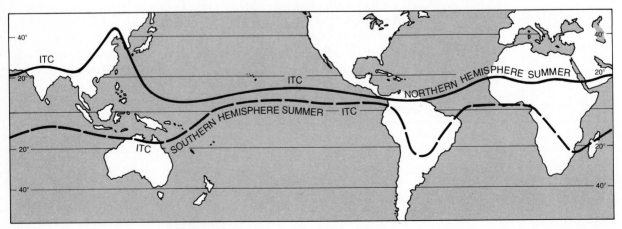

FIGURE 14–6
The seasonal migration of the ITC strongly influences precipitation distribution in the tropics.

TABLE 14—4 Rainfall Regimes and the Movement of the ITC in Africa

	J	F	M	A	M	J	J	A	S	O	N	D
					Maiduguri, Nigeria 11°51′N							
Precip. (mm)	0	0	0	7.6	40.6	68.6	180.3	**220.9**	106.6	17.7	0	0
					Yaounde, Cameroon 3°53′N							
Precip. (mm)	22.8	66.0	147.3	170.1	**195.6**	152.4	73.7	78.7	213.4	**294.6**	116.8	22.9
					Kisangani, Zaire 0°26′N							
Precip. (mm)	53.3	83.8	**177.8**	157.5	137.2	114.3	132.0	165.1	182.9	**218.4**	198.1	83.8
					Luluabourg, Zaire 5°54′S							
Precip. (mm)	137.2	142.2	**195.6**	193.0	83.8	20.3	12.7	58.4	116.8	165.1	**231.1**	226.0
					Zomba, Malawi 15°23′S							
Precip. (mm)	274.3	**289.6**	198.1	76.2	27.9	12.7	5.1	7.6	17.8	17.8	134.6	**279.4**
					Francistown Botswana 21°13′S							
Precip. (mm)	**106.7**	78.7	71.1	17.8	5.1	2.5	0	0	0	22.9	58.4	86.4

lose their leaves and the abundant tall grasses turn brown and wither. The length of the dry season depends primarily on distance from the ITC. Typically the farther an Aw station is from the equator, the shorter the period of ITC control and the longer the locale is under the influence of the stable subtropical high. Consequently, with an increase in latitude, the dry season gets longer and the length of the wet period diminishes.

The importance of the movement of the ITC to an understanding of the rainfall distribution in the tropics cannot be overemphasized. Table 14–4, which shows precipitation data for six African stations, reinforces this point. Notice that at Maiduguri and Francistown there are single rainfall maxima that occur when the ITC reaches its most poleward positions. Between these stations double maximums represent the passage of the ITC on its way to and from these extreme locations. It is important to remember that these statistics represent long-term averages and that on a year-to-year basis the movement of the ITC is far from regular. There is, nevertheless, little doubt that in the tropics rainfall follows the sun.

The Monsoon

Monsoon

In much of India, southeast Asia, and portions of Australia the alternating periods of rainfall and drought characteristic of the Aw precipitation regime are associated with a phenomenon called the **monsoon.** The term is derived from an Arabic word, *mausim*, which means "season" and typically refers to wind systems that have a pronounced seasonal reversal of direction. During the summer, conditions are conducive to rainfall as humid, unstable air moves from the oceans toward the land. In winter a dry wind, having its origins over the continent, blows toward the sea.

The monsoonal circulation system develops partly in response to the differences in annual temperature variations between continents and oceans. In principle, the processes associated with the monsoon are similar to those described in connection with the land and sea breeze (Chapter 7) except that the scales, in both time and space, are much larger. During spring in the northern hemisphere an irregular area of thermally induced low pressure gradually develops over the interior of southern Asia. It is further strengthened by the poleward advance of the ITC. Thus the summertime circulation is from the higher pressure over the ocean toward the lower pressure over the continent. As winter approaches, winds reverse direction as the ITC migrates southward and a deep anticyclone develops over the chilled continent. By midwinter (northern hemisphere) the winds are from the north and converge on Australia and southern Africa.

The Cw Variant

An examination of Figure 14–1 reveals areas adjacent to the wet and dry tropics in southern Africa, South America, northeastern India, and China that are designated as Cw. Although C has been substituted for A, indicating that these regions are subtropical instead of tropical, the Cw climate is nevertheless just a variant of Aw, for the only major difference is somewhat lower temperatures. In Africa and South America Cw climates are highland extensions of Aw. Because they occupy elevated sites, the warmer temperatures of the adjacent wet and dry tropics are reduced. In India and China Cw areas are middle-latitude extensions of the tropical monsoon realm. In some cases, especially in India, the Cw areas are barely poleward enough to have winter temperatures below those of the A climates.

THE DRY CLIMATES

The dry regions of the world cover some 42 million square kilometers, or about 30 percent, of the earth's land surface. No other climatic group covers so large a land area (Figure 14–7).

Although many aspects of the dry climates are discussed in this section, perhaps the most characteristic feature of all, aside from the fact that yearly rainfall totals are relatively meager, is that the amount of precipitation received each year is very unreliable. Generally the smaller the mean annual rainfall, the greater its variability. As a result, yearly averages are often misleading. For example, during one 7-year period, Trujillo, Peru, had an average rainfall of 6.1 centimeters per year. Yet a closer look reveals that during the first 6 years and 11 months of the period the station received a scant 3.5 centimeters (an annual average of slightly

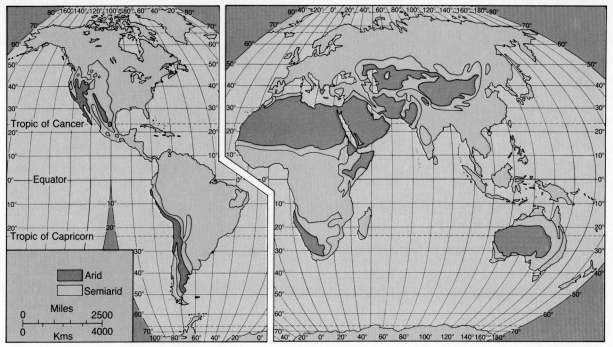

FIGURE 14–7
Arid (desert) and semiarid (steppe) regions cover about 30 percent of the earth's land area. These dry (B) climates constitute the single largest climatic group.

more than 0.5 centimeter). Then during the twelfth month of the seventh year 39 centimeters of rain fell, 23 centimeters of it during a 3-day span. Although perhaps an extreme case, it nevertheless illustrates a common feature of most dry regions: the irregularity of rainfall. It is also significant to note that there are usually more years when rainfall totals are below the average than years when they are above because, as the foregoing example showed, the occasional wet period tends to lift the average.

A number of common misconceptions about deserts continue to persist in the minds of many people, especially those living in more humid regions. One common fallacy about deserts is that they are lifeless or almost lifeless. Although reduced in amount and different in character, plant and animal life are indeed present (Figure 14–8). Desert plants may differ widely from one part of the world to another, but all have one characteristic in common: they have developed adaptations that make them highly tolerant of drought. Many have waxy leaves, stems, or branches or a thickened cuticle (outer-most protective layer) to reduce water loss. Others have very small leaves or no leaves at all. Also, the roots of some species often extend to great depths in order to tap the moisture found there, whereas others produce a shallow but widespread root system that enables them to absorb great amounts of moisture quickly from the infrequent desert downpours. Often the stems of these plants are thickened by a spongy tissue that can store enough water to sustain the plant until the next rainfall comes. Thus although widely dispersed and providing little ground cover, plants of many kinds flourish in the desert.

FIGURE 14—8
One common misconception is that deserts are lifeless or
nearly so. As this scene in Arizona's Sonoran desert illustrates,
these environments are often far from lifeless. (Photo by
R. Scott Dunham)

FIGURE 14—9
Although sand accumulations such as this may be striking features, they represent
only a small percentage of the total desert area. (Photo by Stephen Trimble)

A second widely held belief about the world's dry lands is that they are always hot. As we shall see, it is not true in every case. Unlike any other climatic group, the B climates are defined by precipitation criteria instead of temperature. Therefore dry climates are found from the tropics poleward to the high middle latitudes. Consequently, although tropical deserts lack a cold season, deserts in the middle latitudes do experience seasonal temperature changes.

The last two commonly held misconceptions are more geologic than climatic. One mistaken assumption about the world's deserts is that they consist of mile after mile of drifting sand (Figure 14–9). It is true that sand accumulations do exist in some areas and may be striking features, but they represent only a small percentage of the total desert area. In the Sahara, the world's largest desert, accumulations of sand cover only one-tenth of its area. The sandiest of all deserts is the Arabian, one-third of which consists of sand. The final mistaken assumption is the seemingly logical idea that wind is the most important agent of erosion in deserts. Although wind is relatively more significant in dry areas than anywhere else, most desert landforms are created by running water. When the rains come, they usually take the form of thunderstorms. Because the heavy rain associated with these storms cannot all soak in, rapid runoff results. Without a thick vegetative cover to protect the ground, erosion is great.

What Is Meant by *"Dry"*?

It is important to realize that the concept of dryness is a relative one and refers to any situation in which a water deficiency exists. Thus climatologists define a *dry climate* as one in which the yearly precipitation is not as great as the potential loss of water by evaporation. Dryness, then, is not only related to annual rainfall totals, but it is also a function of evaporation, which, in turn, depends closely on temperature. As temperatures climb, potential evaporation also increases. Twenty-five centimeters of precipitation may be sufficient to support forests in northern Scandinavia, where evaporation into the cool, humid air is slight and a surplus of water remains in the soil. However, the same amount of rain falling on Nevada or Iran supports only a sparse vegetative cover because evaporation into the hot, dry air is great. So clearly no specific amount of precipitation can serve as a universal boundary for dry climates.

To establish the boundary between dry and humid climates, the Koeppen classification uses formulas that involve three variables: (1) average annual precipitation, (2) average annual temperature, and (3) seasonal distribution of precipitation. The use of average annual temperature reflects its importance as an index of evaporation. The amount of rainfall defining the humid–dry boundary will be larger where mean annual temperatures are high and smaller where temperatures are low. The use of seasonal precipitation distribution as a variable is also related to this idea. If rain is concentrated in the warmest months, loss to evaporation is greater than if the precipitation is concentrated in the cooler months. Thus there are considerable differences in the precipitation amounts received at various stations in the B climates. Table 14–5 summarizes these differences. Notice that if a station with an annual mean of 20°C and a summer rainfall maximum does not receive

TABLE 14–5 Average Annual Precipitation (mm) at BS-Humid Boundary

Ave. Ann. Temp. (°C)	Summer Dry Season	Even Distribution	Winter Dry Season
5	100	240	380
10	200	340	480
15	300	440	580
20	400	540	680
25	500	640	780
30	600	740	880

more than 680 millimeters of precipitation per year, it is classified as dry. If the rain falls primarily in winter, however, the station must receive only 400 millimeters or more to be considered humid. If the precipitation is more evenly distributed, the figure defining the humid–dry boundary is between the other two.

Desert

Steppe

Within the regions defined by a general water deficiency there are two climatic types: **arid** or **desert** (BW) and **semiarid** or **steppe** (BS). These two groups have many features in common; their differences are primarily a matter of degree. The semiarid is a marginal and more humid variant of the arid and represents a transition zone that surrounds the desert and separates it from the bordering humid climates. The arid–semiarid boundary is commonly (and arbitrarily) set at one-half the annual precipitation separating dry regions from humid. Thus if the humid–dry boundary happens to be 40 centimeters, the steppe–desert boundary will be 20 centimeters.

Tropical Desert and Steppe

The heart of the low-latitude dry climates lies in the vicinity of the Tropic of Cancer and the Tropic of Capricorn. A glance at Figure 14–7 reveals a virtually unbroken desert environment stretching for more than 9300 kilometers, from the Atlantic coast of North Africa to the dry lands of northwestern India. In addition to this single great expanse, the northern hemisphere contains another much smaller area of tropical desert and steppe in northern Mexico and the southwestern United States. In the southern hemisphere dry climates dominate Australia. Almost 40 percent of the continent is desert, and much of the remainder is steppe. In addition, arid and semiarid areas are found in southern Africa and make a rather limited appearance in coastal Chile and Peru. The existence and distribution of this dry tropical realm are primarily a consequence of the subsidence and marked stability of the subtropical anticyclones.

Within tropical deserts, the scanty precipitation that falls is not only infrequent but also erratic. Indeed, no well-defined seasonal precipitation regime could be termed characteristic of the tropical deserts. These areas are simply located too far poleward to be influenced by the intertropical convergence zone and too far equatorward to benefit from the frontal and cyclonic precipitation of the middle latitudes. Even during the summer months, when extensive daytime heating pro-

TABLE 14–6 Data for Tropical Steppe and Desert Stations

	J	F	M	A	M	J	J	A	S	O	N	D	Yr
				Marrakech, Morocco 31°37′N; 458 m									
Temp. (°C)	11.5	13.4	16.1	18.6	21.3	24.8	28.7	28.7	25.4	21.2	16.5	12.5	19.9
Precip. (mm)	28	28	33	30	18	8	3	3	10	20	28	33	242
				Dakar, Senegal 14°44′N; 23 m									
Temp. (°C)	21.1	20.4	20.9	21.7	23.0	26.0	27.3	27.3	27.5	27.5	26.0	25.2	24.49
Precip. (mm)	0	2	0	0	1	15	88	249	163	49	5	6	578
				Alice Springs, Australia 23°38′S; 570 m									
Temp. (°C)	28.6	27.8	24.7	19.7	15.3	12.2	11.7	14.4	18.3	22.8	25.8	27.8	20.8
Precip. (mm)	43	33	28	10	15	13	8	8	8	18	30	38	252

duces a steep environmental lapse rate and considerable convective motion, clear skies are still the rule. In this case, subsidence aloft prevents the lower air with its modest moisture content from rising high enough to penetrate the condensation level.

The situation is different in the semiarid transitional belts surrounding the desert, for here a seasonal pattern of rainfall becomes better defined. As shown by the data for Dakar in Table 14–6, stations located on the equatorward side of low-latitude deserts have a brief period of relatively heavy rainfall during the summer, when the ITC is farthest poleward. The rainfall regime should look familiar, for it is similar to that found in the adjacent wet and dry tropics except that the amount is less and the period of drought is longer. For steppe areas on the poleward margins of the tropical deserts, the precipitation regime is reversed. As the data for Marrakech illustrate (Table 14–6), the cool season is the period when nearly all precipitation falls. At this time of year middle-latitude cyclones often take more equatorward routes and so bring occasional periods of rain.

To understand temperatures in the desert environment, two factors, humidity and cloud cover, are important to consider. The cloudless sky and low humidity allow an abundance of solar radiation to reach the ground during the day and permit the rapid exit of terrestrial radiation at night. As would be expected, relative humidities are low throughout the year. Relative humidities of from 10 to 30 percent are typical at midday for interior locations. In regard to cloud cover, desert skies are almost always clear (Figure 14–10). In the Sonoran desert region of Mexico and the United States, for example, most stations receive nearly 85 percent of the possible sunshine. Yuma, Arizona, averages 91 percent for the year, with a low of 83 percent in January and a high of 98 percent in June. The Sahara has an average winter cloud cover of about 10 percent, which in summer drops to a mere 3 percent.

During the summer season the desert surface heats rapidly after sunrise, for as we saw in preceding examples, the clear skies permit almost all the solar

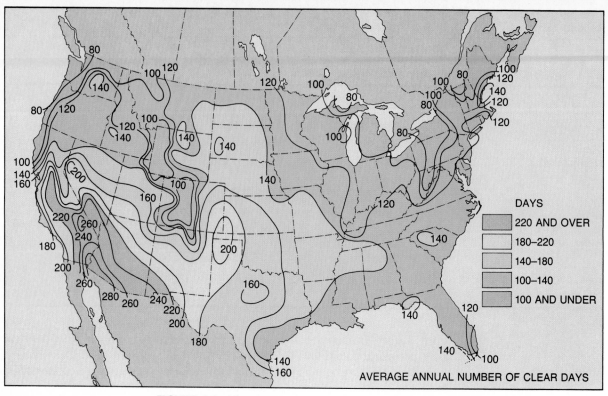

FIGURE 14–10
With few exceptions, desert skies are typically cloudless and hence receive a very high percentage of the possible sunlight. These points are strikingly illustrated when the desert Southwest is examined.

energy to reach the surface. By midafternoon ground-surface temperatures may approach 90°C. Under such circumstances it is not surprising that the world's highest temperatures are recorded in the tropical deserts and that the daily maximums at many stations during the hot season are consistently close to the absolute maximum (the highest temperature ever recorded at a station). At Abadán, Iran, the average daily maximum in July is a scorching 44.7°C, or only 8.3°C lower than the record high. Phoenix, Arizona, is little better, recording an average July maximum of 40.5°C compared with an absolute maximum of 47.7°C. A contributing factor to the high ground and air temperatures is the fact that because very little insolation is used in evaporation, almost all the energy goes to heating the surface. In contrast, humid regions are not as likely to have such extreme ground and air temperatures, for more solar radiation is used to evaporate water and so not as much remains to heat the ground.

At night temperatures typically drop rapidly, partly because the water vapor content of the air is fairly low. The temperature of the ground surface is also a factor, however. Recall from the discussion of radiation in Chapter 2 that the higher the temperature of a radiating body, the more rapidly it loses heat. Thus when applied to a desert setting, this means that not only do such environments heat up quickly by day, but at night they also cool rapidly. Consequently, low-latitude deserts in the interior of continents have the greatest daily temperature ranges on earth. Daily ranges between 15° to 25°C are common and occasionally they may reach even higher values. The highest daily temperature range ever recorded was at In Salah, Algeria, in the Sahara. On October 13, 1927, this station experienced a 24-hour range of 55.5°C, from 52.2° to −3.3°C.

Because most areas of BWh and BSh are located poleward of the A climates, annual temperature ranges are the highest among the tropical climates. During the low sun period, averages are below those in other parts of the tropics, with monthly means of from 16° to 24°C being typical. Still, temperatures during the summer are higher than those in the humid tropics; consequently, annual means at many tropical desert and steppe stations are similar to those in the A climates.

West Coast Tropical Deserts

Where tropical deserts are found along the west coasts of continents, cold ocean currents have a dramatic influence on the climate. The principal west coast deserts are the Atacama in Peru and Chile and the Namib in southern and southwestern Africa. Other areas include portions of the Sonoran desert in Baja California and coastal areas of the Sahara in northwestern Africa. These areas deviate considerably from the general image we have of tropical deserts. Among the most obvious effects of the cold currents are the reduced temperatures, as exemplified by the data for Lima, Peru, and Port Nolloth, South Africa (Table 14–7). Compared with stations at similar latitudes, these places have lower annual means and subdued annual and daily ranges. Port Nolloth, for example, has an annual mean of only 14°C and an annual range of just 4°C compared with Durban, on the opposite side of South Africa, which has a yearly mean of 20°C and an annual range that is twice that at Port Nolloth.

Although these stations are adjacent to the oceans, their yearly rainfall totals are among the lowest in the world. The aridity along these coasts is intensified because the lower air is chilled by the cold offshore waters and hence further stabilized. In addition, the presence of the cold currents causes temperatures to approach, and often reach, the dew point. As a result, these areas are characterized by high relative humidities and much advection fog and dense stratus cloud cover. So it should be remembered that not all tropical deserts are sunny and hot places with low humidities and little cloud cover. Indeed, the presence of cold currents causes west coast tropical deserts to be relatively cool and humid places that are often shrouded by low clouds or fog.

TABLE 14–7 Data for West Coast Tropical Desert Stations

	J	F	M	A	M	J	J	A	S	O	N	D	Yr
Port Nolloth, South Africa 29°14′S; 7 m													
Temp. (°C)	15	16	15	14	14	13	12	12	13	13	15	15	14
Precip. (mm)	2.5	2.5	5.1	5.1	10.2	7.6	10.2	7.6	5.1	2.5	2.5	2.5	63.4
Lima, Peru 12°02′S; 155 m													
Temp. (°C)	22	23	23	21	19	17	16	16	16	17	19	21	19
Precip. (mm)	2.5	T	T	T	5.1	5.1	7.6	7.6	7.6	2.5	2.5	T	40.5

TABLE 14–8 Data for Middle-Latitude Steppe and Desert Stations

	J	F	M	A	M	J	J	A	S	O	N	D	Yr
Ulan Bator, Mongolia 47°55′N; 1311 m													
Temp. (°C)	−26	−21	−13	−1	6	14	16	14	9	−1	−13	−22	−3
Precip. (mm)	1	2	3	5	10	28	76	51	23	7	4	3	213
Denver, Colorado 39°32′N; 1588 m													
Temp. (°C)	0	1	4	9	14	20	24	23	18	12	5	2	11
Precip. (mm)	12	16	27	47	61	32	31	28	23	24	16	10	327

Middle-Latitude Desert and Steppe (BWk and BSk)

Unlike their low-latitude counterparts, middle-latitude deserts and steppes are not controlled by the subsiding air masses of the subtropical anticyclones. Instead these dry lands exist principally because of their position in the deep interiors of large landmasses far removed from oceans. In addition, the presence of high mountains across the paths of the prevailing winds further acts to separate these areas from water-bearing maritime air masses. In North America the Coast ranges, Sierra Nevada, and Cascades are the foremost barriers; in Asia the great Himalayan chain prevents the summertime monsoon flow of moist air from the Indian Ocean from reaching far into the interior. A glance at Figure 14–7 reveals that middle-latitude desert and steppe climates are most widespread in North America and Eurasia. Because of the lack of extensive land areas in the middle latitudes, the southern hemisphere has a much smaller area of BWk and BSk. It is found only at the southern tip of South America in the rain shadow of the towering Andes.

Like tropical deserts and steppes, the dry regions of the middle latitudes have meager and unreliable precipitation. Unlike the dry lands of the low latitudes, however, these more poleward regions have much lower winter temperatures and hence lower annual means and higher annual ranges of temperature. The data in Table 14–8 illustrate this point nicely. The data also reveal that rainfall is most abundant during the warm months. Although not all BWk and BSk stations have a summer precipitation maximum, most do because in winter high pressure and cold temperatures tend to dominate the continents. Both factors oppose precipitation. In summer, however, conditions are somewhat more conducive to cloud formation and precipitation because the anticyclone disappears over the heated continent and higher surface temperatures and greater mixing ratios prevail.

THE HUMID SUBTROPICAL CLIMATE

Humid Subtropical Climate

Located on the eastern sides of the continents, in the 25- to 40-degree latitude range, the **humid subtropical climate** dominates the southeastern United States as well as other similarly situated areas around the world: all of Uruguay and portions of Argentina and southern Brazil in South America, eastern China and southern Japan in Asia, and the eastern coast of Australia.

In the summer a visitor to the humid subtropics would experience hot, sultry weather of the type expected in the rainy tropics. Daytime temperatures are generally in the lower thirties (C°), but it is not uncommon for the thermometer to reach into the upper thirties or even forty on many afternoons. Because the mixing ratio and relative humidity are high, the night brings little relief. An afternoon or evening thunderstorm is also a possibility, for these areas experience such storms on an average of from 40 to 100 days each year, the majority during summer months (Figure 14–11). The primary reason for the tropical summer weather in Cfa regions is the dominating influence of maritime tropical air masses. During the summer months this warm, moist, and unstable air moves inland from the western portions of the oceanic subtropical anticyclone. As the mT air passes over the heated continent, it is made increasingly unstable, giving rise to the common convectional showers and thunderstorms.

As summer turns to autumn, the humid subtropics lose their similarity to the rainy tropics. Although winters are best described as mild, frosts are common

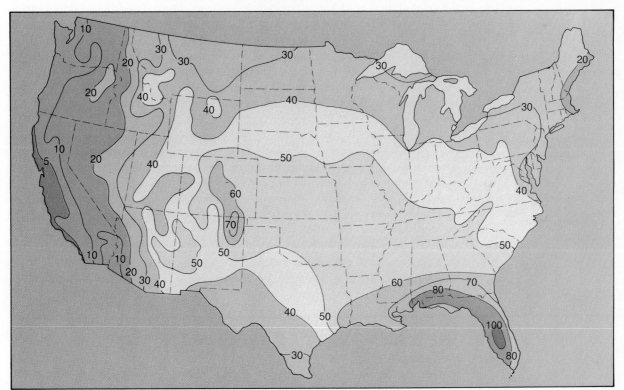

FIGURE 14–11
Average number of days per year with thunderstorms. The humid subtropical climate which dominates the southeastern United States receives much of its precipitation in the form of thunderstorms. Most of the Southeast averages 50 or more days per year with thunderstorms. (Source: Environmental Data Service, NOAA)

TABLE 14–9 Data for Humid Subtropical Stations

	J	F	M	A	M	J	J	A	S	O	N	D	Yr
New Orleans, Louisiana 29°59′N; 1 m													
Temp. (°C)	12	13	16	19	23	26	27	27	25	21	15	13	20
Precip. (mm)	98	101	136	116	111	113	171	136	128	72	85	104	1371
Buenos Aires, Argentina 34°35′S; 27 m													
Temp. (°C)	24	23	21	17	14	11	10	12	14	16	20	22	17
Precip. (mm)	104	82	122	90	79	68	61	68	80	100	90	83	1027

in higher-latitude Cfa areas and occasionally plague the tropical margins as well. The winter precipitation is also different in character from summer. Some is in the form of snow and most is generated along fronts of the frequent middle-latitude cyclones that sweep over these regions. Because the land surface is colder than the maritime air that now arrives less frequently from tropical source regions, the air is chilled in its lower layers as it moves poleward over the ground. Consequently, convectional showers are rare, for the stabilized mT air masses produce clouds and precipitation only when forced to rise.

The data for two humid subtropical stations in Table 14–9 serve to summarize the general characteristics of the Cfa climate. Yearly precipitation totals are usually in excess of 100 centimeters, and the rainfall is well distributed throughout the year. Although summer is ordinarily the season of greatest precipitation, there is considerable variation. In the United States, for example, precipitation in the Gulf states is very evenly distributed. As one moves poleward or toward the drier western margins, however, a much higher percentage falls in summer. Some coastal stations, especially along the equatorward margins, have rainfall maximums in late summer or autumn as tropical cyclones or their remnants visit the area. In Asia the well-developed monsoon circulation favors a summer precipitation maximum. An examination of temperature figures shows that summer temperatures are comparable to the tropics and that winter values are markedly lower. This situation, of course, is to be expected, for the higher latitudinal position of the subtropics results in a wider variation in sun angle and length of daylight as well as in occasional or even frequent invasions of cP air masses during winter.

THE MARINE WEST COAST CLIMATE

Marine West Coast Climate

Situated on the western (windward) side of continents from about 40° to 65° N and S latitude is a climatic region dominated by the onshore flow of oceanic air. The prevalence of maritime air masses means that mild winters and cool summers are the rule, as is an ample amount of rainfall throughout the year. In North America the **marine west coast climate** extends from near the United States–Canadian border northward as a narrow belt into southern Alaska. A similar slender strip occurs in South America along the coast of Chile poleward of about 40°S latitude. In both instances, high mountains parallel the coast and prevent the marine climate from penetrating far inland. The largest area of Cfb climate is found in Europe, for here there is no mountain barrier blocking the movement of cool maritime air from the North Atlantic. Other locations include most of New Zealand as well as tiny slivers of South Africa and Australia.

The data for representative marine west coast stations in Table 14–10 reveal that although there is no pronounced dry period, there is a drop in monthly precipitation totals during the summer. The reduced summer rainfall is due to the poleward migration of the oceanic subtropical highs. Although the areas of marine west coast climate are situated too far poleward to be dominated by these dry anticyclones, their influence is sufficient to cause a decrease in warm-season rainfall. A comparison of precipitation data for London and Vancouver also demonstrates that the presence of coastal mountains has a considerable influence on yearly rainfall totals. Vancouver's total is about 2½ times that of London. In settings like Vancouver's the precipitation totals are higher not only because of orographic uplift but also because the mountains act to slow the passage of cyclonic storms, thus allowing them to drop a greater quantity of water than would otherwise be the case.

Because of the nearness to the ocean, winters are mild and summers relatively cool. Therefore a low annual temperature range is characteristic in the marine west coast climate. Because cP air masses generally drift eastward in the zone of the westerlies, periods of severe winter cold are uncommon. The western edge of North America is especially sheltered from incursions of frigid continental air by the high mountains that intervene between the coast and the source regions for cP air masses. Because of the lack of such a mountain barrier in Europe, cold waves in that region are somewhat more frequent.

The dominance of the ocean in controlling temperatures can be further demonstrated by a look at temperature gradients (changes in temperature per unit distance). Although this climate encompasses a wide latitudinal span, temperatures change much more abruptly moving inland from the coast than they do in a north-south direction. The transport of heat from the oceans more than offsets the latitudinal variation in the receipt of solar energy. In both January and July,

TABLE 14–10 Data for Marine West Coast Stations

	J	F	M	A	M	J	J	A	S	O	N	D	Yr
Vancouver, British Columbia 49°11'N; 0 m													
Temp. (°C)	2	4	6	9	13	15	18	17	14	10	6	4	10
Precip. (mm)	139	121	96	60	48	51	26	36	56	117	142	156	1048
London, U.K. 51°28'N; 5 m													
Temp. (°C)	4	4	7	9	12	16	18	17	15	11	7	5	10
Precip. (mm)	54	40	37	38	46	46	56	59	50	57	64	48	595

for example, the temperature change from coastal Seattle to more inland Spokane, a distance of about 375 kilometers, is equal to the variation between Seattle and Juneau, Alaska. Juneau is about 11 degrees of latitude, or roughly 1200 kilometers, north of Seattle.

THE DRY-SUMMER SUBTROPICAL CLIMATE

Dry-Summer Subtropical Climate

The **dry-summer subtropical climate** is typically located along the west sides of continents between latitudes 30 and 45 degrees. Situated between the marine west coast climate on the poleward side and the tropical steppes on the equatorward side, this climatic region is best described as transitional in character. It is unique for the fact that it is the only humid climate that has a strong winter rainfall maximum, a feature that reflects its intermediate position. In summer the region is dominated by the stable eastern side of the oceanic subtropical highs. In winter, as the wind and pressure systems follow the sun equatorward, it is within range of the cyclonic storms of the polar front. Thus during the course of a year these areas alternate between being a part of the dry tropics and an extension of the humid middle latitudes. Although middle-latitude changeability characterizes the winter, tropical constancy describes the summer.

As was the case for the marine west coast climate, mountain ranges limit the dry-summer subtropics to a relatively narrow coastal zone in both North and South America. Because Australia and southern Africa barely reach to the latitudes where dry-summer climates exist, the development of this climatic type is limited on these continents as well. Consequently, because of the arrangement of the continents, and of their mountain ranges, inland development occurs only in the Mediterranean basin. Here the zone of subsidence extends far to the east in summer; in winter the sea is a major route of cyclonic disturbances. Because the dry-summer climate is particularly extensive in this region, the name **Mediterranean climate** is often used as a synonym.

Mediterranean Climate

Two types of Mediterranean climate are recognized and are based primarily on summertime temperatures. The cool summer type (Csb), as exemplified by San Francisco, California, and Santiago, Chile (Table 14–11), is limited to coastal areas. Here the cooler summer temperatures one expects on a windward coast are further intensified by cold ocean currents. The data for Izmir and Sacramento illustrate the features of the warm summer type (Csa). At both places winter temperatures are not very different from those in the Csb type. But Sacramento in the Central Valley of California is removed from the coast and Izmir is bordered by the warm waters of the Mediterranean; consequently, summer temperatures are noticeably higher. As a result, annual temperature ranges are also higher in Csa areas.

Yearly precipitation totals within the dry-summer subtropics range between about 40 and 80 centimeters. In many areas such amounts mean that a station barely escapes being classified as semiarid. Consequently, some climatologists refer to the dry-summer climate as subhumid instead of humid. This is especially

TABLE 14–11 Data for Dry-Summer Subtropical Stations

	J	F	M	A	M	J	J	A	S	O	N	D	Yr
San Francisco, California 37°37'N; 5 m													
Temp. (°C)	9	11	12	13	15	16	17	17	18	16	13	10	14
Precip. (mm)	102	88	68	33	12	3	0	1	5	19	40	104	475
Sacramento, California 38°35'N; 13 m													
Temp. (°C)	8	10	12	16	19	22	25	24	23	18	12	9	17
Precip. (mm)	81	76	60	36	15	3	0	1	5	20	37	82	416
Izmir, Turkey 38°26'N; 25 m													
Temp. (°C)	9	9	11	15	20	25	28	27	23	19	14	10	18
Precip. (mm)	141	100	72	43	39	8	3	3	11	41	93	141	695
Santiago, Chile 33°27'S; 512 m													
Temp. (°C)	19	19	17	13	11	8	8	9	11	13	16	19	14
Precip. (mm)	3	3	5	13	64	84	76	56	30	13	8	5	360

true along the equatorward margins because rainfall totals increase in the poleward direction. Los Angeles, for example, receives 38 centimeters of precipitation annually, whereas San Francisco, 400 kilometers to the north, receives 51 centimeters per year. Still further north, at Portland, Oregon, the yearly rainfall average is over 90 centimeters.

THE HUMID CONTINENTAL CLIMATE

Situated in the zone dominated by the polar front, this climatic region is a battleground for tropical and polar air masses. In no other climate are rapid nonperiodic changes in the weather more pronounced. Cold waves, heat waves, blizzards, and heavy downpours are all yearly events in the humid continental realm.

Humid Continental Climate

The **humid continental climate,** as its name implies, is a land-controlled climate, the result of broad continents located in the middle latitudes. Because continentality is a basic feature, this climate is not found in the southern hemisphere, where the middle-latitude zone is dominated by the oceans. Instead it is confined to the central and eastern portions of North America and Eurasia in the latitude range between approximately 40° and 50°N latitude. It may at first seem unusual that a continental climate should extend eastward to the margins of the ocean. However, because the prevailing atmospheric circulation is from the west, deep and persistent incursions of maritime air from the east are not likely to occur.

Both winter and summer temperatures in the humid continental climate may be characterized as relatively severe. Consequently, annual temperature ranges

TABLE 14–12 Data for Humid Continental Stations

	J	F	M	A	M	J	J	A	S	O	N	D	Yr
Omaha, Nebraska 41°18′N; 330 m													
Temp. (°C)	−6	−4	3	11	17	22	25	24	19	12	4	−3	10
Precip. (mm)	20	23	30	51	76	102	79	81	86	48	33	23	652
New York City 40°47′N; 40 m													
Temp. (°C)	−1	−1	3	9	15	21	23	22	19	13	7	1	11
Precip. (mm)	84	84	86	84	86	86	104	109	86	86	86	84	1065
Winnipeg, Canada 49°54′N; 240 m													
Temp. (°C)	−18	−16	−8	3	11	17	20	19	13	6	−5	−13	3
Precip. (mm)	26	21	27	30	50	81	69	70	55	37	29	22	517
Harbin, Manchuria 45°45′N; 143 m													
Temp. (°C)	−20	−16	−6	6	14	20	23	22	14	6	−7	−17	3
Precip. (mm)	4	6	17	23	44	92	167	119	52	36	12	5	557

are high throughout the climate. A comparison of the stations whose data are shown in Table 14–12 illustrates the temperature variations within the humid continental realm. July means are generally near and often above 20°C. Thus summertime temperatures, although somewhat lower in the north than in the south, are not markedly different. This situation is also illustrated by the isothermal map of the eastern United States in Figure 14–12(a). Notice that there are only a few widely spaced isotherms, indicating a weak summer temperature gradient. In January, however, the monthly means, although consistently cold, are more variable. The decrease in midwinter values with increasing latitude is appreciable (Figure 14–12b). A glance at Table 14–12 reveals that the temperature change between Omaha and Winnipeg is more than twice as great in winter as in summer. Because of the steeper winter temperature gradient, shifts in wind direction during the cold season often result in sudden large temperature changes. Such is not the case in summer, for temperatures throughout the region are more uniform.

Annual temperature ranges also vary within this climate, generally increasing from south to north as well as from the coast toward the interior. A comparison of the data for Omaha and Winnipeg illustrates the first situation, and a comparison of New York City and Omaha the second.

Records of the four stations in Table 14–12 reveal the general pattern of precipitation within the regions having humid continental climates. A summer maximum occurs at each station, but it is weakly defined at New York City because the East Coast is more accessible to maritime air masses throughout the year. For the same reason, New York City also has the highest total of the four stations. Harbin, Manchuria, on the other hand, shows the most pronounced summer maximum of all, followed by a winter drought. This condition is characteristic of most east Asian stations in the middle latitudes and reflects the powerful control

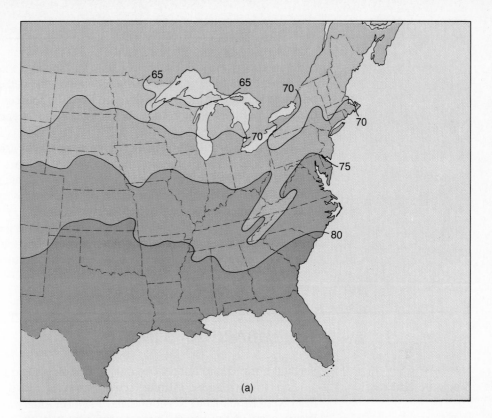

(a)

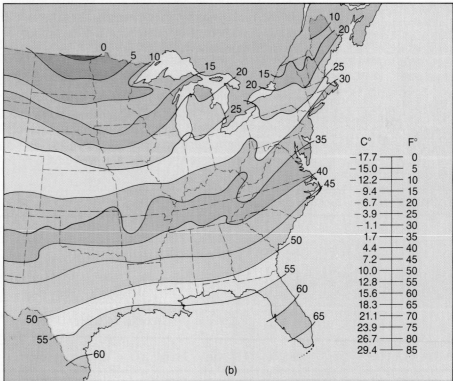

C°	F°
−17.7	0
−15.0	5
−12.2	10
−9.4	15
−6.7	20
−3.9	25
−1.1	30
1.7	35
4.4	40
7.2	45
10.0	50
12.8	55
15.6	60
18.3	65
21.1	70
23.9	75
26.7	80
29.4	85

(b)

FIGURE 14–12
During the summer months north–south temperature variations in the eastern United States are small; that is, the temperature gradient is weak (a). In winter, however, north–south temperature contrasts are sharp (b).

of the monsoon. Another pattern revealed by the data is that precipitation generally decreases toward the interior of the continents as well as from south to north, primarily because of increasing distance from the sources of mT air. Furthermore, the more northerly stations are also influenced for a greater part of the year by drier polar air masses.

Wintertime precipitation in these climates is chiefly associated with the passage of fronts connected with traveling middle-latitude cyclones. Part of this precipitation is in the form of snow, the proportion increasing with latitude. Although precipitation is often considerably less during the cold season, it is usually more conspicuous than the greater amounts that fall during summer. An obvious reason is that snow remains on the ground, often for extended periods, and rain, of course, does not. Moreover, whereas summer rains are often in the form of relatively short convective showers, winter snows generally occur over a more prolonged period.

THE SUBARCTIC CLIMATE

Subarctic Climate

Taiga

Situated north of the humid continental climate and south of the polar tundra is an extensive **subarctic climate** region covering broad, uninterrupted expanses from western Alaska to Newfoundland in North America and from Norway to the Pacific coast of the Soviet Union in Eurasia. It is often referred to as the **taiga** climate, for its extent closely corresponds to the northern coniferous forests region of the same name. Although scrawny, the spruce, fir, larch, and birch trees in the taiga represent the largest stretch of continuous forests on the surface of the earth.

The Subarctic is well illustrated by the data for Yakutsk and Dawson (Table 14–13). Here in the source regions of continental polar air masses the outstanding feature is certainly the dominance of winter. Not only is it long, but temperatures are also bitterly cold. Winter minimum temperatures are among the lowest ever recorded outside the ice caps of Greenland and Antarctica. In fact, for many years the world's coldest temperature was attributed to Verkhoyansk in east-central

TABLE 14–13 Data for Subarctic Stations

	J	F	M	A	M	J	J	A	S	O	N	D	Yr
Yakutsk, U.S.S.R. 62°05′N; 103 m													
Temp. (°C)	−43	−37	−23	−7	7	16	20	16	6	−8	−28	−40	−10
Precip. (mm)	7	6	5	7	16	31	43	38	22	16	13	9	213
Dawson, Canada 64°03′N; 315 m													
Temp. (°C)	−30	−24	−16	−2	8	14	15	12	6	−4	−17	−25	−5
Precip. (mm)	20	20	13	18	23	33	41	41	43	33	33	28	346

Siberia, where the temperature dropped to −68°C on February 5 and 7, 1892. Over a 23-year period, this same station had an average monthly minimum of −62°C during January. Although exceptional temperatures, they illustrate the extreme cold that envelopes the taiga in winter.

In contrast, summers in the Subarctic are remarkably warm despite their short duration. When compared to regions farther south, however, this short season must be characterized as cool; for despite the many hours of daylight, the sun never rises very high in the sky and so solar radiation is not intense. The extremely cold winters and the relatively warm summers of the taiga combine to produce the highest annual temperature ranges on earth. Yakutsk holds the distinction of having the greatest average annual temperature range in the world— 63°C. As the data for Dawson show, the North American Subarctic is not so severe.

Because these far northerly continental interiors are the source regions of cP air masses, only limited moisture is available throughout the year. Precipitation totals are therefore small, seldom exceeding 50 centimeters annually. By far the greatest quantity of precipitation comes in the form of rain from scattered summer convectional showers. Snowfall is not as heavy as in the humid continental climate to the south. Still, there is often the illusion of more. No melting occurs for months at a time, which means that the entire winter accumulation (up to 1 meter) is visible all at once. Furthermore, during blizzards the dry powdery snow is easily picked up and whirled about by the high winds, creating high drifts and giving the impression that more snow is falling than is actually the case. So although snowfall is not excessive, a visitor to this region could leave with that impression.

THE POLAR CLIMATES

Polar Climates

According to the Koeppen classification, **polar climates** are those in which the mean temperature of the warmest month is below 10°C. Therefore just as the tropics are defined by their year-round warmth, so the polar realm is known for its enduring cold. These regions hold the distinction of having the lowest annual means for any part of the planet. Because winters are periods of perpetual night, or nearly so, temperatures at most polar locations are understandably bitter. During the summer months temperatures remain cool despite the long days, for the sun is so low in the sky that its oblique rays are not effective enough to produce a genuine warming. In addition, much solar radiation is reflected by the ice and snow or used in melting the snow cover. In either case, energy that could have warmed the land is lost or consumed. Although cool, summer temperatures are still much higher than those experienced during the extremely severe winter months. Consequently, annual temperature ranges are usually large.

Although polar climates are classified as humid, precipitation is generally meager, with many nonmarine stations receiving less than 25 centimeters annually. Evaporation, of course, is also limited. The scanty precipitation totals are easily understood in view of the temperature characteristics of the region. The amount of water vapor in the air is always small because low mixing ratios must accompany low temperatures. In addition, steep lapse rates are not possible under polar temperature conditions. Usually precipitation is most abundant during the warmer summer months when the air's moisture content is highest.

The Tundra Climate

Tundra Climate

The **tundra climate** on land is found almost exclusively in the northern hemisphere. Here it occupies the coastal fringes of the Arctic Ocean as well as many Arctic islands and the ice-free shores of northern Iceland and southern Greenland. In the southern hemisphere no extensive land areas exist in the latitudes where tundra climates prevail. Consequently, except for some small islands in the southern oceans, the ET climate occupies only the southwestern tip of South America and the northern portion of the Palmer Peninsula in Antarctica.

The 10°C summer isotherm that marks the equatorward limit of the tundra also marks the poleward limit of tree growth. Thus it is essentially a treeless region of grasses, sedges, mosses, and lichens. During the long cold season plant life is dormant, but once the short cool summer commences, these plants mature and produce seeds with great rapidity. Because summers are cool and short, the frozen soils of the tundra generally thaw to depths of less than a meter; consequently, the subsoil remains permanently frozen. This condition, known as **permafrost,**

Permafrost

blocks the downward loss of water and results in poorly drained, boggy soils that make the summer tundra landscape difficult to traverse (Figure 14–13).

The data for Point Barrow, Alaska (Table 14–14), on the shores of the frozen Arctic Ocean, exemplify the most common type of ET station—that is, one in which continentality prevails. Because of the combination of high latitude and continentality, winters are severe, summers are cool, and annual temperature ranges are high. Also, yearly precipitation is small, with a modest summertime maximum.

While Point Barrow represents the most common type of tundra setting, the data for Angmagssalik, Greenland (Table 14–14), reveal that some ET stations are different. Although summer temperatures at both stations are roughly equivalent, winters at Angmagssalik are much warmer and the annual precipitation is eight times greater than at Point Barrow. The reason for the contrast is Angmagssalik's location on the southeast coast of Greenland, where there is considerable marine influence. The warm North Atlantic Drift keeps winter temperatures relatively high, and mP air masses supply moisture throughout the year. Because winters are not as severe at stations like Angmagssalik, annual temperature ranges are much smaller than at ET stations like Point Barrow, where continentality is a major control.

Finally, it should be pointed out that tundra climates are not entirely confined to the high latitudes. The summer coolness of this climate is also found at increasingly higher elevations as one moves equatorward. So even in the tropics you can find ET climates if you go high enough. When compared with the Arctic tundra, however, winter temperatures in these lower-latitude counterparts become milder and less distinct from summer, as the data for Cruz Loma, Ecuador (Table 14–14), illustrate.

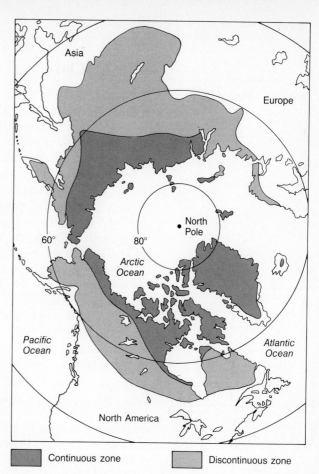

FIGURE 14–13

In the northern hemisphere, permafrost decreases in thickness progressively from north to south. Two major zones are distinguished: a northern zone in which permafrost forms a continuous layer at shallow depths and, to the south, a discontinuous zone in which there are scattered permafrost-free areas. Land underlain by continuous permafrost almost completely circumscribes the Arctic Ocean. (After U.S. Geological Survey)

■ Continuous zone   ▨ Discontinuous zone

TABLE 14–14 Data for Polar Stations

	J	F	M	A	M	J	J	A	S	O	N	D	Yr
Eismitte, Greenland 70°53′N; 2953 m													
Temp. (°C)	−42	−47	−40	−32	−24	−17	−12	−11	−11	−36	−43	−38	−29
Precip. (mm)	15	5	8	5	3	3	3	10	8	13	13	25	111
Angmagssalik, Greenland 65°36′N; 29 m													
Temp. (°C)	−7	−7	−6	−3	2	6	7	7	4	0	−3	−5	0
Precip. (mm)	57	81	57	55	52	45	28	70	72	96	87	75	775
Barrow, Alaska 71°18′N; 9 m													
Temp. (°C)	−28	−28	−26	−17	−8	0	4	3	−1	−9	−18	−24	−12
Precip. (mm)	5	5	3	3	3	10	20	23	15	13	5	5	110
Cruz Loma, Ecuador 0°08′S; 3888 m													
Temp. (°C)	6.1	6.6	6.6	6.6	6.6	6.1	6.1	6.1	6.1	6.1	6.6	6.6	6.4
Precip. (mm)	198	185	241	236	221	122	36	23	86	147	124	160	1779

Ice Cap Climate

The **ice cap climate,** designated by Koeppen as EF, does not have a single monthly mean above 0°C. Consequently, because the average temperature for all months is below freezing, the growth of vegetation is prohibited and the landscape is one of permanent ice and snow. This climate of perpetual frost covers a surprisingly large area of more than 15.5 million square kilometers or about 9 percent of the earth's land area, and aside from scattered occurrences in high mountain areas, it is confined to the ice caps of Greenland and Antarctica.

Average annual temperatures are extremely low; for example, the annual mean at Eismitte, Greenland (Table 14–14), is −29°C; at Byrd Station, Antarctica, −21°C; and at Vostok, the USSR Antarctic Meteorological Station, −57°C. Vostok has also experienced the lowest temperature ever recorded, −88.3°C on August 24, 1960. In addition to the effect of latitude, the primary reason for the very low temperatures is undoubtedly the presence of permanent ice. Ice has a very high albedo, reflecting up to 80 percent of the insolation striking it. Also, the energy that is not reflected is used largely to melt the ice and so is not available for raising the temperature of the air. An additional factor that contributes to the very low temperatures at many EF stations is elevation. Eismitte, at the center of the Greenland ice cap, rests at an elevation of almost 3000 meters, and much of Antarctica is even higher. Thus because of the permanent ice and often high elevations, the already low temperatures of the polar realm are further reduced.

The intense chilling of the air close to the ice cap means that strong surface temperature inversions are common. Temperatures near the surface may be as much as 30°C colder than air just a few hundred meters above. Often this cold, dense air flows downslope because of the force of gravity and produces strong winds and blizzard conditions. Such air movements, called *katabatic winds*, are an important aspect of ice cap weather at many locations. In some places, where the slope is sufficient, these gravity-induced air movements are sometimes strong enough to flow in a direction that is opposite the pressure gradient.

REVIEW

1. Why is classification often a helpful or even necessary task in science?

2. What climatic data are needed in order to classify a climate using the Koeppen scheme?

3. Should climatic boundaries, such as those shown on the world map in Figure 14–1, be regarded as fixed? Explain.

4. List the major climatic controls and briefly describe their influence.

5. How does the selva differ from a typical middle-latitude forest?

6. Distinguish between jungle and selva.

7. Explain each of the following characteristics of the wet tropics.
   a. This climate is restricted to elevations below 1000 meters.
   b. Mean monthly and annual temperatures are high and the annual temperature range is low.
   c. This climate is rainy throughout the year or nearly so.

8. Why are the wet tropics considered oppressive and monotonous?

9. What is the difference between Af areas and Am areas?

10. **a.** What primary factor distinguishes Aw climates from Af and Am?
    b. How is this difference reflected in the vegetation?

11. Describe the influence of the ITC and the subtropical high on the precipitation regime in the Aw climate.

12. In which of the following climates is the annual rainfall likely to be more consistent from year to year: BSh, Aw, BWh, Af? In which of these climates is the annual rainfall most variable from year to year? Explain your answers.

13. In the dry (B) climates there are usually more years when rainfall totals are below the average than above. Explain and give an example.

14. List and briefly discuss the four common misconceptions about deserts.

15. Why is the amount of precipitation that defines the humid–dry boundary variable?

16. What is the primary reason (control) for the existence of the dry tropical realm (BWh and BSh)?

17. **a.** Describe and explain the seasonal distribution of precipitation for a BSh station located on the poleward side of a tropical desert and a BSh station located on the equatorward side of the desert.
    b. If both stations barely meet the requirements for steppe climates (that is, with only a little more rainfall both stations would be considered humid), which station would probably have the lowest rainfall total? Explain.

18. Why do ground and air temperatures reach such high values in tropical deserts?

19. Tropical deserts, such as the Atacama and Namib, deviate considerably from the general image that we have of tropical deserts. In what ways are these deserts not "typical" and why?

20. What is the primary cause for the existence of the middle-latitude deserts and steppes?

21. Why are desert and steppe areas uncommon in the middle latitudes of the southern hemisphere?

22. Describe and explain the differences between summertime and wintertime precipitation in the humid subtropics (Cfa).

23. Why is the marine west coast climate (Cfb and Cfc) represented by only slender strips of land in North and South America and why is it very extensive in western Europe?

24. How do temperature gradients (north-south versus east-west) reveal the strong oceanic influence along the west coast of North America?

25. In this chapter the dry-summer subtropics were described as transitional. Explain why this statement is true.

26. What other name is given to the dry-summer subtropical climate?

27. Why are summer temperatures cooler at San Francisco than at Sacramento (Table 14–11)?

28. Why is the humid continental climate confined to the northern hemisphere?

29. Why do coastal stations like New York City experience primarily continental climatic conditions?

30. Using the four stations whose data are shown in Table 14–12, describe the general pattern of precipitation in the humid continental climates.

31. In the dry-summer subtropics precipitation totals increase with an increase in latitude, but in the humid continental climates the reverse is true. Explain.

32. Although generally characterized by small precipitation totals, subarctic and polar climates are considered humid. Explain.

33. Although snowfall in the subarctic climate is relatively small, a wintertime visitor to this region might leave with the impression that snowfall is great. Explain.

34. Describe and explain the annual temperature range one should expect in the realm of the taiga.

35. Although polar regions experience extended periods of almost perpetual sunlight in the summer, temperatures remain cool. Explain.

36. What is the significance of the 10°C summer isotherm?

37. Why is the tundra landscape characterized by poorly drained, boggy soils?

38. Why are winter temperatures higher and the annual precipitation greater at Angmagssalik, Greenland, than at Point Barrow, Alaska?

39. The tundra climate is not confined solely to high latitudes. Under what circumstances might the ET climate be found in more equatorward locations?

40. Where are EF climates most extensively developed?

41. Besides the effect of latitude, what other factor(s) contributes to the extremely low temperatures that characterize the EF climate? Explain.

VOCABULARY REVIEW

Koeppen classification

marine climate

continental climate

tropical rain forest (selva)

jungle

intertropical convergence zone (ITC)

tropical wet and dry

savanna

monsoon

desert (arid)

steppe (semiarid)

humid subtropical climate

marine west coast climate

dry-summer subtropical climate

Mediterranean climate

humid continental climate

subarctic climate

taiga

polar climates

tundra climate

permafrost

ice cap climate

Metric Units

TABLE A–1 The International System of Units (SI)

I. Basic Units

Quantity	Unit	SI Symbol
Length	meter	m
Mass	kilogram	kg
Time	second	s
Electric current	ampere	A
Thermodynamic temperature	kelvin	K
Amount of substance	mole	mol
Luminous intensity	candela	cd

TABLE A–1 (Continued)

II. Prefixes

Prefix	Factor by Which Unit Is Multiplied	Symbol
tera	10^{12}	T
giga	10^{9}	G
mega	10^{6}	M
kilo	10^{3}	k
hecto	10^{2}	h
deka	10	da
deci	10^{-1}	d
centi	10^{-2}	c
milli	10^{-3}	m
micro	10^{-6}	μ
nano	10^{-9}	n
pico	10^{-12}	p
femto	10^{-15}	f
atto	10^{-18}	a

III. Derived Units

Quantity	Units	Expression
Area	square meter	m^2
Volume	cubic meter	m^3
Frequency	hertz (Hz)	s^{-1}
Density	kilogram per cubic meter	kg/m^3
Velocity	meter per second	m/s
Angular velocity	radian per second	rad/s
Acceleration	meter per second squared	m/s^2
Angular acceleration	radian per second squared	rad/s^2
Force	newton (N)	$kg \cdot m/s^2$
Pressure	newton per square meter	N/m^2
Work, energy, quantity of heat	joule (J)	$N \cdot m$
Power	watt (W)	J/s
Electric charge	coulomb (C)	$A \cdot s$
Voltage, potential difference, electromotive force	volt (V)	W/A
Luminance	candela per square meter	cd/m^2

TABLE A–2 Metric–English Conversion

When You Want to Convert:	Multiply by:	To Find:
Length		
inches	2.54	centimeters
centimeters	0.39	inches
feet	0.30	meters
meters	3.28	feet
yards	0.91	meters
meters	1.09	yards
miles	1.61	kilometers
kilometers	0.62	miles
Area		
square inches	6.45	square centimeters
square centimeters	0.15	square inches
square feet	0.09	square meters
square meters	10.76	square feet
square miles	2.59	square kilometers
square kilometers	0.39	square miles
Volume		
cubic inches	16.38	cubic centimeters
cubic centimeters	0.06	cubic inches
cubic feet	0.028	cubic meters
cubic meters	35.3	cubic feet
cubic miles	4.17	cubic kilometers
cubic kilometers	0.24	cubic miles
liters	1.06	quarts
liters	0.26	gallons
gallons	3.78	liters
Masses and Weights		
ounces	20.33	grams
grams	0.035	ounces
pounds	0.45	kilograms
kilograms	2.205	pounds

Temperature

When you want to convert degrees Fahrenheit (°F) to degrees Celsius (°C), subtract 32 degrees and divide by 1.8 (also see Table A–3).

When you want to convert degrees Celsius (°C) to degrees Fahrenheit (°F), multiply by 1.8 and add 32 degrees.

When you want to convert degrees Celsius (°C) to degrees Kelvin (K), delete the degree symbol and add 273.

When you want to convert degrees Kelvin (K) to degrees Celsius (°C), add the degree symbol and subtract 273.

TABLE A-3 Temperature Conversion Table. (To find either the Celsius or the Fahrenheit equivalent, locate the known temperature in the center column. Then read the desired equivalent value from the appropriate column.)

°C		°F	°C		°F	°C		°F	°C		°F
−40.0	−40	−40	−17.2	+1	33.8	5.0	41	105.8	27.2	81	177.8
−39.4	−39	−38.2	−16.7	2	35.6	5.6	42	107.6	27.8	82	179.6
−38.9	−38	−36.4	−16.1	3	37.4	6.1	43	109.4	28.3	83	181.4
−38.3	−37	−34.6	−15.4	4	39.2	6.7	44	111.2	28.9	84	183.2
−37.8	−36	−32.8	−15.0	5	41.0	7.2	45	113.0	29.4	85	185.0
−37.2	−35	−31.0	−14.4	6	42.8	7.8	46	114.8	30.0	86	186.8
−36.7	−34	−29.2	−13.9	7	44.6	8.3	47	116.6	30.6	87	188.6
−36.1	−33	−27.4	−13.3	8	46.4	8.9	48	118.4	31.1	88	190.4
−35.6	−32	−25.6	−12.8	9	48.2	9.4	49	120.2	31.7	89	192.2
−35.0	−31	−23.8	−12.2	10	50.0	10.0	50	122.0	32.2	90	194.0
−34.4	−30	−22.0	−11.7	11	51.8	10.6	51	123.8	32.8	91	195.8
−33.9	−29	−20.2	−11.1	12	53.6	11.1	52	125.6	33.3	92	197.6
−33.3	−28	−18.4	−10.6	13	55.4	11.7	53	127.4	33.9	93	199.4
−32.8	−27	−16.6	−10.0	14	57.2	12.2	54	129.2	34.4	94	201.2
−32.2	−26	−14.8	−9.4	15	59.0	12.8	55	131.0	35.0	95	203.0
−31.7	−25	−13.0	−8.9	16	60.8	13.3	56	132.8	35.6	96	204.8
−31.1	−24	−11.2	−8.3	17	62.6	13.9	57	134.6	36.1	97	206.6
−30.6	−23	−9.4	−7.8	18	64.4	14.4	58	136.4	36.7	98	208.4
−30.0	−22	−7.6	−7.2	19	66.2	15.0	59	138.2	37.2	99	210.2
−29.4	−21	−5.8	−6.7	20	68.0	15.6	60	140.0	37.8	100	212.0
−28.9	−20	−4.0	−6.1	21	69.8	16.1	61	141.8	38.3	101	213.8
−28.3	−19	−2.2	−5.6	22	71.6	16.7	62	143.6	38.9	102	215.6
−27.8	−18	−0.4	−5.0	23	73.4	17.2	63	145.4	39.4	103	217.4
−27.2	−17	+1.4	−4.4	24	75.2	17.8	64	147.2	40.0	104	219.2
−26.7	−16	3.2	−3.9	25	77.0	18.3	65	149.0	40.6	105	221.0
−26.1	−15	5.0	−3.3	26	78.8	18.9	66	150.8	41.1	106	222.8
−25.6	−14	6.8	−2.8	27	80.6	19.4	67	152.6	41.7	107	224.6
−25.0	−13	8.6	−2.2	28	82.4	20.0	68	154.4	42.2	108	226.4
−24.4	−12	10.4	−1.7	29	84.2	20.6	69	156.2	42.8	109	228.2
−23.9	−11	12.2	−1.1	30	86.0	21.1	70	158.0	43.3	110	230.0
−23.3	−10	14.0	−0.6	31	87.8	21.7	71	159.8	43.9	111	231.8
−22.8	−9	15.8	0.0	32	89.6	22.2	72	161.6	44.4	112	233.6
−22.2	−8	17.6	+0.6	33	91.4	22.8	73	163.4	45.0	113	235.4
−21.7	−7	19.4	1.1	34	93.2	23.3	74	165.2	45.6	114	237.2
−21.1	−6	21.2	1.7	35	95.0	23.9	75	167.0	46.1	115	239.0
−20.6	−5	23.0	2.2	36	96.8	24.4	76	168.8	46.7	116	240.8
−20.0	−4	24.8	2.8	37	98.6	25.0	77	170.6	47.2	117	242.6
−19.4	−3	26.6	3.3	38	100.4	25.6	78	172.4	47.8	118	244.4
−18.9	−2	28.4	3.9	39	102.2	26.1	79	174.2	48.3	119	246.2
−18.3	−1	30.2	4.4	40	104.0	26.7	80	176.0	48.9	120	248.0
−17.8	0	32.0									

TABLE A–4 Wind Conversion Table (Wind Speed Units: 1 mile per hour = 0.868391 knot = 1.609344 km/hr = 0.44704 m/sec)

Miles per Hour	Knots	Meters per Second	Kilometers per Hour	Miles per Hour	Knots	Meters per Second	Kilometers per Hour
1	0.9	0.4	1.6	41	35.6	18.3	66.0
2	1.7	0.9	3.2	42	36.5	18.8	67.6
3	2.6	1.3	4.8	43	37.3	19.2	69.2
4	3.5	1.8	6.4	44	38.2	19.7	70.8
5	4.3	2.2	8.0	45	39.1	20.1	72.4
6	5.2	2.7	9.7	46	39.9	20.6	74.0
7	6.1	3.1	11.3	47	40.8	21.0	75.6
8	6.9	3.6	12.9	48	41.7	21.5	77.2
9	7.8	4.0	14.5	49	42.6	21.9	78.9
10	8.7	4.5	16.1	50	43.4	22.4	80.5
11	9.6	4.9	17.7	51	44.3	22.8	82.1
12	10.4	5.4	19.3	52	45.2	23.2	83.7
13	11.3	5.8	20.9	53	46.0	23.7	85.3
14	12.2	6.3	22.5	54	46.9	24.1	86.9
15	13.0	6.7	24.1	55	47.8	24.6	88.5
16	13.9	7.2	25.7	56	48.6	25.0	90.1
17	14.8	7.6	27.4	57	49.5	25.5	91.7
18	15.6	8.0	29.0	58	50.4	25.9	93.3
19	16.5	8.5	30.6	59	51.2	26.4	95.0
20	17.4	8.9	32.2	60	52.1	26.8	96.6
21	18.2	9.4	33.8	61	53.0	27.3	98.2
22	19.1	9.8	35.4	62	53.8	27.7	99.8
23	20.0	10.3	37.0	63	54.7	28.2	101.4
24	20.8	10.7	38.6	64	55.6	28.6	103.0
25	21.7	11.2	40.2	65	56.4	29.1	104.6
26	22.6	11.6	41.8	66	57.3	29.5	106.2
27	23.4	12.1	43.5	67	58.2	30.0	107.8
28	24.3	12.5	45.1	68	59.1	30.4	109.4
29	25.2	13.0	46.7	69	59.9	30.8	111.0
30	26.1	13.4	48.3	70	60.8	31.3	112.7
31	26.9	13.9	49.9	71	61.7	31.7	114.3
32	27.8	14.3	51.5	72	62.5	32.2	115.9
33	28.7	14.8	53.1	73	63.4	32.6	117.5
34	29.5	15.2	54.7	74	64.3	33.1	119.1
35	30.4	15.6	56.3	75	65.1	33.5	120.7
36	31.3	16.1	57.9	76	66.0	34.0	122.3
37	32.1	16.5	59.5	77	66.9	34.4	123.9
38	33.0	17.0	61.2	78	67.7	34.9	125.5
39	33.9	17.4	62.8	79	68.6	35.3	127.1
40	34.7	17.9	64.4	80	69.5	35.8	128.7

(continued)

TABLE A—4 (Continued)

Miles per Hour	Knots	Meters per Second	Kilometers per Hour	Miles per Hour	Knots	Meters per Second	Kilometers per Hour
81	70.3	36.2	130.4	91	79.0	40.7	146.5
82	71.2	36.7	132.0	92	79.9	41.1	148.1
83	72.1	37.1	133.6	93	80.8	41.6	149.7
84	72.9	37.6	135.2	94	81.6	42.0	151.3
85	73.8	38.0	136.8	95	82.5	42.5	152.9
86	74.7	38.4	138.4	96	83.4	42.9	154.5
87	75.5	38.9	140.0	97	84.2	43.4	156.1
88	76.4	39.3	141.6	98	85.1	43.8	157.7
89	77.3	39.8	143.2	99	86.0	44.3	159.3
90	78.2	40.2	144.8	100	86.8	44.7	160.9

Explanation and Decoding of the Daily Weather Map

Weather maps showing the development and movement of weather systems are among the most important tools used by the weather forecaster. Of the several types of maps used, some portray conditions near the surface of the earth and others depict conditions at various heights in the atmosphere. Some cover the entire northern hemisphere and others cover only local areas as required for special purposes. The maps used for daily forecasting by the National Weather Service are similar in many respects to the printed Daily Weather Map. At Weather Service offices maps showing conditions at the earth's surface are drawn four times daily. Maps of upper-level temperature, pressure, and humidity are prepared twice each day.

Principal Surface Weather Map

To prepare the surface map and present the information quickly and pictorially, two actions are necessary: (1) weather observers at many places must go to their posts at regular times each day to observe the weather and send the information to the offices where the maps are drawn; (2) the information must be quickly transcribed to the maps. In order that the necessary speed and economy of space and transmission time may be realized, codes have been devised for sending the information and for plotting it on the maps.

Codes and Map Plotting

A great deal of information is contained in a brief coded weather message. If each item were named and described in plain language, a very lengthy message would be required, one confusing to read and difficult to transfer to a map. A code permits the message to be condensed to a few five-figure numeral groups, each figure of which has a meaning, depending on its position in the message. People trained in the use of the code can read the message as easily as plain language.

The location of the reporting station is printed on the map as a small circle (the station circle). A definite arrangement of the data around the station circle, called the *station model*, is used. When the report is plotted in these fixed positions around the station circle on the weather map, many code figures are transcribed exactly as sent. Entries in the station model that are not made in code figures or actual values found in the message are usually in the form of symbols that graphically represent the element concerned. In some cases, certain of the data may or may not be reported by the observer, depending on local weather conditions. Precipitation and clouds are examples. In such cases, the absence of an entry on the map is interpreted as nonoccurrence or nonobservance of the phenomena. The letter M is entered where data are normally observed but not received.

Both the code and the station model are based on international agreements. These standardized numerals and symbols enable a meteorologist of one country to use the weather reports and weather maps of another country even though that person does not understand the language. Weather codes are, in effect, an international language that permits complete interchange and use of worldwide weather reports so essential in present-day activities.

The boundary between two different air masses is called a *front*. Important changes in weather, temperature, wind direction, and clouds often occur with the passage of a front. Half circles or triangular symbols or both are placed on the lines representing fronts to indicate the kind of front. The side on which the symbols are placed indicates the direction of frontal movement. The boundary of relatively cold air of polar origin advancing into an area occupied by warmer air, often of tropical origin, is called a *cold front*. The boundary of relatively warm air advancing into an area occupied by colder air is called a *warm front*. The line along which a cold front has overtaken a warm front at the ground is called an *occluded front*. A boundary between two air masses, which shows at the time of observation little tendency to advance into either the warm or cold areas, is called a *stationary front*. Air-mass boundaries are known as *surface fronts* when they intersect the ground and as *upper-air fronts* when they do not. Surface fronts are drawn in solid black, fronts aloft are drawn in outline only. Front symbols are given in Table B–1.

A front that is disappearing or weak and decreasing in intensity is labeled *frontolysis*. A front that is forming is labeled *frontogenesis*. A *squall line* is a line

TABLE B–1 Weather Map Symbols

Symbol	Explanation
▲▲▲	Cold front (surface)
●●●	Warm front (surface)
▲●▲●	Occluded front (surface)
●▼●▼	Stationary front (surface)
△△△	Cold front (aloft)
⌒⌒⌒	Warm front (aloft)
—..—..—..	Squall line
⟶ ⟶	Path of low-pressure center
⊠	Location of low pressure at 6-hour intervals
— — — —	32°F isotherm
—..—.—.—	0°F isotherm

of thunderstorms or squalls usually accompanied by heavy showers and shifting winds (Table B–1).

The paths followed by individual disturbances are called *storm tracks* and are shown by arrows (Table B–1). A symbol (a box containing an x) indicates past positions of a low-pressure center at 6-hour intervals. HIGH (H) and LOW (L) indicate the centers of high- and low-barometric pressure. Solid lines are isobars and connect points of equal sea-level barometric pressure. The spacing and orientation of these lines on weather maps are indications of speed and direction of windflow. In general, wind direction is parallel to these lines with low pressure to the left of an observer looking downwind. Speed is directly proportional to the closeness of the lines (termed *pressure gradient*). Isobars are labeled in millibars.

Isotherms are lines connecting points of equal temperature. Two isotherms are drawn on the large surface weather map when applicable. The freezing, or 32°F, isotherm is drawn as a dashed line, and the 0°F isotherm is drawn as a dash-dot line (Table B–1). Areas where precipitation is occurring at the time of observation are shaded.

AUXILIARY MAPS

500-Millibar Map

Contour lines, isotherms, and wind arrows are shown on the insert map for the 500-millibar contour level. Solid lines are drawn to show height above sea level and are labeled in feet. Dashed lines are drawn at 5° intervals of temperature and are labeled in degrees Celsius. True wind direction is shown by "arrows" that are plotted as flying with the wind. The wind speed is shown by flags and feathers. Each flag represents 50 knots, each full feather represents 10 knots, and each half-feather represents 5 knots.

Temperature Map (Highest and Lowest)

Temperature data are entered from selected weather stations in the United States. The figures entered above the station dot denote maximum temperatures reported from these stations during the 24 hours ending 1:00 A.M. EST; the figures entered below the station dot denote minimum temperature during the 24 hours ending at 1:00 P.M. EST of the previous day. The letter M denotes missing data.

Precipitation Map

Precipitation data are entered from selected weather stations in the United States. When precipitation has occurred at any of these stations in the 24-hour period ending at 1:00 A.M. EST, the total amount, in inches and hundredths, is entered above the station dot. When the figures for total precipitation have been compiled from incomplete data and entered on the map, the amount is underlined. T indicates a trace of precipitation and the letter M denotes missing data. The geographical areas where precipitation has fallen during the 24 hours ending at 1:00 A.M. EST are shaded. Dashed lines show depth of snow on ground in inches as of 7:00 A.M. EST of the previous day.

SYMBOLIC STATION MODEL	SAMPLE PLOTTED REPORT
	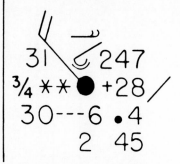

EXPLANATION OF STATION SYMBOLS AND MAP ENTRIES

N Sky coverage (total amount)—see block **6**

dd True direction *from* which wind is blowing

ff Wind speed in knots or miles per hour—see block **9**

VV Visibility in miles

ww Present weather—see block **8**

W Past weather—see block **11**

PPP Barometric pressure in millibars reduced to sea level. To decode barometric pressure, follow the steps listed here.

1. Place a decimal point to the left of the last number. For example, in the sample plotted report above, 247 becomes 24.7.

2. Next, place either a 9 or a 10 in front (to the left of the first digit). Thus 24.7 becomes either 924.7 mb or 1024.7 mb.

3. To determine whether to place a 9 or a 10 in front of the first digit, follow the rule stated next.

   If the number falls between 56.0 and 99.9, place a 9 before the first digit.

   If the number falls between 00.0 and 55.9, place a 10 before the first digit.

   Thus 24.7 would be 1024.7 mb.

TT Current air temperature in Fahrenheit

N_h Fraction of sky covered by low or middle clouds—see block **7**

C_L Low clouds or clouds of vertical development—see block **3**

h Height in feet of the base of the lowest clouds—see block **5**

C_M Middle clouds—see block **3**

C_H High clouds—see block **3**

T_dT_d Dew-point temperature in Fahrenheit

a Pressure tendency—see block **10**

pp Pressure change (in millibars) in preceding 3 hours ($+28 = +2.8$ mb)

RR Amount of precipitation ($45 = 0.45$ inch)

R_t Time precipitation began or ended—see block **4**

439

C_L

CLOUD ABBREVIATION	C_L Code	DESCRIPTION (Abridged From W M O Code)
St or Fs – Stratus or Fractostratus	1	Cu of fair weather, little vertical development and seemingly flattened
Ci – Cirrus	2	Cu of considerable development, generally towering, with or without other Cu or Sc bases all at same level
Cs – Cirrostratus	3	Cb with tops lacking clear-cut outlines, but distinctly not cirriform or anvil-shaped, with or without Cu, Sc, or St
Cc – Cirrocumulus	4	Sc formed by spreading out of Cu, Cu often present also
Ac – Altocumulus	5	Sc not formed by spreading out of Cu
As – Altostratus	6	St or Fs or both, but no Fs of bad weather
Sc – Stratocumulus	7	Fs and/or Fc of bad weather (scud)
Ns – Nimbostratus	8	Cu and Sc (not formed by spreading out of Cu) with bases at different levels
Cu or Fc – Cumulus or Fractocumulus	9	Cb having a clearly fibrous (cirriform) top, often anvil-shaped, with or without Cu, Sc, St, or scud
Cb – Cumulonimbus		

C_M

C_M Code	DESCRIPTION (Abridged From W M O Code)
1	Thin As (most of cloud layer semi-transparent)
2	Thick As, greater part sufficiently dense to hide sun (or moon), or Ns
3	Thin Ac, mostly semi-transparent; cloud elements not changing much and at a single level
4	Thin Ac in patches; cloud elements continually changing and/or occurring at more than one level
5	Thin Ac in bands or in a layer gradually spreading over sky and usually thickening as a whole
6	Ac formed by the spreading out of Cu
7	Double-layered Ac, or a thick layer of Ac, not increasing, or Ac with As and/or Ns
8	Ac in the form of Cu-shaped tufts or Ac with turrets
9	Ac of a chaotic sky, usually at different levels; patches of dense Ci are usually present also

C_H

C_H Code	DESCRIPTION (Abridged From W M O Code)
1	Filaments of Ci, or "mares tails," scattered and not increasing
2	Dense Ci in patches or twisted sheaves, usually not increasing, sometimes like remains of Cb; or towers or tufts
3	Dense Ci, often anvil-shaped, derived from or associated with Cb
4	Ci, often hook-shaped, gradually spreading over the sky and usually thickening as a whole
5	Ci and Cs, often in converging bands, or Cs alone, generally overspreading and growing denser; the continuous layer not reaching 45° altitude
6	Ci and Cs, often in converging bands, or Cs alone, generally overspreading and growing denser; the continuous layer exceeding 45° altitude
7	Veil of Cs covering the entire sky
8	Cs not increasing and not covering entire sky
9	Cc alone or Cc with some Ci or Cs, but the Cc being the main cirriform cloud

Rt (4) / h	TIME OF PRECIPITATION	HEIGHT IN FEET (Rounded Off)	HEIGHT IN METERS (Approximate) (5)	N	SKY COVERAGE (Total Amount) (6)	Nh	SKY COVERAGE (Low And/Or Middle Clouds) (7)
0	No Precipitation	0 - 149	0 - 49	○	No clouds	0	No clouds
1	Less than 1 hour ago	150 - 299	50 - 99	⊖	Less than one-tenth or one-tenth	1	Less than one-tenth or one-tenth
2	1 to 2 hours ago	300 - 599	100 - 199	◑	Two-tenths or three-tenths	2	Two-tenths or three-tenths
3	2 to 3 hours ago	600 - 999	200 - 299	◕	Four-tenths	3	Four-tenths
4	3 to 4 hours ago	1,000 - 1,999	300 - 599	◕	Five-tenths	4	Five-tenths
5	4 to 5 hours ago	2,000 - 3,499	600 - 999	◕	Six-tenths	5	Six-tenths
6	5 to 6 hours ago	3,500 - 4,999	1,000 - 1,499	◕	Seven-tenths or eight-tenths	6	Seven-tenths or eight-tenths
7	6 to 12 hours ago	5,000 - 6,499	1,500 - 1,999	◗	Nine-tenths or overcast with openings	7	Nine-tenths or overcast with openings
8	More than 12 hours ago	6,500 - 7,999	2,000 - 2,499	●	Completely overcast	8	Completely overcast
9	Unknown	At or above 8,000, or no clouds	At or above 2,500, or no clouds	⊗	Sky obscured	9	Sky obscured

WW PRESENT WEATHER (Descriptions

	0	1	2	3	4
00	Cloud development NOT observed or NOT observable during past hour	Clouds generally dissolving or becoming less developed during past hour	State of sky on the whole unchanged during past hour	Clouds generally forming or developing during past hour	Visibility reduced by smoke
10	Light fog	Patches of shallow fog at station, NOT deeper than 6 feet on land	More or less continuous shallow fog at station, NOT deeper than 6 feet on land	Lightning visible, no thunder heard	Precipitation within sight, but NOT reaching the ground
20	Drizzle (NOT freezing and NOT falling as showers) during past hour, but NOT at time of observation	Rain (NOT freezing and NOT falling as showers) during past hour, but NOT at time of observation	Snow (NOT falling as showers) during past hour, but NOT at time of observation	Rain and snow (NOT falling as showers) during past hour, but NOT at time of observation	Freezing drizzle or freezing rain (NOT falling as showers) during past hour, but NOT at time of observation
30	Slight or moderate dust storm or sand storm, has decreased during past hour	Slight or moderate dust storm or sand storm, no appreciable change during past hour	Slight or moderate dust storm or sand storm, has increased during past hour	Severe dust storm or sand storm, has decreased during past hour	Severe dust storm or sand storm, no appreciable change during past hour
40	Fog at distance at time of observation, but NOT at station during past hour	Fog in patches	Fog, sky discernible, has become thinner during past hour	Fog, sky NOT discernible, has become thinner during past hour	Fog, sky discernible, no appreciable change during past hour
50	Intermittent drizzle (NOT freezing) slight at time of observation	Continuous drizzle (NOT freezing) slight at time of observation	Intermittent drizzle (NOT freezing) moderate at time of observation	Continuous drizzle (NOT freezing), moderate at time of observation	Intermittent drizzle (NOT freezing), thick at time of observation
60	Intermittent rain (NOT freezing), slight at time of observation	Continuous rain (NOT freezing), slight at time of observation	Intermittent rain (NOT freezing) moderate at time of obs.	Continuous rain (NOT freezing), moderate at time of observation	Intermittent rain (NOT freezing), heavy at time of observation
70	Intermittent fall of snow flakes, slight at time of observation	Continuous fall of snowflakes, slight at time of observation	Intermittent fall of snowflakes, moderate at time of observation	Continuous fall of snowflakes, moderate at time of observation	Intermittent fall of snowflakes, heavy at time of observation
80	Slight rain shower(s)	Moderate or heavy rain shower(s)	Violent rain shower(s)	Slight shower(s) of rain and snow mixed	Moderate or heavy shower(s) of rain and snow mixed
90	Moderate or heavy shower(s) of hail, with or without rain or rain and snow mixed, not associated with thunder	Slight rain at time of observation, thunderstorm during past hour, but NOT at time of observation	Moderate or heavy rain at time of observation, thunderstorm during past hour, but NOT at time of observation	Slight snow or rain and snow mixed or hail at time of observation; thunderstorm during past hour, but not at time of observation	Moderate or heavy snow, or rain and snow mixed or hail at time of observation, thunderstorm during past hour, but NOT at time of obs.

Abridged from W. M. O. Code)

8

5	6	7	8	9
Haze	Widespread dust in suspension in the air, NOT raised by wind, at time of observation	Dust or sand raised by wind, at time of observation	Well developed dust devil(s) within past hour	Dust storm or sand storm within sight of or at station during past hour
Precipitation within sight, reaching the ground, but distant from station	Precipitation within sight, reaching the ground, near to but NOT at station	Thunder heard, but no precipitation at the station	Fog during past hour, but NOT at time of observation	Funnel cloud(s) within sight during past hour
Showers of rain during past hour, but NOT at time of observation	Showers of snow, or of rain and snow, during past hour, but NOT at time of observation	Showers of hail, or of hail and rain, during past hour, but NOT at time of observation	Slight or moderate drifting snow, generally high	Thunderstorm (with or without precipitation) during past hour, but NOT at time of obs.
Severe dust storm or sand storm, has increased during past hour	Slight or moderate drifting snow, generally low	Heavy drifting snow, generally low	Fog, depositing rime, sky discernible	Heavy drifting snow, generally high
Fog, sky NOT discernible, no appreciable change during past hour	Fog, sky discernible, has begun or become thicker during past hour	Fog, sky NOT discernible, has begun or become thicker during past hour	Drizzle and rain, slight	Fog, depositing rime, sky NOT discernible
Continuous drizzle (NOT freezing), thick at time of observation	Slight freezing drizzle	Moderate or thick freezing drizzle	Drizzle and rain, slight	Drizzle and rain, moderate or heavy
Continuous rain (NOT freezing), heavy at time of observation	Slight freezing rain	Moderate or heavy freezing rain	Rain or drizzle and snow, slight	Rain or drizzle and snow, moderate or heavy
Continuous fall of snowflakes, heavy at time of observation	Ice needles (with or without fog)	Granular snow (with or without fog)	Isolated starlike snow crystals (with or without fog)	Ice pellets (sleet, U. S. definition)
Slight snow shower(s)	Moderate or heavy snow shower(s)	Slight shower(s) of soft or small hail with or without rain, or rain and snow mixed	Moderate or heavy shower(s) of soft or small hail with or without rain, or rain and snow mixed	Slight shower(s) of hail, with or without rain or rain and snow mixed, not associated with thunder
Slight or moderate thunderstorm without hail, but with rain and/or snow at time of obs.	Slight or moderate thunderstorm, with hail at time of observation	Heavy thunderstorm, without hail, but with rain and/or snow at time of observation	Thunderstorm combined with dust storm or sand storm at time of obs.	Heavy thunderstorm with hail at time of observation

ff	(MILES) (Statute) Per Hour	KNOTS	Code Number ⑨	a	BAROMETRIC TENDENCY ⑩
◎	Calm	Calm	0	╱	Rising, then falling
—	1 - 2	1 - 2	1	╱	Rising, then steady; or rising, then rising more slowly
⌐	3 - 8	3 - 7	2	╱	Rising steadily, or unsteadily
⌐	9 - 14	8 - 12	3	✓	Falling or steady, then rising; or rising, then rising more quickly
⌐	15 - 20	13 - 17	4	—	Steady, same as 3 hours ago
⌐	21 - 25	18 - 22	5	╲	Falling, then rising, same or lower than 3 hours ago
⌐	26 - 31	23 - 27	6	╲	Falling, then steady; or falling, then falling more slowly
⌐	32 - 37	28 - 32	7	╲	Falling steadily, or unsteadily
⌐	38 - 43	33 - 37	8	╲	Steady or rising, then falling; or falling, then falling more quickly

Barometer now higher than 3 hours ago (codes 1, 2, 3)
Barometer now lower than 3 hours ago (codes 6, 7, 8)

ff	(MILES) (Statute) Per Hour	KNOTS	Code Number ⑪	W	PAST WEATHER
⌐	44 - 49	38 - 42	0		Clear or few clouds
⌐	50 - 54	43 - 47	1		Partly cloudy (scattered) or variable sky
∟	55 - 60	48 - 52	2		Cloudy (broken) or overcast
⌐	61 - 66	53 - 57	3	S/⇉	Sandstorm or dust-storm, or drifting or blowing snow
⌐	67 - 71	58 - 62	4	≡	Fog, or smoke, or thick dust haze
⌐	72 - 77	63 - 67	5	'	Drizzle
⌐	78 - 83	68 - 72	6	●	Rain
⌐	84 - 89	73 - 77	7	✳	Snow, or rain and snow mixed, or ice pellets (sleet)
⌐	119 - 123	103 - 107	8	▽	Shower(s)
			9	℞	Thunderstorm, with or without precipitation

Codes 0, 1, 2 under Past Weather: Not Plotted

Correcting Mercurial Barometer Readings

When barometric readings taken from separate locations are being compared, corrections must be made to ensure that the true differences in atmospheric pressure are shown. In general, this step is done by reducing these barometric readings to normal sea-level conditions. So a correction is made for temperature, elevation, and latitude (gravity correction).

Temperature Correction

This correction accounts for the expansion and contraction of the mercury column and brass scale caused by temperature variations.

Elevation Correction

Because pressure decreases with increases in elevation, barometers not located at sea level will show a reduced pressure. The elevation correction accounts for the position of the barometer as well as for the temperature of the assumed air column extending down to sea level.

Latitude Correction

The earth is not a perfect sphere; it has a shorter polar diameter than an equatorial diameter. Consequently, the pull of gravity on the mercury column is greater at the poles and lesser at the equator. Because the adjustment needed to correct this latitudinal variation in gravity is usually negligible compared with the other corrections, we have not included it in the following procedure for correcting the barometer.

To reduce a mercurial barometer reading to sea level, take the following steps.

1. Read the height of the mercury column on the millimeter scale.
2. Correct this reading for the temperature of the barometer by using Table C–1 and subtracting the obtained value from your reading.
3. Using the out-of-doors air temperature and the elevation (altitude) of the barometer, obtain the temperature-altitude factor from Table C–2.
4. Using the temperature-altitude factor from step 3, the temperature-corrected barometric reading from step 2, and Table C–3, find the sea-level correction.
5. Add the sea-level correction to the temperature-corrected barometer reading. This sum gives the desired barometric reading reduced to sea level.

TABLE C–1 Temperature Correction

Temp. °C	Observed Height in Millimeters																	
	620	630	640	650	660	670	680	690	700	710	720	730	740	750	760	770	780	790
0	0.00	0.00	0.00	0.00	0.00	0.00	0.00	0.00	0.00	0.00	0.00	0.00	0.00	0.00	0.00	0.00	0.00	0.00
1	.10	.10	.10	.11	.11	.11	.11	.11	.11	.12	.12	.12	.12	.12	.12	.13	.13	.13
2	.20	.21	.21	.21	.22	.22	.22	.23	.23	.23	.24	.24	.24	.25	.25	.25	.25	.26
3	.30	.31	.31	.32	.32	.33	.33	.34	.34	.35	.35	.36	.36	.37	.37	.38	.38	.39
4	.40	.41	.42	.42	.43	.44	.44	.45	.46	.46	.47	.48	.48	.49	.50	.50	.51	.52
5	0.51	0.51	0.52	0.53	0.54	0.55	0.56	0.56	0.57	0.58	0.59	0.60	0.60	0.61	0.62	0.63	0.64	0.64
6	.61	.62	.63	.64	.65	.66	.67	.68	.69	.70	.71	.71	.72	.73	.74	.75	.76	.77
7	.71	.72	.73	.74	.75	.77	.78	.79	.80	.81	.82	.83	.85	.86	.87	.88	.89	.90
8	.81	.82	.84	.85	.86	.87	.89	.90	.91	.93	94	.95	.97	.98	99	1.01	1.02	1.03
9	.91	.92	.94	.95	.97	.98	1.00	1.01	1.03	1.04	1.06	1.07	1.09	1.10	1.12	1.13	1.15	1.16
10	1.01	1.03	1.04	1.06	1.08	1.09	1.11	1.13	1.14	1.16	1.17	1.19	1.21	1.22	1.24	1.26	1.27	1.29
11	1.11	1.13	1.15	1.17	1.18	1.20	1.22	1.24	1.26	1.27	1.29	1.31	1.33	1.35	1.36	1.38	1.40	1.42
12	1.21	1.23	1.25	1.27	1.29	1.31	1.33	1.35	1.37	1.39	1.41	1.43	1.45	1.47	1.49	1.51	1.53	1.55
13	1.31	1.34	1.36	1.38	1.40	1.42	1.44	1.46	1.48	1.50	1.53	1.55	1.57	1.59	1.61	1.63	1.65	1.67
14	1.41	1.44	1.46	1.48	1.51	1.53	1.55	1.57	1.60	1.62	1.64	1.67	1.69	1.71	1.73	1.76	1.78	1.80
15	1.52	1.54	1.56	1.59	1.61	1.64	1.66	1.69	1.71	1.74	1.76	1.78	1.81	1.83	1.86	1.88	1.91	1.93
16	1.62	1.64	1.67	1.69	1.72	1.75	1.77	1.80	1.82	1.85	1.88	1.90	1.93	1.96	1.98	2.01	2.03	2.06
17	1.72	1.74	1.77	1.80	1.83	1.86	1.88	1.91	1.94	1.97	1.99	2.02	2.05	2.08	2.10	2.13	2.16	2.19
18	1.82	1.85	1.88	1.91	1.93	1.96	1.99	2.02	2.05	2.08	2.11	2.14	2.17	2.20	2.23	2.26	2.29	2.32
19	1.92	1.95	1.98	2.01	2.04	2.07	2.10	2.13	2.17	2.20	2.23	2.26	2.29	2.32	2.35	2.38	2.41	2.44
20	2.02	2.05	2.08	2.12	2.15	2.18	2.21	2.25	2.28	2.31	2.34	2.38	2.41	2.44	2.47	2.51	2.54	2.57
21	2.12	2.15	2.19	2.22	2.26	2.29	2.32	2.36	2.39	2.43	2.46	2.50	2.53	2.56	2.60	2.63	2.67	2.70
22	2.22	2.26	2.29	2.33	2.36	2.40	2.43	2.47	2.51	2.54	2.58	2.61	2.65	2.69	2.72	2.76	2.79	2.83
23	2.32	2.36	2.40	2.43	2.47	2.51	2.54	2.58	2.62	2.66	2.69	2.73	2.77	2.81	2.84	2.88	2.92	2.96
24	2.42	2.46	2.50	2.54	2.58	2.62	2.66	2.69	2.73	2.77	2.81	2.85	2.89	2.93	2.97	3.01	3.05	3.08
25	2.52	2.56	2.60	2.64	2.68	2.72	2.77	2.81	2.85	2.89	2.93	2.97	3.01	3.05	3.09	3.13	3.17	3.21
26	2.62	2.66	2.71	2.75	2.79	2.83	2.88	2.92	2.96	3.00	3.04	3.09	3.13	3.17	3.21	3.26	3.30	3.34
27	2.72	2.77	2.81	2.85	2.90	2.94	2.99	3.03	3.07	3.12	3.16	3.20	3.25	3.29	3.34	3.38	3.42	3.47
28	2.82	2.87	2.91	2.96	3.00	3.05	3.10	3.14	3.19	3.23	3.28	3.32	3.37	3.41	3.46	3.51	3.55	3.60
29	2.92	2.97	3.02	3.06	3.11	3.16	3.21	3.25	3.30	3.35	3.39	3.44	3.49	3.54	3.58	3.63	3.68	3.72
30	3.02	3.07	3.12	3.17	3.22	3.27	3.32	3.36	3.41	3.46	3.51	3.56	3.61	3.66	3.71	3.75	3.80	3.85
31	3.12	3.17	3.22	3.27	3.32	3.37	3.43	3.48	3.53	3.58	3.63	3.68	3.73	3.78	3.83	3.88	3.93	3.98
32	3.22	3.28	3.33	3.38	3.43	3.48	3.54	3.59	3.64	3.69	3.74	3.79	3.85	3.90	3.95	4.00	4.05	4.11
33	3.32	3.38	3.43	3.48	3.54	3.59	3.64	3.70	3.75	3.81	3.86	3.91	3.97	4.02	4.07	4.13	4.18	4.23
34	3.42	3.48	3.53	3.59	3.64	3.70	3.75	3.81	3.87	3.92	3.98	4.03	4.09	4.14	4.20	4.25	4.31	4.36
35	3.52	3.58	3.64	3.69	3.75	3.81	3.86	3.92	3.98	4.03	4.09	4.15	4.21	4.26	4.32	4.38	4.43	4.49

TABLE C–2 Temperature-Altitude Factor Values (Enter in Table C–3)

Altitude in Meters	Assumed Temperature of Air Column °C									
	−16°	−8°	0°	+4°	+8°	+12°	+16°	+20°	+24°	+28°
10	1.2	1.1	1.1	1.1	1.0	1.0	1.0	1.0	1.0	1.0
50	5.8	5.6	5.4	5.3	5.2	5.2	5.1	5.0	4.9	4.9
100	11.5	11.2	10.8	10.7	10.5	10.3	10.2	10.0	9.9	9.7
150	17.3	16.7	16.2	16.0	15.7	15.5	15.3	15.0	14.8	14.6
200	23.0	22.3	21.6	21.3	21.0	20.7	20.3	20.0	19.7	19.5
250	28.8	27.9	27.0	26.6	26.2	25.8	25.4	25.0	24.7	24.3
300	34.5	33.5	32.5	32.0	31.5	31.0	30.5	30.1	29.6	29.2
350	40.3	39.0	37.9	37.3	36.7	36.2	35.6	35.1	34.6	34.0
400	46.0	44.6	43.3	42.6	42.0	41.3	40.7	40.1	39.5	38.9
450	51.8	50.2	48.7	47.9	47.2	46.5	45.8	45.1	44.4	43.8
500	57.5	55.8	54.1	53.3	52.4	51.6	50.9	50.1	49.4	48.6
550	63.3	61.4	59.5	58.6	57.7	56.8	55.9	55.1	54.3	53.5
600	69.0	66.9	64.9	63.9	62.9	62.0	61.0	60.1	59.2	58.3
650	74.8	72.5	70.3	69.2	68.2	67.1	66.1	65.1	64.2	63.2
700	80.6	78.1	75.7	74.6	73.4	72.3	71.2	70.1	69.1	68.1
750	86.3	83.7	81.1	79.9	78.7	77.5	76.3	75.1	74.0	72.9
800	92.1	89.2	86.5	85.2	83.9	82.6	81.4	80.1	79.0	77.8
850	97.8	94.8	92.0	90.5	89.2	87.8	86.4	85.2	83.9	82.7
900	103.6	100.4	97.4	95.9	94.4	93.0	91.5	90.2	88.8	87.5
950	109.3	106.0	102.8	101.2	99.6	98.1	96.6	95.2	93.8	92.4
1000	115.1	111.5	108.2	106.5	104.9	103.3	101.7	100.2	98.7	97.3
1050	120.8	117.1	113.6	111.8	110.1	108.4	106.8	105.2	103.6	102.1
1100	126.6	122.7	119.0	117.2	115.4	113.6	111.9	110.2	108.6	107.0
1150	132.3	128.3	124.4	122.5	120.6	118.8	117.0	115.2	113.5	111.8
1200	138.1	133.8	129.8	127.8	125.9	123.9	122.0	120.2	118.4	116.7
1250	143.8	139.4	135.2	133.1	131.1	129.1	127.1	125.2	123.4	121.6
1300	149.6	145.0	140.6	138.5	136.3	134.3	132.2	130.2	128.3	126.4
1350	155.3	150.6	146.0	143.8	141.6	139.4	137.3	135.2	133.2	131.3
1400	161.1	156.2	151.4	149.1	146.8	144.6	142.4	140.2	138.2	136.2
1450	166.8	161.7	156.8	154.5	152.1	149.7	147.5	145.3	143.1	141.0
1500	172.6	167.3	162.3	159.8	157.3	154.9	152.5	150.3	148.0	145.9
1550	178.3	172.9	167.7	165.1	162.6	160.1	157.6	155.3	153.0	150.7
1600	184.1	178.5	173.1	170.4	167.8	165.2	162.7	160.3	157.9	155.6
1650	189.8	184.0	178.5	175.7	173.0	170.4	167.8	165.3	162.8	160.5
1700	195.6	189.6	183.9	181.1	178.3	175.6	172.9	170.3	167.8	165.3
1750	201.4	195.2	189.3	186.4	183.5	180.7	178.0	175.3	172.7	170.2
1800	207.1	200.8	194.7	191.7	188.8	185.9	183.1	180.3	177.6	175.0
1850	212.9	206.3	200.1	197.0	194.0	191.0	188.1	185.3	182.6	179.9
1900	218.6	211.9	205.5	202.4	199.3	196.2	193.2	190.3	187.5	184.8
1950	224.4	217.5	210.9	207.7	204.5	201.4	198.3	195.3	192.4	189.6
2000	230.1	223.0	216.3	213.0	209.7	206.5	203.4	200.3	197.4	194.5
2050	235.9	228.6	221.7	218.3	215.0	211.7	208.5	205.3	202.3	199.3
2100	241.6	234.2	227.1	223.7	220.2	216.8	213.5	210.4	207.2	204.2
2150	247.4	239.8	232.5	229.0	225.5	222.0	218.6	215.4	212.2	209.1
2200	253.1	245.4	237.9	234.3	230.7	227.2	223.7	220.4	217.1	213.9
2250	258.9	250.9	243.4	239.6	235.9	232.3	228.8	225.4	222.0	218.8

(continued)

TABLE C–2 (*Continued*)

Altitude in Meters	Assumed Temperature of Air Column °C									
	−16°	−8°	0°	+4°	+8°	+12°	+16°	+20°	+24°	+28°
2300	264.6	256.5	248.8	245.0	241.2	237.5	233.9	230.4	227.0	223.6
2350	270.4	262.1	254.2	250.3	246.4	242.7	239.0	235.4	231.9	228.5
2400	276.1	267.7	259.6	255.6	251.7	247.8	244.0	240.4	236.8	233.4
2450	281.9	273.2	265.0	260.9	256.9	253.0	249.1	245.4	241.8	238.2
2500	287.6	278.8	270.4	266.2	262.2	258.1	254.2	250.4	246.7	243.1
2550	293.4	284.4	275.8	271.6	267.4	263.3	259.3	255.4	251.6	247.9
2600	299.1	290.0	281.2	276.9	272.6	268.5	264.4	260.4	256.6	252.8
2650	304.9	295.5	286.6	282.2	277.9	273.6	269.5	265.4	261.5	257.7
2700	310.6	301.1	292.0	287.5	283.1	278.8	274.5	270.4	266.4	262.5
2750	316.4	306.7	297.4	292.9	288.4	283.9	279.6	275.4	271.4	267.4
2800	322.1	312.3	302.8	298.2	293.6	289.1	284.7	280.4	276.3	272.2
2850	327.9	317.8	308.2	303.5	298.8	294.3	289.8	285.4	281.2	277.1
2900	333.6	323.4	313.6	308.8	304.1	299.4	294.9	290.4	286.2	282.0
2950	339.4	329.0	319.0	314.2	309.3	304.6	299.9	295.5	291.1	286.8
3000	345.1	334.5	324.4	319.5	314.6	309.7	305.0	300.5	296.0	291.7

TABLE C–3 Barometric Correction Values

Temp.–Alt. Factor	Barometer Reading				
	780 mm	760 mm	740 mm	720 mm	700 mm
1	0.9	0.9	0.9	0.8	0.8
5	4.5	4.4	4.3	4.2	4.0
10	9.0	8.8	8.6	8.3	8.1
15	13.6	13.2	12.9	12.5	12.2
20	18.2	17.7	17.2	16.8	16.3
25	22.8	22.2	21.6	21.0	20.4
30	27.4	26.7	26.0	25.3	24.6
35	......	31.2	30.4	29.6	28.8

Temp.–Alt. Factor	760 mm	740 mm	720 mm	700 mm	680 mm	660 mm
40	35.8	34.9	33.9	33.0	32.0	31.1
45	40.4	39.3	38.3	37.2	36.2	35.1
50	45.0	43.8	42.7	41.5	40.3	39.1
55	49.7	48.4	47.1	45.8	44.5	43.1
60	......	52.9	51.5	50.1	48.6	47.2
65	......	57.5	55.9	54.4	52.8	51.3
70	......	62.1	60.4	58.7	57.1	55.4
75	......	66.7	64.9	63.1	61.3	59.5

Temp.–Alt. Factor	720 mm	700 mm	680 mm	660 mm	640 mm
80	69.5	67.5	65.6	63.7	61.7
85	74.0	72.0	69.9	67.9	65.8
90	78.6	76.4	74.2	72.1	69.9
95	83.2	80.9	78.6	76.3	74.0
100	87.9	85.4	83.0	80.5	78.1
105	......	89.9	87.4	84.8	82.2
110	......	94.5	91.8	89.1	86.4
115	......	99.1	96.3	93.4	90.6
120	......	103.7	100.7	97.8	94.8
125	......	108.3	105.3	102.2	99.1

Temp.–Alt. Factor	680 mm	660 mm	640 mm	620 mm	600 mm
125	105.3	102.2	99.1	96.0	92.9
130	109.8	106.6	103.3	100.1	96.9
135	114.3	111.0	107.6	104.3	100.9
140	118.9	115.4	111.9	108.4	104.9
145	123.5	119.9	116.3	112.6	109.0
150	128.2	124.4	120.6	116.9	113.1
155	......	128.9	125.0	121.1	117.2
160	......	133.5	129.4	125.4	121.4
165	......	138.1	133.9	129.7	125.5
170	......	142.7	138.4	134.0	129.7

Temp.–Alt. Factor	Barometer Reading				
	640 mm	620 mm	600 mm	580 mm	560 mm
170	138.4	134.0	129.7	125.4	121.1
175	142.9	138.4	133.9	129.5	125.0
180	147.4	142.8	138.2	133.6	129.0
185	151.9	147.2	142.4	137.7	132.9
190	156.5	151.6	146.7	141.8	136.9
195	161.1	156.1	151.0	146.0	141.0
200	165.7	160.5	155.4	150.2	145.0
205	170.4	165.0	159.7	154.4	149.1
210	......	169.6	164.1	158.6	153.2
215	......	174.1	168.5	162.9	157.3

Temp.–Alt. Factor	620 mm	600 mm	580 mm	560 mm	540 mm
215	174.1	168.5	162.9	157.3	151.7
220	178.7	172.9	167.2	161.4	155.7
225	183.3	177.4	171.5	165.6	159.7
230	188.0	181.9	175.8	169.8	163.7
235	192.6	186.4	180.2	174.0	167.8
240	......	191.0	184.6	178.2	171.9
245	......	195.5	189.0	182.5	176.0
250	......	200.1	193.4	186.8	180.1
255	......	204.7	197.9	191.1	184.3
260	......	209.4	202.4	195.4	188.4

Temp.–Alt. Factor	580 mm	560 mm	540 mm	520 mm
260	202.4	195.4	188.4	181.5
265	206.9	199.8	192.6	185.5
270	211.5	204.2	196.9	189.6
275	216.0	208.6	201.1	193.7
280	220.6	213.0	205.4	197.8
285	225.2	217.5	209.7	201.9
290	229.9	222.0	214.0	206.1
295	......	226.5	218.4	210.3
300	......	231.0	222.8	214.5

Temp.–Alt. Factor	560 mm	540 mm	520 mm	500 mm	480 mm
305	235.6	227.2	218.8	210.3	201.9
310	240.2	231.6	223.0	214.4	205.9
315	244.8	236.0	227.3	218.6	209.8
320	249.4	240.5	231.6	222.7	213.8
325	254.1	245.0	236.0	226.9	217.8
330	......	249.6	240.3	231.1	221.8
335	......	254.1	244.7	235.3	225.9
340	......	258.7	249.1	239.6	230.0
345	......	263.3	253.6	243.8	234.1

Appendix **D**

Forces
and Air Motions

Newton's Second Law

When motions in the atmosphere are being considered, Newton's laws relate the forces that are involved to acceleration. In particular, Newton's second law states: The acceleration of a body is directly proportional to the net force acting on the body and is inversely proportional to the mass of the body. Stated mathematically,

$$A = \frac{F}{M}$$

where A is the acceleration that a force F will produce on mass M. In the metric system force is measured in newtons, where 1 newton of force will accelerate 1 kilogram of matter 1 meter per second squared. Two forces frequently encountered in basic meteorology are the Coriolis force and the pressure gradient force.

Coriolis Force

The magnitude of the Coriolis force can be expressed as

$$F_c = 2v\Omega \sin \phi$$

where F_c is the Coriolis force per unit mass of air, v is the wind speed, Ω is the earth's rate of rotation (angular velocity), which is 7.29×10^{-5} radian per second, and ϕ is the latitude. Note that sin ϕ is a trigonometric function that equals zero for an angle of 0 degrees (equator), 0.64 when $\phi = 40$ degrees, and 1 when $\phi = 90$ degrees (poles). Because the earth's rate of rotation remains rather constant, we can see from the preceding expression that the magnitude of the Coriolis force depends on wind speed and latitude. When this formula is used to determine the Coriolis force for a specific latitude and wind speed, it is common practice to use the letter f to represent 2Ω sin ϕ, which has the following values at various latitudes:

ϕ	0	10	20	30	40	50	60	70	80	90
$f \cdot 10^5$ sec^{-1}	0	2.5	5.0	7.3	9.4	11.2	12.6	13.7	14.4	14.6

Using this information, we can calculate the following: At 30 degrees latitude with a wind speed of 10 meters per second, the Coriolis force would equal 0.073 centimeter per second squared, whereas with the same wind speed at 60 degrees latitude the Coriolis force would equal 0.126 centimeter per second squared.

Pressure Gradient Force

The magnitude of the pressure gradient force can be expressed as

$$F_p = \frac{1}{d} \frac{\Delta p}{\Delta n}$$

where F_p is the pressure gradient force per unit mass, d is the density of air, Δp is the pressure difference obtained from the isobar spacing, and Δn is the distance between isobars. Using this formula and the standard sea-level density of 1.293×10^{-3} gram per cubic centimeter, we can calculate the pressure gradient force for any isobaric surface. For example, suppose that 4-millibar isobars are spaced 200 kilometers apart. Then

$$F_p = \frac{1 \times 4 \times 10^3}{1.293 \times 10^{-3} \times 1 \times 10^7} = 0.155 \text{ centimeter per second squared}$$

Comparing this value with that obtained for the Coriolis force calculated earlier suggests that a balance between these forces is very possible at reasonable wind speeds.

Appendix E

Laws Relating to Gases

Kinetic Energy

All moving objects, by virtue of their motion, are capable of doing work. We call this energy of motion, or *kinetic energy*. The kinetic energy of a moving object is equal to one-half its mass multiplied by its velocity squared. Stated mathematically,

$$\text{Kinetic energy} = \frac{1}{2} Mv^2$$

So we can see that by doubling the velocity of a moving object, the object's kinetic energy will increase four times.

First Law of Thermodynamics

According to the kinetic theory, the temperature of a gas is proportional to the kinetic energy of the moving molecules. When a gas is heated, its kinetic energy increases because of an increase in molecular motion. Further, when a gas is compressed, the kinetic energy will also be increased and the temperature of the gas will rise. These relationships are expressed in the first law of thermodynamics, which states: The temperature of a gas may be changed by the addition or subtraction of heat, or by changing the pressure (compression or expansion), or by a combination of both. It is easy to understand how the atmosphere is heated or cooled by the gain or loss of heat. However, when we consider rising and sinking air, the relationships between temperature and pressure become more important. Here an increase in temperature is brought about by performing work on the gas and not by the addition of heat. This phenomenon is called the *adiabatic form* of the first law of thermodynamics.

Boyle's Law

About 1600 the Englishman Robert Boyle showed that if the temperature is kept constant when the pressure exerted on a gas is increased, the volume decreases. This principle, called Boyle's law, states: At a constant temperature the volume of a given mass of gas varies inversely with the pressure. Stated mathematically,

$$P_1 V_1 = P_2 V_2$$

The symbols P_1 and V_1 refer to the original pressure and volume, and P_2 and V_2 indicate the new pressure and volume after a change occurs. Boyle's law shows that if a given volume of gas is compressed so that the volume is reduced by one-half, the pressure exerted by the gas is doubled. This increase in pressure can be explained by the kinetic theory, which predicts that when the volume of the gas is reduced by one-half, the molecules collide with the walls of the container twice as often. Because density is defined as the mass per unit volume, an increase in pressure results in increased density.

Charles's Law

The relationships between temperature and volume (hence density) of a gas were recognized about 1787 by the French scientist Jacques Charles, and were stated formally by J. Gay-Lussac in 1802. Charles's law states: At a constant pressure the volume of a given mass is directly proportional to the absolute temperature. In other words, when a quantity of gas is kept at a constant pressure, an increase in temperature results in an increase in volume and vice versa. Stated mathematically,

$$\frac{V_1}{V_2} = \frac{T_1}{T_2}$$

where V_1 and T_1 represent the original volume and temperature, respectively, and V_2 and T_2 represent the final volume and temperature, respectively. This law explains the fact that a gas expands when it is heated. According to the kinetic theory, when heated, particles move more rapidly and therefore collide more often.

The Ideal Gas Law or Equation of State

In describing the atmosphere, three variable quantities must be considered: pressure, temperature, and density (mass per unit volume). The relationships between these variables can be found by combining in a single statement the laws of Boyle and Charles as follows:

$$PV = RT \qquad \text{or} \qquad P = \rho RT$$

where P is pressure, V the volume, R the constant of proportionality, T the absolute temperature, and ρ the density. This law, called the ideal gas law, states:

1. When the volume is kept constant, the pressure of a gas is directly proportional to its absolute temperature.
2. When the temperature is kept constant, the pressure of a gas is proportional to its density and inversely proportional to its volume.
3. When the pressure is kept constant, the absolute temperature of a gas is proportional to its volume and inversely proportional to its density.

Worldwide Extremes of Temperature and Precipitation Recorded by Continental Area

TABLE F–1 Temperature Extremes

Key No.	Area	Highest (°C)	Place	Elevation (meters)	Date
1	Africa	57.8	Azizia, Libya	114	Sept. 13, 1922
2	North America	56.7	Death Valley, Calif.	−53	July 10, 1913
3	Asia	53.9	Tirat Tsvi, Israel	−217	June 21, 1942
4	Australia	53.3	Cloncurry, Queensland	187	Jan. 16, 1889
5	Europe	50.0	Seville, Spain	8	Aug. 4, 1881
6	South America	48.9	Rivadavia, Argentina	203	Dec. 11, 1905
7	Oceania	42.2	Tuguegarao, Philippines	22	Apr. 29, 1912
8	Antarctica	14.4	Esperanza, Palmer Peninsula	8	Oct. 20, 1956

Key No.	Area	Lowest (°C)	Place	Elevation (meters)	Date
9	Antarctica	−88.3	Vostok	3366	Aug. 24, 1960
10	Asia	−67.8	Oymykon, U.S.S.R.	788	Feb. 6, 1933
11	Greenland	−66.1	Northice	2307	Jan. 9, 1954
12	North America	−62.8	Snag, Yukon, Canada	578	Feb. 3, 1947
13	Europe	−55.0	Ust'Shchugor, U.S.S.R.	84	January*
14	South America	−32.8	Sarmiento, Argentina	264	June 1, 1907
15	Africa	−23.9	Ifrane, Morocco	1609	Feb. 11, 1935
16	Australia	−22.2	Charlotte Pass, N.S.W.	‡	July 22, 1947†
17	Oceania	−10.0	Haleakala Summit, Hawaii	293	Jan. 2, 1961

* Exact data unknown; lowest in 15-year period.
† And earlier date.
‡ Elevation unknown.

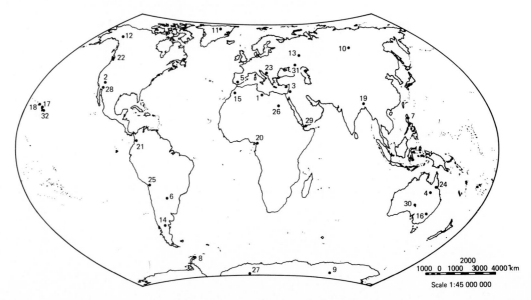

● Key numbers correspond to data entries.

TABLE F–2 Extremes of Average Annual Precipitation

Key No.	Area	Greatest Amount (mm)	Place	Elevation (meters)	Years of Record
18	Oceania	11,684	Mt. Waialeale, Kauai, Hawaii	1,523	32
19	Asia	11,430	Cherrapunji, India	1,313	74
20	Africa	10,277	Debundscha, Cameroon	9	32
21	South America	8,989	Quibdo, Colombia	72	10–16
22	North America	6,657	Henderson Lake, B.C., Canada	4	14
23	Europe	4,643	Crkvica, Yugoslavia	1,001	22
24	Australia	4,554	Tully, Queensland	—	31

Key No.	Area	Least Amount (mm)	Place	Elevation (meters)	Years of Record
25	South America	0.76	Arica, Chile	29	59
26	Africa	<2.54	Wadi Halfa, Sudan	123	39
27	Antarctica	20.32*	South Pole Station	2,756	10
28	North America	30.48	Batagues, Mexico	5	14
29	Asia	45.72	Aden, Arabia	7	50
30	Australia	102.87	Mulka, South Australia	—	34
31	Europe	162.56	Astrakhan, U.S.S.R.	14	25
32	Oceania	226.82	Puako, Hawaii	2	13

* The value given is the average amount of solid snow accumulating in 1 year as indicated by snow markers. The liquid content of the snow is undetermined.

Appendix G

Climatic Data

The following climatic data are for 51 representative stations from around the world. Temperatures are given in degrees Celsius and precipitation in millimeters. Names and locations are given in Table G–2, along with the elevation (in meters) of each station and its Koeppen classification. This format was selected so that you can use the data in exercises to reinforce your understanding of climatic controls and classification. For example, after classifying a station using Table 14–1, write out a likely location based on such items as mean annual temperature, annual temperature range, total precipitation, and seasonal precipitation distribution. Your location need not be a specific city; it could be a description of the station's setting, such as "middle-latitude continental" or "subtropical with a strong monsoon influence." It would also be a good idea to list the reasons for your selection. You may then check your answer by examining the list of stations in Table G–2.

If you simply wish to examine the data for a specific place or data for a specific climatic type, consult the list at the end of this appendix.

TABLE G—1

	J	F	M	A	M	J	J	A	S	O	N	D	Yr.
1.	1.7	4.4	7.9	13.2	18.4	23.8	25.8	24.8	21.4	14.7	6.7	2.8	13.8
	10	10	13	13	20	15	30	32	23	18	10	13	207
2.	−10.4	−8.3	−4.6	3.4	9.4	12.8	16.6	14.9	10.8	5.5	−2.3	−6.4	3.5
	18	25	25	30	51	89	64	71	33	20	18	15	459
3.	10.2	10.8	13.7	17.9	22.2	25.7	26.7	26.5	24.2	19.0	13.3	10.0	18.4
	66	84	99	74	91	127	196	168	147	71	53	71	1247
4.	−23.9	−17.5	−12.5	−2.7	8.4	14.8	15.6	12.8	6.4	−3.1	−15.8	−21.9	−3.3
	23	13	18	8	15	33	48	53	33	20	18	15	297
5.	−17.8	−15.3	−9.2	−4.4	4.7	10.9	16.4	14.4	10.3	3.3	−3.6	−12.8	−0.3
	58	58	61	48	53	61	81	71	58	61	64	64	738
6.	−8.2	−7.1	−2.4	5.4	11.2	14.7	18.9	17.4	12.7	7.4	−0.4	−4.6	5.4
	21	23	29	35	52	74	39	40	35	29	26	21	424
7.	12.8	13.9	15.0	15.0	17.8	20.0	21.1	22.8	22.2	18.3	17.2	15.0	17.6
	69	74	46	28	3	3	0	0	5	10	28	61	327
8.	18.9	20.0	21.1	22.8	25.0	26.7	27.2	27.8	27.2	25.0	21.1	20.0	23.6
	51	48	58	99	163	188	172	178	241	208	71	43	1520
9.	−4.4	−2.2	4.4	10.6	16.7	21.7	23.9	22.7	18.3	11.7	3.8	−2.2	10.4
	46	51	69	84	99	97	97	81	97	61	61	51	894
10.	−5.6	−4.4	0.0	6.1	11.7	16.7	20.0	18.9	15.5	10.0	3.3	−2.2	7.5
	112	96	109	94	86	81	74	61	89	81	107	99	1089
11.	−2.1	0.9	4.7	9.9	14.7	19.4	24.7	23.6	18.3	11.5	3.4	−0.2	10.7
	34	30	40	45	36	25	15	22	13	29	33	31	353
12.	12.8	13.9	15.0	16.1	17.2	18.8	19.4	22.2	21.1	18.8	16.1	13.9	15.9
	53	56	41	20	5	0	0	2	5	13	23	51	269
13.	−0.1	1.8	6.2	13.0	18.7	24.2	26.4	25.4	21.1	14.9	6.7	1.6	13.3
	50	52	78	94	95	109	84	77	70	73	65	50	897
14.	2.7	3.2	7.1	13.2	18.8	23.4	25.7	24.7	20.9	15.0	8.7	3.4	13.9
	77	63	82	80	105	82	105	124	97	78	72	71	1036
15.	12.8	15.0	18.9	21.1	26.1	31.1	32.7	33.9	31.1	22.2	17.7	13.9	23.0
	10	9	6	2	0	0	6	13	10	10	3	8	77
16.	25.6	25.6	24.4	25.0	24.4	23.3	23.3	24.4	24.4	25.0	25.6	25.6	24.7
	259	249	310	165	254	188	168	117	221	183	213	292	2619
17.	25.9	25.8	25.8	25.9	26.4	26.6	26.9	27.5	27.9	27.7	27.3	26.7	26.7
	365	326	383	404	185	132	68	43	96	99	189	143	2433
18.	13.3	13.3	13.3	13.3	13.9	13.3	13.3	13.3	13.9	13.3	13.3	13.9	13.5
	99	112	142	175	137	43	20	30	69	112	97	79	1115
19.	25.9	26.1	25.2	23.9	22.3	21.3	20.8	21.1	21.5	22.3	23.1	24.4	23.2
	137	137	143	116	73	43	43	43	53	74	97	127	1086
20.	13.8	13.5	11.4	8.0	3.7	1.2	1.4	2.9	5.5	9.2	11.4	12.9	7.9
	21	16	18	13	25	15	15	17	12	7	15	18	171
21.	1.5	1.3	3.1	5.8	10.2	12.6	15.0	14.7	12.0	8.3	5.5	3.3	7.8
	179	139	109	140	83	126	141	167	228	236	207	203	1958
22.	−0.5	0.2	3.9	9.0	14.3	17.7	19.4	18.8	15.0	9.6	4.7	1.2	9.5
	41	37	30	39	44	60	67	65	45	45	44	39	556
23.	6.1	5.8	7.8	9.2	11.6	14.4	15.6	16.0	14.7	12.0	9.0	7.0	10.8
	133	96	83	69	68	56	62	80	87	104	138	150	1126

TABLE G–1 (Continued)

	J	F	M	A	M	J	J	A	S	O	N	D	Yr.
24.	10.8	11.6	13.6	15.6	17.2	20.1	22.2	22.5	21.2	18.2	14.4	11.5	16.6
	111	76	109	54	44	16	3	4	33	62	93	103	708
25.	−9.9	− 9.5	−4.2	4.7	11.9	16.8	19.0	17.1	11.2	4.5	−1.9	−6.8	4.4
	31	28	33	35	52	67	74	74	58	51	36	36	575
26.	8.0	9.0	10.9	13.7	17.5	21.6	24.4	24.2	21.5	17.2	12.7	9.5	15.9
	83	73	52	50	48	18	9	18	70	110	113	105	749
27.	−9.0	−9.0	−6.6	−4.1	0.4	3.6	5.6	5.5	3.5	−0.6	−4.5	−7.6	−1.9
	202	180	164	166	197	249	302	278	208	183	190	169	2488
28.	−2.9	−3.1	−0.7	4.4	10.1	14.9	17.8	16.6	12.2	7.1	2.8	0.1	6.6
	43	30	26	31	34	45	61	76	60	48	53	48	555
29.	12.8	13.9	17.2	18.9	22.2	23.9	25.5	26.1	25.5	23.9	18.9	15.0	20.3
	66	41	20	5	3	0	0	0	3	18	46	66	268
30.	24.6	24.9	25.0	24.9	25.0	24.2	23.7	23.8	23.9	24.2	24.2	24.7	24.4
	81	102	155	140	133	119	99	109	206	213	196	122	1675
31.	21.1	20.4	20.9	21.7	23.0	26.0	27.3	27.3	27.5	27.5	26.0	25.2	24.3
	0	2	0	0	1	15	88	249	163	49	5	6	578
32.	20.4	22.7	27.0	30.6	33.8	34.2	33.6	32.7	32.6	30.5	25.5	21.3	28.7
	0	0	0	0	0	2	1	11	2	0	0	0	16
33.	17.8	18.1	18.8	18.8	17.8	16.2	14.9	15.5	16.8	18.6	18.3	17.8	17.5
	46	51	102	206	160	46	18	25	25	53	109	81	922
34.	20.6	20.7	19.9	19.2	16.7	13.9	13.9	16.3	19.1	21.8	21.4	20.9	18.7
	236	168	86	46	13	8	0	3	8	38	94	201	901
35.	11.7	13.3	16.7	18.6	19.2	20.0	20.3	20.5	20.5	19.1	15.9	12.9	17.4
	20	41	179	605	1705	2875	2455	1827	1231	447	47	5	11437
36.	26.2	26.3	27.1	27.2	27.3	27.0	26.7	27.0	27.4	27.4	26.9	26.6	26.9
	335	241	201	141	116	97	61	50	78	91	151	193	1755
37.	−0.8	2.6	5.3	8.5	13.1	17.0	17.2	17.3	15.3	11.5	5.7	0.3	9.4
	0	0	1	1	18	72	157	151	68	4	1	0	473
38.	24.5	25.8	27.9	30.5	32.7	32.5	30.7	30.1	29.7	28.1	25.9	24.6	28.6
	24	7	15	25	52	53	83	124	118	267	308	157	1233
39.	−18.7	−18.1	−16.7	−11.7	−5.0	0.6	5.3	5.8	1.4	−4.2	−12.3	−15.8	−7.5
	8	8	8	8	15	20	36	43	43	33	13	12	247
40.	−21.9	−18.6	−12.5	−5.0	9.7	15.6	18.3	16.1	10.3	0.8	−10.6	−18.4	−1.4
	15	8	8	13	30	51	51	51	28	25	18	20	318
41.	−4.7	−1.9	4.8	13.7	20.1	24.7	26.1	24.9	19.9	12.8	3.8	−2.7	11.8
	4	5	8	17	35	78	243	141	58	16	10	3	623
42.	24.3	25.2	27.2	29.8	29.5	27.8	27.6	27.1	27.6	28.3	27.7	25.0	27.3
	8	5	6	17	260	524	492	574	398	208	34	3	2530
43.	25.8	26.3	27.8	28.8	28.2	27.4	27.1	27.1	26.7	26.5	26.1	25.7	27.0
	6	13	12	65	196	285	242	277	292	259	122	37	1808
44.	3.7	4.3	7.6	13.1	17.6	21.1	25.1	26.4	22.8	16.7	11.3	6.1	14.7
	48	73	101	135	131	182	146	147	217	220	101	60	1563
45.	−15.8	−13.6	−4.0	8.5	17.7	21.5	23.9	21.9	16.7	6.1	−6.2	−13.0	5.3
	8	15	15	33	25	33	16	35	15	47	22	11	276
46.	−46.8	−43.1	−30.2	−13.5	2.7	12.9	15.7	11.4	2.7	−14.3	−35.7	−44.5	−15.2
	7	5	5	4	5	25	33	30	13	11	10	7	155

TABLE G–1 (*Continued*)

	J	F	M	A	M	J	J	A	S	O	N	D	Yr.
47.	19.2	19.6	18.4	16.4	13.8	11.8	10.8	11.3	12.6	16.3	15.9	17.7	15.2
	84	104	71	109	122	140	140	109	97	106	81	79	1242
48.	28.2	27.9	28.3	28.2	26.8	25.4	25.1	25.8	27.7	29.1	29.2	28.7	27.6
	341	338	274	121	9	1	2	5	17	66	156	233	1562
49.	21.9	21.9	21.2	18.3	15.7	13.1	12.3	13.4	15.3	17.6	19.4	21.0	17.6
	104	125	129	101	115	141	94	83	72	80	77	86	1205
50.	−7.2	−7.2	−4.4	−0.6	4.4	8.3	10.0	8.3	5.0	1.1	−3.3	−6.1	0.7
	84	66	86	62	89	81	79	94	150	145	117	79	1132
51.	−4.4	−8.9	−15.5	−22.8	−23.9	−24.4	−26.1	−26.1	−24.4	−18.8	−10.0	−3.9	−17.4
	13	18	10	10	10	8	5	8	10	5	5	8	110

TABLE G–2

Station No.	City	Location	Elevation	Koeppen Classification
		North America		
1.	Albuquerque, N.M.	lat. 35°05′N long. 106°40′W	1593 m	BWk
2.	Calgary, Canada	lat. 51°03′N long. 114°05′W	1062 m	Dfb
3.	Charleston, S.C.	lat. 32°47′N long. 79°56′W	18 m	Cfa
4.	Fairbanks, Alaska	lat. 64°50′N long. 147°48′W	134 m	Dfc
5.	Goose Bay, Canada	lat. 53°19′N long. 60°33′W	45 m	Dfb
6.	Lethbridge, Canada	lat. 49°40′N long. 112°39′W	920 m	Dfb
7.	Los Angeles, Calif.	lat. 34°00′N long. 118°15′W	29 m	BSk
8.	Miami, Fla.	lat. 25°45′N long. 80°11′W	2 m	Am
9.	Peoria, Ill.	lat. 40°45′N long. 89°35′W	180 m	Dfa
10.	Portland, Me.	lat. 43°40′N long. 70°16′W	14 m	Dfb
11.	Salt Lake City, Utah	lat. 40°46′N long. 111°52′W	1288 m	BSk
12.	San Diego, Calif.	lat. 32°43′N long. 117°10′W	26 m	BSk
13.	St. Louis, Mo.	lat. 38°39′N long. 90°15′W	172 m	Cfa
14.	Washington, D.C.	lat. 38°50′N long. 77°00′W	20 m	Cfa
15.	Yuma, Ariz.	lat. 32°40′N long. 114°40′W	62 m	BWh

TABLE G–2 (Continued)

Station No.	City	Location	Elevation	Koeppen Classification
		South America		
16.	Iquitos, Peru	lat. 3°39′S long. 73°18′W	115 m	Af
17.	Manaus, Brazil	lat. 3°01′S long. 60°00′W	60 m	Am
18.	Quito, Ecuador	lat. 0°17′S long. 78°32′W	2766 m	Cfb
19.	Rio de Janeiro, Brazil	lat. 22°50′S long. 43°20′W	26 m	Aw
20.	Santa Cruz, Argentina	lat. 50°01′S long. 60°30′W	111 m	BSk
		Europe		
21.	Bergen, Norway	lat. 60°24′N long. 5°20′E	44 m .	Cfb
22.	Berlin, Germany	lat. 52°28′N long. 13°26′E	50 m	Cfb
23.	Brest, France	lat. 48°24′N long. 4°30′W	103 m	Cfb
24.	Lisbon, Portugal	lat. 38°43′N long. 9°05′W	93 m	Csa
25.	Moscow, U.S.S.R.	lat. 55°45′N long. 37°37′E	156 m	Dfb
26.	Rome, Italy	lat. 41°52′N long. 12°37′E	3 m	Csa
27.	Santis, Switzerland	lat. 47°15′N long. 9°21′E	2496 m	ET
28.	Stockholm, Sweden	lat. 59°21′N long. 18°00′E	52 m	Dfb
		Africa		
29.	Benghazi, Libya	lat. 32°06′N long. 20°06′E	25 m	BSh
30.	Coquilhatville, Zaire	lat. 0°01′N long. 18°17′E	21 m	Af
31.	Dakar, Senegal	lat. 14°40′N long. 17°28′W	23 m	BSh
32.	Faya, Chad	lat. 18°00′N long. 21°18′E	251 m	BWh
33.	Nairobi, Kenya	lat. 1°16′S long. 36°47′E	1791 m	Csb
34.	Salisbury, Rhodesia	lat. 17°50′S long. 30°52′E	1449 m	Cwb
		Asia		
35.	Cherrapunji, India	lat. 25°15′N long. 91°44′E	1313 m	Cwb
36.	Djakarta, Indonesia	lat. 6°11′S long. 106°45′E	8 m	Am

TABLE G–2 (*Continued*)

Station No.	City	Location	Elevation	Koeppen Classification
37.	Lhasa, Tibet	lat. 29°40′N long. 91°07′E	3685 m	Cwb
38.	Madras, India	lat. 13°00′N long. 80°11′E	16 m	Aw
39.	Novaya Zemlya, U.S.S.R.	lat. 72°23′N long. 54°46′E	15 m	ET
40.	Omsk, U.S.S.R.	lat. 54°48′N long. 73°19′E	85 m	Dfb
41.	Peking, China	lat. 39°57′N long. 116°23′E	52 m	Dwa
42.	Rangoon, Burma	lat. 16°46′N long. 96°10′E	23 m	Am
43.	Saigon, Viet Nam	lat. 10°49′N long. 106°40′E	10 m	Aw
44.	Tokyo, Japan	lat. 35°41′N long. 139°46′E	6 m	Cfa
45.	Urumchi, China	lat. 43°47′N long. 87°43′E	912 m	Dfa
46.	Verkhoyansk, U.S.S.R.	lat. 67°33′N long. 133°23′E	137 m	Dfd
	Australia and New Zealand			
47.	Auckland, New Zealand	lat. 37°43′S long. 174°53′E	49 m	Csb
48.	Darwin, Australia	lat. 12°26′S long. 131°00′E	27 m	Aw
49.	Sydney, Australia	lat. 33°52′S long. 151°17′E	42 m	Cfb
	Greenland			
50.	Ivigtut, Greenland	lat. 61°12′N long. 48°10′W	29 m	ET
	Antarctica			
51.	McMurdo Station, Antarctica	lat. 77°53′S long. 167°00′E	2 m	EF

Glossary

Absolute Humidity: The weight of water vapor per volume of air (usually expressed as grams of water vapor per cubic meter of air).

Absolute Instability: The condition of air that has an environmental lapse rate that is greater than the dry adiabatic rate (1°C per 100 meters).

Absolute Stability: The condition of air that has an environmental lapse rate that is less than the wet adiabatic rate.

Absolute Zero: The zero point on the Kelvin temperature scale, representing the temperature at which all molecular motion is presumed to cease.

Acid Precipitation: Rain or snow with pH values of less than 5.6

Adiabatic Temperature Change: The cooling or warming of air caused when air is allowed to expand or is compressed, not because heat is added or subtracted.

Advection: Horizontal convective motion, such as wind.

Advection Fog: Fog formed when warm, moist air is blown over a cool surface and chilled below the dew point.

Aerovane: A device that resembles a wind vane with a propeller at one end. Used to indicate wind speed and direction.

Air: A mixture of many discrete gases, of which nitrogen and oxygen are most abundant, in which varying quantities of tiny solid and liquid particles are suspended.

Air Mass: A large body of air, usually 1600 kilometers or more across, that is characterized by homogeneous physical properties at any given altitude.

Air-Mass Thunderstorm: A localized thunderstorm that forms in a warm, moist, unstable air mass. Most frequent in the afternoon in spring and summer.

Air-Mass Weather: The conditions experienced in an area as an air mass passes over it. Because air masses are large and relatively homogeneous, air-mass weather will be very fairly constant and may last for several days.

Air Pressure: The force exerted by the weight of a column of air above a given point.

Albedo: The reflectivity of a substance, usually expressed as a percentage of the incident radiation reflected.

Aleutian Low: A large cell of low pressure centered over the Aleutian Islands of the North Pacific during the winter.

Altimeter: An aneroid barometer calibrated to indicate altitude instead of pressure.

Altitude (of the sun): The angle of the sun above the horizon.

Analog Method: A statistical approach to weather forecasting in which current conditions are matched with records of similar past weather events with the idea that the succession of events in the past will be paralleled by current conditions.

Anemometer: An instrument used to determine wind speed.

Aneroid Barometer: An instrument for measuring air pressure that consists of evacuated metal chambers that are very sensitive to variations in air pressure.

Annual March of Temperature: The annual cycle of air temperature changes. Usually the lowest monthly mean occurs a month, or on occasion 2 months, after the winter solstice, whereas the highest monthly mean commonly occurs a month or two after the summer solstice. January and July generally represent the extreme months.

Annual Mean: An average of the 12 monthly means.

Annual Temperature Range: The difference between the warmest and coldest monthly means.

Anticyclone: An area of high atmospheric pressure characterized by diverging and rotating winds and subsiding air aloft.

Anticyclonic Flow: Winds blow out and clockwise about an anticyclone (high) in the northern hemisphere, and they blow out and counterclockwise about an anticyclone in the southern hemisphere.

Aphelion: The point in the orbit of a planet that is farthest from the sun.

Apparent Temperature: The air temperature perceived by a person.

Arctic (A) Air Mass: A bitterly cold air mass that forms over the frozen Arctic Ocean.

Arctic Sea Smoke: A dense and often extensive steam fog that occurs over high-latitude ocean areas in winter.

Arid: *See* Desert.

Astronomical Theory: A theory of climatic change first developed by the Yugoslavian astronomer Milankovitch. It is based on changes in the shape of the earth's orbit, variations in the obliquity of the earth's axis, and the wobbling of the earth's axis.

Atmosphere: The gaseous portion of a planet; the planet's envelope of air; one of the traditional subdivisions of the earth's physical environment.

Atmospheric Window: Refers to the fact that the troposphere is transparent to (i.e., does not absorb) terrestrial radiation between 8 and 11 micrometers in length.

Aurora: A bright display of ever-changing light caused by solar radiation interacting with the upper atmosphere in the region of the poles. Termed aurora borealis in the northern hemisphere and aurora australis in the southern hemisphere.

Autumnal Equinox: *See* Equinox.

Backing Wind Shift: A wind shift in a counterclockwise direction, such as a shift from east to north.

Barograph: A recording barometer.

Barometric Tendency: *See* Pressure Tendency.

Beaufort Scale: A scale that may be used for estimating wind speed when an anemometer is not available.

Bergeron Process: A theory that relates the formation of precipitation to supercooled clouds, freezing nuclei, and the different saturation levels of ice and liquid water.

Bimetal Strip: A thermometer consisting of two thin strips of metal that are welded together and have widely different coefficients of thermal expansion. When temperature changes, the two metals expand or contract unequally and cause changes in the curvature of the element. Commonly used in thermographs.

Biosphere: The totality of life-forms on earth.

Blackbody: A material that is able to absorb 100 percent of the radiation that strikes it.

Blizzard: A violent and extremely cold wind laden with dry snow picked up from the ground.

Bora: In the region of the eastern shore of the Adriatic Sea, a cold, dry northeasterly wind that blows down from the mountains.

Buys Ballott's Law: With your back to the wind in the northern hemisphere, low pressure will be to your left and high pressure to your right. The reverse is true in the southern hemisphere.

Calorie: The amount of heat required to raise the temperature of 1 gram of water 1°C.

Ceiling: The height ascribed to the lowest layer of clouds or obscuring phenomena when the sky is reported as broken, overcast, or obscured and the clouds are not classified "thin" or "partial." The ceiling is termed unlimited when the foregoing conditions are not present.

Celsius Scale: A temperature scale (at one time called the centigrade scale) devised by Anders Celsius in 1742 and used where the metric system is used. For water at sea level, 0° is designated the ice point and 100° the steam point.

Centrifugal Force: The tendency of a particle to move in a straight line when rotated creates this imaginary outward force.

Chinook: The name applied to a foehn wind in the Rocky Mountains.

Circle of Illumination: The line (great circle) separating daylight from darkness on the earth.

Cirrus: One of three basic cloud forms; also one of the three high cloud types. They are thin, delicate ice crystal clouds often appearing as veil-like patches or thin, wispy fibers.

Climate: A description of aggregate weather conditions; the sum of all statistical weather information that helps describe a place or region.

Climatic Change: A study dealing with variations in climate on many different time scales, from decades to millions of years, and the possible causes of such variations.

Climatic-Feedback Mechanisms: Because the atmosphere is a complex interactive physical system, several different possible outcomes may result when one of the system's elements is altered. These various possibilities are climatic feedback mechanisms.

Cloud: A form of condensation best described as a dense concentration of suspended water droplets or tiny ice crystals.

Cloud Seeding: The introduction into clouds of particles (most commonly dry ice or silver iodide) for the purpose of altering the cloud's natural development.

Clouds of Vertical Development: A cloud that has its base in the low height range but extends upward into the middle or high altitudes.

Cold Front: The discontinuity at the forward edge of an advancing cold air mass that is displacing warmer air in its path.

Cold-Type Occluded Front: A front that forms when the air behind the cold front is colder than the air underlying the warm front it is overtaking.

Cold Wave: A rapid and marked fall of temperature. The Weather Service applies this term to a fall of temperature in 24 hours equaling or exceeding a specified number of degrees and reaching a specified minimum temperature or lower. These specifications vary for different parts of the country and for different periods of the year.

Collision–Coalescence Process: A theory of raindrop formation in warm clouds (above 0°C) in which large cloud droplets ("giants") collide and join together with smaller droplets to form a raindrop. Opposite electrical charges may bind the cloud droplets together.

Condensation: The change of state from a gas to a liquid.

Condensation Nuclei: Microscopic particles that serve as surfaces on which water vapor condenses.

Conditional Instability: The condition of moist air with an environmental lapse rate between the dry and wet adiabatic rates.

Conduction: The transfer of heat through matter by molecular activity. Energy is transferred through collisions from one molecule to another.

Constant Pressure Surface: A surface along which the atmospheric pressure is everywhere equal at any given moment.

Continental (c) Air Mass: An air mass that forms over land; it is normally relatively dry.

Continental Climate: A climate lacking marine influence and characterized by more extreme temperatures than in marine climates; therefore it has a relatively high annual temperature range for its latitude.

Contrail: A cloudlike streamer frequently observed behind aircraft flying in clear, cold, humid air and caused by the addition to the atmosphere of water vapor from engine exhaust gases.

Controls of Temperature: Those factors that cause variations in temperature from place to place, such as latitude and altitude.

Convection: The transfer of heat by the movement of a mass or substance. It can only take place in fluids.

Convection Cell: Circulation that results from the uneven heating of a fluid; the warmer parts of the fluid expand and rise because of their buoyancy and the cooler parts sink.

Convergence: The condition that exists when the distribution of winds within a given area results in a net horizontal inflow of air into the area. Since convergence at lower levels is associated with an upward movement of air, areas of convergent winds are regions favorable to cloud formation and precipitation.

Cooling Degree Days: Each degree of temperature of the daily mean above 65°F is counted as one cooling degree day. The amount of energy required to maintain a certain temperature in a building is proportional to the cooling-degree days total.

Coriolis Effect: The deflective effect of the earth's rotation on all free-moving objects, including the atmosphere and oceans. Deflection is to the right in the northern hemisphere and to the left in the southern hemisphere.

Corona: A bright whitish disk centered on the moon or sun that results from defraction when the objects are veiled by a thin cloud layer.

Country Breeze: A circulation pattern characterized by a light wind blowing into a city from the surrounding countryside. It is best developed on clear and otherwise calm nights when the urban heat island is most pronounced.

Cumulus: One of three basic cloud forms; also the name given one of the clouds of vertical development. Cumulus are billowy individual cloud masses that often have flat bases.

Cumulus Stage: The initial stage in thunderstorm development in which the growing cumulonimbus is dominated by strong updrafts.

Cup Anemometer: *See* Anemometer.

Cyclogenesis: The process that creates or develops a new cyclone; also the process that produces an intensification of a preexisting cyclone.

Cyclone: An area of low atmospheric pressure characterized by rotating and converging winds and ascending air.

Cyclonic Flow: Winds blow in and counterclockwise about a cyclone (low) in the northern hemisphere and in and clockwise about a cyclone in the southern hemisphere.

Daily March of Temperature: The daily cycle of air temperature changes. Commonly the minimum temperature occurs near sunrise and the maximum temperature occurs during midafternoon.

Daily Mean: The mean temperature for a day that is determined by averaging the hourly readings or, more commonly, by averaging the maximum and minimum temperatures for a day.

Daily Range: The difference between the maximum and minimum temperatures for a day.

Dart Leader: *See* Leader.

Degassing: The release of gases dissolved in molten rock.

Deposition: The process whereby water vapor changes directly to ice without going through the liquid state.

Desert: One of the two types of dry climate; the driest of the dry climates.

Dew Point: The temperature to which air has to be cooled in order to reach saturation.

Diffraction: The slight bending of light as it passes sharp edges.

Diffused Light: Solar energy is scattered and reflected in the atmosphere and reaches the earth's surface in the form of diffuse blue light from the sky.

Directional Divergence: The spreading out of an airstream that typically occurs slightly downwind of the axis of an upper-air trough.

Discontinuity: A zone characterized by a comparatively rapid transition of meteorological elements.

Dispersion: The separation of colors by refraction.

Dissipating Stage: The final stage of a thunderstorm that is dominated by downdrafts and entrainment that leads to the evaporation of the cloud structure.

Diurnal: Daily, especially pertaining to actions that are completed within 24 hours and that recur every 24 hours.

Divergence: The condition that exists when the distribution of winds within a given area results in a net horizontal outflow of air from the region. In divergence at lower levels the resulting deficit is compensated for by a downward movement of air from aloft; hence, areas of divergent winds are unfavorable to cloud formation and precipitation.

Doldrums: The equatorial belt of calms or light variable winds lying between the two trade wind belts.

Doppler Radar: A type of radar that has the capability of detecting motion directly.

Drizzle: Precipitation from stratus clouds consisting of numerous tiny droplets.

Dry Adiabatic Rate: The rate of adiabatic cooling or warming in unsaturated air. The rate of temperature change is 1°C per 100 meters.

Dry Climate: A climate in which yearly precipitation is not as great as the potential loss of water by evaporation.

Dry-Summer Subtropical Climate: A climate located on the west sides of continents between latitudes 30 degrees and 45 degrees. It is the only humid climate with a strong winter precipitation maximum.

Dynamic Seeding: A type of cloud seeding that uses massive seeding, a process that results in an increase of the release of latent heat and causes the cloud to grow larger.

Eccentricity: The variation of an ellipse from a circle.

Electromagnetic Radiation: *See* Radiation.

Elements (atmospheric): Those quantities or properties of the atmosphere that are measured regularly and that are used to express the nature of weather and climate.

El Niño: The name given to the periodic warming of the ocean that occurs in the central and eastern Pacific. A major El Niño episode can cause extreme weather in many parts of the world.

Entrainment: The influx of cool, dry air from the surrounding environment into the downdraft of a cumulonimbus cloud; a process that acts to intensify the downdraft.

Environmental Lapse Rate: The rate of temperature decrease with height in the troposphere.

Equatorial (E) Air Mass: A warm-to-hot air mass that forms in the tropics near the equator.

Equatorial Low: A quasi-continuous belt of low pressure lying near the equator and between the subtropical highs.

Equinox: The point in time when the vertical rays of the sun are striking the equator. March 21 is the vernal or spring equinox in the northern hemisphere and September 22 or 23 is the autumnal equinox in the northern hemisphere. Length of daylight and darkness is equal at all latitudes at equinox.

Evaporation: The process by which a liquid is transformed into a gas.

Eye: A roughly circular area of relatively light winds and fair weather at the center of a hurricane.

Eye Wall: The doughnut-shaped area of intense cumulonimbus development and very strong winds that surrounds the eye of a hurricane.

Fahrenheit Scale: A temperature scale devised by Gabriel Daniel Fahrenheit in 1714 and used in the English system. For water at sea level, 32° is designated the ice point and 212° the steam point.

Fall Wind: *See* Katabatic Wind.

Fata Morgana: A mirage most frequently observed in coastal areas in which extreme towering occurs.

Fixed Points: Reference points, such as the steam point and the ice point, used in the construction of temperature scales.

Flash: The total discharge of lightning, which is usually perceived as a single flash of light but which actually consists of several strokes (see Strokes).

Foehn: A warm, dry wind on the lee side of a mountain range that owes its relatively high temperature largely to adiabatic heating during descent down mountain slopes.

Fog: A cloud with its base at or very near the earth's surface.

Freezing: The change of state from a liquid to a solid.

Freezing Nuclei: Solid particles that have a crystal form resembling that of ice; they serve as cores for the formation of ice crystals.

Front: A boundary (discontinuity) separating air masses of different densities, one warmer and often higher in moisture content than the other.

Frontal Fog: Fog formed when rain evaporates as it falls through a layer of cool air.

Frontal Wedging: The lifting of air resulting when cool air acts as a barrier over which warmer, lighter air will rise.

Frontogenesis: The beginning or creation of a front.

Frontolysis: The destruction and dying of a front.

Frost: Occurs when the temperature falls to 0°C or below. (See White Frost.)

Fujita Intensity Scale (F-scale): A scale developed by T. Theodore Fujita for classifying the severity of a tornado, based on the correlation of wind speed with the degree of destruction.

Geostationary Satellite: A satellite that remains over a fixed point because its rate of travel corresponds to the earth's rate of rotation. Because the satellite must orbit at distances of about 35,000 kilometers, images from this type of satellite are not as detailed as those from polar satellites.

Geostrophic Wind: A wind, usually above a height of 600 meters, that blows parallel to the isobars.

Glaze: A coating of ice on objects formed when supercooled rain freezes on contact. A storm that produces glaze is called an icing storm.

Global Circulation: The general circulation of the atmosphere; the average flow of air over the entire globe.

Glory: A series of rings of colored light most commonly appearing around the shadow of an airplane that is projected on clouds below.

Gradient Wind: The curved airflow pattern around a pressure center resulting from a balance among pressure gradient force, Coriolis force, and centrifugal force.

Greenhouse Effect: The transmission of shortwave solar radiation by the atmosphere coupled with the selective absorption of longer wavelength terrestrial radiation, especially by water vapor and carbon dioxide.

Growing Degree Days: A practical application of temperature data for determining the approximate date when crops will be ready for harvest.

Gust Front: The boundary separating the cold downdraft from a thunderstorm and the relatively warm, moist surface air. Lifting along this boundary may initiate the development of thunderstorms.

Hadley Cell: The thermally driven circulation system of equatorial and tropical latitudes consisting of two convection cells, one in each hemisphere. The existence of this circulation system was first proposed by George Hadley in 1735 as an explanation for the trade winds.

Hail: Precipitation in the form of hard, round pellets or irregular lumps of ice that may have concentric shells formed by the successive freezing of layers of water.

Halo: A narrow whitish ring of large diameter centered around the sun. The commonly observed *22-degree halo* subtends an angle of 22 degrees from the observer.

Heat: The kinetic energy of random molecular motion.

Heat Budget: The balance of incoming and outgoing radiation.

Heating Degree Day: Each degree of temperature of the daily mean below 65°F is counted as one heating degree day. The amount of heat required to maintain a certain temperature in a building is proportional to the heating-degree days total.

Heterosphere: A zone of the atmosphere beyond about 80 kilometers where the gases are arranged into four roughly spherical shells, each with a distinctive composition.

High Cloud: A cloud that normally has its base above 6000 meters; the base may be lower in winter and at high latitude locations.

Homosphere: A zone of the atmosphere extending from the surface to about 80 kilometers that is uniform in terms of the proportions of its component gases.

Horse Latitudes: A belt of calms or light variable winds and subsiding air located near the center of the subtropical high.

Humid Continental Climate: A relatively severe climate characteristic of broad continents in the middle latitudes between approximately 40 and 50 degrees north latitude. This climate is not found in the southern hemisphere, where the middle latitudes are dominated by the oceans.

Humidity: A general term referring to water vapor in the air.

Humid Subtropical Climate: A climate generally located on the eastern side of a continent and characterized by hot, sultry summers and cool winters.

Hurricane: A tropical cyclonic storm having winds in excess of 115 kilometers per hour; also known as typhoon (western Pacific) and cyclone (Indian Ocean).

Hydrologic Cycle: The continuous movement of water from the oceans to the atmosphere (by evaporation), from the atmosphere to the land (by condensation and precipitation), and from the land back to the sea (via stream flow).

Hydrosphere: The water portion of our planet; one of the traditional subdivisions of the earth's physical environment.

Hygrometer: An instrument designed to measure relative humidity.

Hygroscopic Nuclei: Condensation nuclei having a high affinity for water, such as salt particles.

Ice Cap Climate: A climate that has no monthly means above freezing and supports no vegetative cover except in a few scattered high mountain areas. This climate, with its perpetual ice and snow, is confined largely to the ice caps of Greenland and Antarctica.

Icelandic Low: A large cell of low pressure centered over Iceland and southern Greenland in the North Atlantic during the winter.

Ice Point: The temperature at which ice melts.

Ideal Gas Law: The pressure exerted by a gas is proportional to its density and absolute temperature.

Inclination of the Axis: The tilt of the earth's axis from the perpendicular to the plane of the earth's orbit (plane of the ecliptic). Currently the inclination is about 23½ degrees away from the perpendicular.

Inferior Mirage: A mirage in which the image appears below the true location of the object.

Infrared: Radiation with a wavelength from 0.7 to 200 micrometers.

Interference: Occurs when light rays of different frequencies (i.e., colors) meet. Such interference results in the cancellation or subtraction of some frequencies, which is responsible for the colors associated with coronas.

Internal Reflection: Occurs when light that is traveling through a transparent material, such as water, reaches the opposite surface and is reflected back into the material. This is an important factor in the formation of such optical phenomena as rainbows.

Intertropical Convergence Zone (ITC): The zone of general convergence between the northern and southern hemisphere trade winds.

Ionosphere: A complex atmospheric zone of ionized gases that extends between 80 and 400 kilometers, thus coinciding with the lower thermosphere and heterosphere.

Isobar: A line drawn on a map connecting points of equal barometric pressure, usually corrected to sea level.

Isohyet: A line connecting places having equal rainfall.

Isotherm: A line connecting points of equal air temperature.

ITC: *See* Intertropical Convergence Zone.

Jet Stream: Swift geostrophic airstreams in the upper troposphere that meander in relatively narrow belts.

Jungle: An almost impenetrable growth of tangled vines, shrubs, and short trees that characterizes areas where the tropical rain forest has been cleared.

Katabatic Wind: The flow of cold dense air downslope under the influence of gravity; the direction of flow is controlled largely by topography.

Kelvin Scale: A temperature scale (also called the absolute scale) used primarily for scientific purposes and having intervals equivalent to those on the Celsius scale but beginning at absolute zero.

Kinetic Energy: The energy of motion.

Koeppen Classification: A system for classifying climates devised by Wladimir Koeppen that is based on mean monthly and annual values of temperature and precipitation.

Lake-Effect Snow: Snow showers associated with a cP air mass to which moisture and heat are added from below as it traverses a large and relatively warm lake (such as one of the Great Lakes), rendering the air mass humid.

Land Breeze: A local wind blowing from the land toward the sea during the night in coastal areas.

Lapse Rate: *See* Environmental Lapse Rate and Normal Lapse Rate.

Latent Heat: The energy absorbed or released during a change of state.

Law of Conservation of Angular Momentum: The product of the velocity of an object around a center of rotation (axis) and the distance squared of the object from the axis is constant.

Leader: The conductive path of ionized air that forms near a cloud base prior to a lightning stroke. The initial conductive path is referred to as a *step leader* because it extends itself earthward in short, nearly invisible bursts. A *dart leader,* which is continuous and less branched than a step leader, precedes each subsequent stroke along the same path.

Lifting Condensation Level: The height at which rising air that is cooling at the dry adiabatic rate becomes saturated and condensation begins.

Lightning: A sudden flash of light generated by the flow of electrons between oppositely charged parts of a cumulonimbus cloud or between the cloud and the ground.

Liquid-in-Glass Thermometer: A device for measuring temperature that consists of a tube with a liquid-filled bulb at one end. The expansion or contraction of the fluid indicates temperature.

Lithosphere: The solid portion of the earth; one of the traditional subdivisions of the earth's physical environment.

Long-Range Forecasting: Estimating rainfall and temperatures for a period beyond 3 to 5 days, usually for 7- to 30-day periods. Such forecasts are not as detailed or reliable as those for shorter periods.

Looming: A mirage that allows objects that are below the horizon to be seen.

Low Cloud: A cloud that forms below a height of 2000 meters.

Macroscale Wind Systems: Such phenomena as cyclones and anticyclones that persist for days or weeks and have a horizontal dimension of hundreds to several thousands of kilometers; also features of the atmospheric circulation that persist for weeks or months and have horizontal dimensions of up to 10,000 kilometers.

Marine Climate: A climate dominated by the ocean; because of the moderating effect of water, sites having this climate are considered relatively mild.

Marine West Coast Climate: A climate found on windward coasts from latitudes 40 degrees to 65 degrees and dominated by maritime air masses. Winters are mild and summers are cool.

Maritime (m) Air Mass: An air mass that originates over the ocean. These air masses are relatively humid.

Mature Stage: The second of the three stages of a thunderstorm. This stage is characterized by violent weather as downdrafts exist side by side with updrafts.

Maximum Thermometer: A thermometer that measures the maximum temperature for a given period of time, usually 24 hours. A constriction in the base of the glass tube allows mercury to rise but prevents it from returning to the bulb until the thermometer is shaken or whirled.

Mediterranean Climate: A common name applied to the dry-summer subtropical climate.

Melting: The change of state from a solid to a liquid.

Mercurial Barometer: A mercury-filled glass tube in which the height of the mercury column is a measure of air pressure.

Mesocyclone: An intense rotating wind system in the lower part of a thunderstorm that precedes the development of damaging hail, severe winds, or tornadoes.

Mesopause: The boundary between the mesosphere and the thermosphere.

Mesoscale Winds: Small convective cells that exist for minutes or hours, such as thunderstorms, tornadoes, and the land and sea breeze. Typical horizontal dimensions range from 1 to 100 kilometers.

Mesosphere: The layer of the atmosphere above the stratosphere where temperatures drop fairly rapidly with increasing height.

Microscale Winds: Phenomena, such as turbulence, with life spans of less than a few minutes that affect small areas and are strongly influenced by local conditions of temperature and terrain.

Middle Cloud: A cloud occupying the height range from 2000 to 6000 meters.

Middle-Latitude Cyclone: *See* Wave Cyclone.

Millibar: The standard unit of pressure measurement used by the National Weather Service. One millibar (mb) equals 100 newtons per square meter.

Minimum Thermometer: A thermometer that measures the minimum temperature for a given period of time, usually 24 hours. By checking the small dumbbell-shaped index, the minimum temperature can be read.

Mirage: An optical effect of the atmosphere caused by refraction in which the image of an object appears displaced from its true position.

Mistral: A cold northwest wind that blows into the western Mediterranean basin from higher elevations to the north.

Mixing Depth: The height to which convectional movements extend above the earth's surface. The greater the mixing depth, the better the air quality.

Mixing Ratio: The mass of water vapor in a unit mass of dry air; commonly expressed as grams of water vapor per kilogram of dry air.

Monsoon: The seasonal reversal of wind direction associated with large continents, especially Asia. In winter the wind blows from land to sea; in summer it blows from sea to land.

Monthly Mean: The mean temperature for a month that is calculated by averaging the daily means.

Mountain Breeze: The nightly downslope winds commonly encountered in mountain valleys.

Negative-Feedback Mechanism: As used in climatic change, any effect that is opposite of the initial change and tends to offset it.

Newton: A unit of force used in physics. One newton is the force that is necessary to accelerate 1 kilogram of mass 1 meter per second per second.

Normal: The average value of a meteorological element over any fixed period of years that is recognized as standard for the country and for the element of concern.

Normal Lapse Rate: The average drop in temperature with increasing height in the troposphere; about 6.5°C per kilometer.

Northeaster: The term applied to describe the weather associated with an incursion of mP air into the Northeast from the North Atlantic; strong northeast winds, freezing or near-freezing temperatures, and the possibility of precipitation make this an unwelcome weather event.

Numerical Weather Prediction (NWP): Forecasting the behavior of atmospheric disturbances based upon the solution of the governing fundamental equations of hydrodynamics, subject to the observed initial conditions. Because of the vast number of calculations involved, high-speed computers are always used.

Obliquity: The angle between the planes of the earth's equator and orbit.

Occluded Front: A front formed when a cold front overtakes a warm front.

Occlusion: The overtaking of one front by another.

Orographic Lifting: Mountains or highlands acting as barriers to the flow of air force the air to ascend. The air cools adiabatically and clouds and precipitation may result.

Overrunning: Warm air gliding up a retreating cold air mass.

Oxygen-Isotope Analysis: A method of deciphering past temperatures based on the precise measurement of the ratio between two isotopes of oxygen, ^{16}O and ^{18}O. Analysis is commonly made of sea-floor sediments and cores from ice sheets.

Ozone: A molecule of oxygen containing three oxygen atoms.

Paleosol: An old buried soil that may furnish some evidence of the nature of past climates because climate is the most important factor in soil formation.

Parhelia: *See* Sun Dogs.

Perihelion: The point in the orbit of a planet closest to the sun.

Permafrost: The permanent freezing of the subsoil in tundra regions.

Persistence Forecast: A forecast which assumes that the weather occurring up-stream will persist and move on and will affect the areas in its path in much the same way. Persistence forecasts do not account for changes that might occur in the weather system.

pH Scale: A 0 to 14 scale that is used for expressing the exact degree of acidity or alkalinity of a solution. A pH of 7 signifies a neutral solution. Values below 7 signify an acid solution, and values above 7 signify an alkaline solution.

Photochemical Reaction: A chemical reaction in the atmosphere that is triggered by sunlight, often yielding a secondary pollutant.

Photosynthesis: The production of sugars and starches by plants using air, water, sunlight, and chlorophyll. In the process atmospheric carbon dioxide is changed to organic matter and oxygen is released.

Plate Tectonics Theory: A theory that states that the outer portion of the earth is made up of several individual pieces, called plates, which move in relation to one another upon a partially molten zone below. As plates move, so do continents, which might explain some climatic changes in the geologic past.

Polar (P) Air Mass: A cold air mass that forms in a high-latitude source region.

Polar Climates: Climates in which the mean temperature of the warmest month is below 10°C; climates that are too cold to support the growth of trees.

Polar Easterlies: In the global pattern of prevailing winds, winds that blow from the polar high toward the subpolar low. These winds, however, should not be thought of as persistent winds, such as the trade winds.

Polar Front: The stormy frontal zone separating air masses of polar origin from air masses of tropical origin.

Polar Front Theory: A theory developed by J. Bjerknes and other Scandinavian meteorologists in which the polar front, separating polar and tropical air masses, gives rise to cyclonic disturbances that intensify and move along the front and pass through a succession of stages.

Polar High: Anticyclones that are assumed to occupy the inner polar regions and are believed to be thermally induced, at least in part.

Polar Jet Stream: A midlatitude jet stream that migrates from as far south as

30°N latitude to around 50°N latitude. Outbreaks of thunderstorms and tornadoes follow this seasonal migration.

Polar Satellite: Satellites that orbit the poles at rather low altitudes of a few hundred kilometers and require only 100 minutes per orbit.

Positive-Feedback Mechanism: As used in climatic change, any effect that acts to reinforce the initial change.

Potential Energy: The energy that exists by virtue of a body's position with respect to gravitation.

Precession: The slow migration of the earth's axis that traces out a cone over a period of 26,000 years.

Precipitation Fog: *See* Frontal Fog.

Pressure Gradient: The amount of pressure change occurring over a given distance.

Pressure Tendency: The nature of the change in atmospheric pressure over the past several hours. It can be a useful aid in short-range weather prediction.

Prevailing Westerlies: The dominant west-to-east motion of the atmosphere that characterizes the regions on the poleward side of the subtropical highs.

Prevailing Wind: A wind that consistently blows from one direction more than from any other.

Primary Pollutant: A pollutant emitted directly from an identifiable source.

Psychrometer: Device consisting of two thermometers (wet bulb and dry bulb) that is rapidly whirled and, with the use of tables, yields the relative humidity and dew point.

Radiation: The wavelike energy emitted by any substance that possesses heat. This energy travels through space at 300,000 kilometers per second (the speed of light).

Radiation Fog: Fog resulting from radiation cooling of the ground and adjacent air; primarily a nighttime and early morning phenomenon.

Radiosonde: A lightweight package of weather instruments fitted with a radio transmitter and carried aloft by a balloon.

Rainbow: A luminous arc formed by the refraction and reflection of light in drops of water.

Rain Shadow Desert: A dry area on the lee side of a mountain range.

Rawinsonde: A radiosonde that is tracked by radio-location devices in order to obtain data on upper-air winds.

Reflection, Law of: The angle of incidence (incoming ray) is equal to the angle of reflection (outgoing ray).

Refraction: The bending of light as it passes obliquely from one transparent medium to another.

Relative Humidity: The ratio of the air's water vapor content to its water vapor capacity.

Return Stroke: The term applied to the electric discharge resulting from the downward (earthward) movement of electrons from successively higher levels along the conductive path of lightning.

Revolution: The motion of one body about another, as the earth about the sun.

Ridge: An elongate region of high atmospheric pressure.

Rossby Waves: Upper-air waves in the middle and upper troposphere of the middle-latitudes with wavelengths of from 4000 to 6000 kilometers; named for C. G. Rossby, the meteorologist who developed the equations for parameters governing the waves.

Rotation: The spinning of a body, such as the earth, about its axis.

Santa Ana: The local name given a foehn wind in southern California.

Saturation: The maximum possible quantity of water vapor that the air can hold at any given temperature and pressure.

Saturation Vapor Pressure: The vapor pressure, at a given temperature, wherein the water vapor is in equilibrium with a surface of pure water or ice.

Savanna: A tropical grassland, usually with scattered trees and shrubs.

Sea Breeze: A local wind blowing from the sea during the afternoon in coastal areas.

Secondary Pollutant: A pollutant that is produced in the atmosphere by chemical reactions that occur among primary pollutants.

Selva: *See* Tropical Rain Forest.

Semiarid: *See* Steppe.

Severe Thunderstorm: A thunderstorm that produces frequent lightning, locally damaging wind, or hail that is 2 centimeters or more in diameter. In the middle latitudes, most form along or ahead of cold fronts.

Siberian High: The high-pressure center that forms over the Asian interior in January and produces the dry winter monsoon for much of the continent.

Sleet: Frozen or semifrozen rain formed when raindrops pass through a subfreezing layer of air.

Smog: A word currently used as a synonym for general air pollution. It was originally created by combining the words "smoke" and "fog."

Snow: Precipitation in the form of white or translucent ice crystals, chiefly in complex branched hexagonal form and often clustered into snowflakes.

Solar Constant: The rate at which solar radiation is received outside the earth's atmosphere on a surface perpendicular to the sun's rays when the earth is at an average distance from the sun.

Solstice: The point in time when the vertical rays of the sun are striking either the Tropic of Cancer (summer solstice in northern hemisphere) or the Tropic of Capricorn (winter solstice in northern hemisphere). Solstice represents the longest or shortest day (length of daylight) of the year.

Source Region: The area where an air mass acquires its characteristic properties of temperature and moisture.

Southern Oscillation: The seesaw pattern of atmospheric pressure change that occurs between the eastern and western Pacific. The interaction of this effect and that of El Niño can cause extreme weather events in many parts of the world.

Specific Heat: The amount of heat needed to raise 1 gram of a substance 1°C at sea-level atmospheric pressure.

Specific Humidity: The weight of water vapor per weight of a chosen mass of air, including the water vapor (usually expressed as grams of water vapor per kilogram of air).

Speed Divergence: The divergence of air aloft that results from the variations in velocity that occur along the axis of a jet stream. On passing from a zone of slower wind speed to one of faster speed, air accelerates and therefore experiences divergence.

Squall Line: Any nonfrontal line or narrow band of active thunderstorms.

Stable Air: Air that resists vertical displacement. If it is lifted, adiabatic cooling will cause its temperature to be lower than the surrounding environment; and if it is allowed, it will sink to its original position.

Standard Rain Gauge: Having a diameter of about 20 centimeters, this gauge funnels rain into a cylinder that magnifies precipitation amounts by a factor of 10, allowing for accurate measurement of small amounts.

Static Seeding: The most commonly used technique of cloud seeding; based on the assumption that cumulus clouds are deficient in freezing nuclei and that the addition of nuclei will spur additional precipitation formation.

Stationary Front: A situation in which the surface position of a front does not move; the flow on either side of such a boundary is nearly parallel to the position of the front.

Statistical Methods (in forecasting): Methods in which tables or graphs are prepared from a long series of observations to show the probability of certain weather events under certain conditions of pressure, temperature, or wind direction.

Steam Fog: Fog having the appearance of steam; produced by evaporation from a warm water surface into the cool air above.

Steam Point: The temperature at which water boils.

Step Leader: *See* Leader.

Steppe: One of the two types of dry climate. A marginal and more humid variant of the desert that separates it from bordering humid climates. Steppe also refers to the short-grass vegetation associated with this semiarid climate.

Storm Surge: The abnormal rise of the sea along a shore as a result of strong winds.

Stratopause: The boundary between the stratosphere and the mesosphere.

Stratosphere: The zone of the atmosphere above the troposphere that is characterized at first by isothermal conditions and then a gradual temperature increase. The earth's ozone is concentrated here.

Stratus: One of three basic cloud forms; also the name given one of the low clouds. They are sheets or layers that cover much or all of the sky.

Stroke: One of the individual components that make up a flash of lightning. There are usually three to four strokes per flash, roughly 50 milliseconds apart.

Subarctic Climate: A climate found north of the humid continental climate and south of the polar climate and characterized by bitterly cold winters and short cool summers. Places within this climatic realm experience the highest annual temperature ranges on earth.

Sublimation: The process whereby a solid changes directly to a gas without going through the liquid state.

Subpolar Low: Low pressure located at about the latitudes of the Arctic and Antarctic circles. In the northern hemisphere the low takes the form of individual oceanic cells; in the southern hemisphere there is a deep and continuous trough of low pressure.

Subsidence: An extensive sinking motion of air, most frequently occurring in anticyclones. The subsiding air is warmed by compression and becomes more stable.

Subtropical High: Not a continuous belt of high pressure but rather several semipermanent anticyclonic centers characterized by subsidence and divergence located roughly between latitudes 25 and 35 degress.

Summer Solstice: *See* Solstice.

Sun Dogs: Two bright spots of light, sometimes called mock suns, that sit at a distance of 22 degrees on either side of the sun.

Sun Pillar: Shafts of light caused by reflection from ice crystals that extend upward or, less commonly, downward from the sun when the sun is near the horizon.

Sunspot: A dark area on the sun associated with powerful magnetic storms that extend from the sun's surface deep into the interior.

Supercooled: The condition of water droplets that remain in the liquid state at temperatures well below 0°.

Superior Mirage: A mirage in which the image appears above the true position of the object.

Synoptic Weather Chart: A weather chart describing the state of the atmosphere over a large area at a given moment.

Synoptic Weather Forecasting: A system of forecasting based on careful studies of synoptic weather charts over a period of years; from such studies a set of empirical rules is established to aid the forecaster in estimating the rate and direction of weather system movements.

Taiga: The northern coniferous forest; also a name applied to the subarctic climate.

Temperature: A measure of the degree of hotness or coldness of a substance.

Temperature-Humidity Index (THI): A well-known and often used guide to human comfort or discomfort based on the conditions of temperature and relative humidity.

Temperature Inversion: A layer in the atmosphere of limited depth where the temperature increases rather than decreases with height.

Thermal Low: An area of low atmospheric pressure created by abnormal surface heating.

Thermistor: An electric thermometer consisting of a conductor whose resistance to the flow of current is temperature dependent; commonly used in radiosondes.

Thermocouple: An electric thermometer that operates on the principle that differences in temperature between the junction of two unlike metal wires in a circuit will induce a current to flow.

Thermograph: An instrument that continuously records temperature.

Thermometer: An instrument for measuring temperature; in meteorology, generally used to measure the temperature of the air.

Thermosphere: The zone of the atmosphere beyond the mesosphere in which there is a rapid rise in temperature with height.

Thunder: The sound emitted by rapidly expanding gases along the channel of a lightning discharge.

Thunderstorm: A storm produced by a cumulonimbus cloud and always accompanied by lightning and thunder. It is of relatively short duration and usually accompanied by strong wind gusts, heavy rain, and sometimes hail.

Tipping-Bucket Gauge: A recording rain gauge consisting of two compartments ("buckets") each capable of holding 0.025 centimeter of water. When one compartment fills, it tips and the other compartment takes its place.

TIROS 1: The first weather satellite; launched by the United States on April 1, 1960.

Tornado: A violently rotating column of air attended by a funnel-shaped or tubular cloud extending downward from a cumulonimbus cloud.

Tornado Warning: A warning issued when a tornado has actually been sighted in an area or is indicated by radar.

Tornado Watch: A forecast issued for areas of about 65,000 square kilometers, indicating that conditions are such that tornadoes may develop; they are intended to alert people to the possibility of tornadoes.

Towering: A mirage in which the size of an object is magnified.

Trace of Precipitation: An amount less than 0.025 centimeter.

Trade Winds: Two belts of winds that blow almost constantly from easterly directions and are located on the equatorward sides of the subtropical highs.

Transpiration: The release of water vapor to the atmosphere by plants.

Tropic of Cancer: The parallel of latitude, 23½°N latitude, marking the northern limit of the sun's vertical rays.

Tropic of Capricorn: The parallel of latitude, 23½°S latitude, marking the southern limit of the sun's vertical rays.

Tropical (T) Air Mass: A warm-to-hot air mass that forms in the subtropics.

Tropical Depression: By international agreement, a tropical cyclone with maximum winds that do not exceed 61 kilometers per hour.

Tropical Disturbance: A term used by the U.S. National Weather Service for a cyclonic wind system in the tropics that is in its formative stages.

Tropical Rain Forest (Selva): A luxuriant broadleaf evergreen forest; also the name given the climate associated with this vegetation.

Tropical Storm: By international agreement, a tropical cyclone with maximum winds between 61 and 115 kilometers per hour.

Tropical Wet and Dry: A climate that is transitional between the wet tropics and the tropical steppes.

Tropopause: The boundary between the troposphere and the stratosphere.

Troposphere: The lowermost layer of the atmosphere marked by considerable turbulence and, in general, a decrease in temperature with increasing height.

Trough: An elongate region of low atmospheric pressure.

Tundra Climate: Found almost exclusively in the northern hemisphere or at high altitudes in many mountainous regions. A treeless climatic realm of sedges, grasses, mosses, and lichens that is dominated by a long, bitterly cold winter.

Ultraviolet: Radiation with a wavelength from 0.2 to 0.4 micrometer.

Unstable Air: Air that does not resist vertical displacement. If it is lifted, its temperature will not cool as rapidly as the surrounding environment and so it will continue to rise on its own.

Upslope Fog: Fog created when air moves up a slope and cools adiabatically.

Urban Heat Island: Refers to the fact that temperatures within a city are generally higher than in surrounding rural areas.

Valley Breeze: The daily upslope winds commonly encountered in a mountain valley.

Vapor Pressure: That part of the total atmospheric pressure attributable to its water vapor content.

Veering Wind Shift: A wind shift in a clockwise direction, such as a shift from east to south.

Vernal Equinox: *See* Equinox.

Virga: Wisps or streaks of water or ice particles falling out of a cloud but evaporating before reaching the earth's surface.

Visibility: The greatest distance that prominent objects can be seen and identified by unaided, normal eyes.

Visible Light: Radiation with a wavelength from 0.4 to 0.7 micrometer.

Vorticity: The tendency of air to rotate in either a cyclonic or anticyclonic manner.

Warm Front: The discontinuity at the forward edge of an advancing warm air mass that is displacing cooler air in its path.

Warm-Type Occluded Front: A front that forms when the air behind the cold front is warmer than the air underlying the warm front it is overtaking.

Water Hemisphere: A term used to refer to the southern hemisphere, where the oceans cover 81 percent of the surface (compared to 61 percent in the northern hemisphere).

Wave Cyclone: A cyclone that forms and moves along a front. The circulation around the cyclone tends to produce the wavelike deformation of the front.

Weather: The state of the atmosphere at any given time.

Weather Analysis: The stage prior to developing a weather forecast. This stage involves collecting, compiling, and transmitting observational data.

Weather Forecasting: Predicting the future state of the atmosphere.

Weather Modification: Deliberate human intervention to influence and improve atmospheric processes.

Weather-Stress Index: A means of estimating relative human discomfort and stress attributed to the weather. This index considers temperature, relative humidity, and wind.

Weighing Gauge: A recording precipitation gauge consisting of a cylinder that rests on a spring balance.

Westerlies: *See* Prevailing Westerlies.

Wet Adiabatic Rate: The rate of adiabatic temperature change in saturated air. The rate of temperature change is variable, but it is always less than the dry adiabatic rate.

White Frost: Ice crystals that form on surfaces instead of dew when the dew point is below freezing.

Wind: Air flowing horizontally with respect to the earth's surface.

Wind Chill: A measure of apparent temperature that uses the effects of wind and temperature on the cooling rate of the human body. The wind-chill chart translates the cooling power of the atmosphere with the wind to a temperature under calm conditions.

Wind Vane: An instrument used to determine wind direction.

Winter Solstice: *See* Solstice.

World Meteorological Organization (WMO): Established by the United Nations, the WMO consists of more than 130 nations. The organization is responsible for gathering needed observational data and compiling some general prognostic charts.

Index